Music and Technology

Music and Technology

A Historical Encyclopedia

James E. Perone

BLOOMSBURY ACADEMIC
NEW YORK • LONDON • OXFORD • NEW DELHI • SYDNEY

BLOOMSBURY ACADEMIC
Bloomsbury Publishing Inc
1385 Broadway, New York, NY 10018, USA
50 Bedford Square, London, WC1B 3DP, UK
29 Earlsfort Terrace, Dublin 2, Ireland

BLOOMSBURY, BLOOMSBURY ACADEMIC and the Diana logo
are trademarks of Bloomsbury Publishing Plc

First published in the United States of America by ABC-CLIO 2022
Paperback edition published by Bloomsbury Academic 2025

Copyright © Bloomsbury Publishing Inc, 2025

Cover photos: Vintage jukebox. (Delstudio/Dreamstime.com); Gramophone. (06photo/Dreamstime.com); Vintage reel-to-reel tape recorder. (Grungemaster/Dreamstime.com)

All rights reserved. No part of this publication may be reproduced or transmitted in any form or by any means, electronic or mechanical, including photocopying, recording, or any information storage or retrieval system, without prior permission in writing from the publishers.

Bloomsbury Publishing Inc does not have any control over, or responsibility for, any third-party websites referred to or in this book. All internet addresses given in this book were correct at the time of going to press. The author and publisher regret any inconvenience caused if addresses have changed or sites have ceased to exist, but can accept no responsibility for any such changes.

Library of Congress Cataloging-in-Publication Data
Names: Perone, James E., author.
Title: Music and technology : a historical encyclopedia / James E. Perone.
Description: [1st edition.] | Santa Barbara : Greenwood, 2022. | Includes bibliographical references and index.
Identifiers: LCCN 2021054210 (print) | LCCN 2021054211 (ebook) | ISBN 9781440878299 (hardcover) | ISBN 9781440878305 (ebook)
Subjects: LCSH: Music and technology—Encyclopedias. | Sound—Recording and reproducing— Encyclopedias. | Electronic musical instruments—Encyclopedias.
Classification: LCC ML102.T44 P37 2022 (print) | LCC ML102.T44 (ebook) | DDC 780.3—dc23
LC record available at https://lccn.loc.gov/2021054210
LC ebook record available at https://lccn.loc.gov/2021054211

ISBN: HB: 978-1-4408-7829-9
PB: 979-8-2162-7013-3
ePDF: 978-1-4408-7830-5
eBook: 979-8-2161-2029-2

To find out more about our authors and books visit www.bloomsbury.com
and sign up for our newsletters.

Contents

Preface and Acknowledgments   ix

Introduction   xiii

Chronology   xxiii

An A–Z of Music Technology through the Years

Amplification   1

Analog Synthesizers   5

Artificial Intelligence   7

Auto-Tune   10

Boomboxes   13

Bose, Amar G. and the Bose Corporation   16

Brass Instruments   19

Cassette Tapes   23

Collaboration Software   27

Compact Discs (CDs)   30

Computer-Composed Music   33

Computer-Generated Music   36

Desktop and Laptop Computers   39

Digital Audio Tape (DAT)   41

Digital Recording   42

Digital Synthesizers   44

Distortion   46

DJ-ing   49

Dolby Noise Reduction 51

Drum Machines 53

Edison Phonograph 59

Effects Pedals 61

8-Track Tapes 64

Electric Guitars 67

Electrical Recording 70

Electronic Keyboard Instruments 73

Electronic Tuners 77

Electronic Wind Instruments (EWI) 80

45-rpm Records 83

GarageBand 87

Generative Music 89

Gramophones and Victrolas 90

Headphones 95

Instructional Software 99

Internet Radio 101

iPads and Other Tablets 103

iPods and Other Portable Digital Music Players 105

iTunes 108

Jukeboxes 111

Karaoke 115

Leslie Speaker 119

Loops 121

Loudspeakers 123

Mechanical Organs 127

Mellotrons 130

Metronomes 132

Microphones 135

MIDI 138

Mixing Boards 140

Moog Synthesizers   143

Motion Picture Soundtracks   147

MP3s   150

MTV and Music Videos   152

Multi-Track Recording   155

Music Boxes   158

Music Notation Software   161

Music Printing with Movable Type   164

Musical Notation   167

Muzak   170

Ondes Martenots   173

Orchestral String Instruments   174

Organs   177

PA Systems   181

Pandora and the Music Genome Project   182

Patches   184

Pedal Steel Guitars   185

Peer-to-Peer File Sharing   188

Percussion Instruments   191

Player Pianos   192

Quadraphonic and Surround Sound   197

Radio   201

Reel-to-Reel Tape   206

Resonator Guitars   209

Reverb   212

Sampling   215

Satellite Radio   217

Sequencers   219

78-rpm Records   220

Sheet Music   223

Smartphones   226

Social Media   229

Sony Walkman and Discman   231

Stereophonic Sound   234

Streaming Audio Services   236

Tablature   239

Tape Manipulation   241

Theater Organs   244

Theremins   247

Transistor Radios   250

Video Games   253

Vinyl Albums   256

Virtual Accompanists   259

Virtual Studio Software   263

Virtual Synthesizers   264

Vocoders and Talk Boxes   266

Wire Recording   269

Woodwind Instruments   270

Yamaha Disklavier   275

Yamaha DX7 Synthesizer   277

YouTube   279

Glossary   283

Bibliography   287

Index   307

Preface and Acknowledgments

The topic of "music and technology" can suggest many different things to different people. As discussed in this book's Introduction, this volume is based on a broad definition of the topic—broad in the sense of the time period covered and in the types of technologies that have influenced, affected, and, in some cases, suddenly changed the way we create, distribute, and consume music.

The Chronology chapter provides brief overviews of significant connections between music and technology over the course of hundreds of years, with the caveat that the rate of change has dramatically accelerated, particularly with the development of radio, electronic musical instruments, and amplification beginning in the 1920s, and with the advent of sound synthesis and the linkage of computers and music during the 1950s through the present. The next section, the Introduction, provides more detail on the major technological changes connected with music over the years, along with discussion of some of the broader sociological significance of these developments.

The alphabetically arranged entries on music technology through the years focus on 100 significant examples of technological changes that have impacted the creation, production, dissemination, recording, and/or consumption of music. Again, I have taken a historically broad approach to selecting these 100 examples. As one might reasonably expect, Robert Moog's analog monophonic synthesizer of the 1960s is included, as are streaming audio services, and MIDI (Musical Instrument Digital Interface); however, I have also included some examples of the connection between music and technology that, important as they might be, are easy to be overlooked by the general public. These include the development of music printing from movable type to the development and now widespread use of music education software. The recent lockdowns, stay-at-home orders, and so on, associated with the COVID-19 pandemic, have also helped to expose more of us around the world to some technologies and music-sharing and music-collaboration applications that, although some may have been around for several years, suddenly became commonplace in the year 2020. These entries include discussion of some of these technologies, particularly from the standpoint of their sociological significance during the pandemic.

Throughout this book, I have attempted to avoid the use of technical terms without explanation, as well as the discussion of technological details that require the knowledge base of a musical or technology expert. Still, some terms and concepts related to music and/or technology are necessary in order to adequately cover the topics. Therefore, this volume includes a Glossary, in which I provide brief explanations of some of the terms that might require more explanation than that which will fit in the other sections of the book. For other musical or technological terms that are unfamiliar, I suggest using one of the concise and reputable online dictionaries, such as *Encyclopædia Britannica* (brittanica.com), which is an easily accessible source for basic definitions.

I have included annotations for some of the entries in the Bibliography; however, these generally are limited to entries in which the names of the author and source alone might not make it clear exactly what can be gleaned from that particular source.

As one might imagine, technological advances related to music that occurred over the course of hundreds of years include some that are more notable, more influential, and more widely known than others. I have included 22 sidebars throughout the volume to provide moderately sized chunks of information about individuals and technological innovations that either expand upon the main entries in the book or that provide a look at some of the connections of music and technology that were not quite up to the level of meriting an entire 1,000-plus word entry in the A–Z section. Included among the sidebars are a few that decidedly are not well known and that may at first appear to be quirky outliers on the spectrum of music-technology connections. I have included these as examples of some of the oddities that have occurred or that have been invented over the years that might broaden our understanding of what has been tried, even if sometimes with little success or little lasting impact.

ACKNOWLEDGMENTS

I wish to thank the entire staff at ABC-CLIO, as well as the copyeditors and others with whom they subcontract, for all of their help in getting this volume from the concept to the published stage. I am especially indebted to Catherine Lafuente, who first approached me with the idea for this volume. I wish to thank my wife and best friend, Karen Perone, for all the support that she continues to give me on all my writing projects. As I worked on this project, I drew particularly strong inspiration from one of my graduate school professors, the late Dr. Lejaren Hiller. Dr. Hiller was acknowledged as one of America's early computer music experts, particularly because of his work in the 1950s at the University of Illinois. One of the most unusual and most memorable courses I took in graduate school was Dr. Hiller's course on 20th-century music composition techniques. It is because Dr. Hiller began our study course by looking back to such things as medieval isorhythmic motets, Johann Sebastian Bach's use of numerology, Wolfgang Amadeus Mozart's extensive tonicization of various keys in his *Symphony No. 40 in G minor*, K. 550, and Robert Schumann's unusual key motion by the interval of a third in the *Symphony No. 1*, Op. 38, that I was inspired to offer a wide definition of "music technology" in selecting the topics for this volume. I also wish to thank composer Charles Ames who also broadened my knowledge of computer applications in music during my graduate student years at the State University of New York at Buffalo. It was because of Charles's friendship and influence that my first publication came in the form of a computer-based BINGO caller, programmed in BASIC, that appeared years and years ago in *Antic: The Atari Resource*.

Introduction

The topic of "music and technology" probably conjures up images of computers, artificial intelligence, the latest in recording studio and editing techniques, the latest in notation software, the latest apps for tablets and smartphones, and so on. The danger in focusing on what is viewed as cutting edge today is that by the time this volume is published, something newer, better, faster, and more useful might well have taken the place of at least some of these.

This volume takes a broader view of music and technology. Although it includes information about such things as current physical studio techniques, music-related computer software for sound synthesis, virtual studios, computer-based music generating programs, smartphone apps, and so on, it also examines connections between music and earlier forms of technology. So, we will take a look at such topics as Theobald Boehm's development of the Boehm system flute—the principles of which as still used today—back in the 1840s; the development of the plastic 45-rpm record, which undoubtedly played a significant role in the popularizing of rock and roll in the 1950s; music boxes; player pianos; analog synthesizers; early electrical recording; the now-almost-extinct 8-track tape; and other examples of music technology that are easy to overlook in an age in which it seems that virtually everything associated with the word "technology" involves a computer.

Of course, if one opens up consideration of a topic on one end of the historical timeline, the question will always arise as to where to begin. Because of documentation—or lack thereof—and because it is one of the connections of music and technology that still impact us today, let us begin with the work of the ancient Greek writer and mathematician Pythagoras. Using a monochord, a one-stringed musical instrument, Pythagoras described and discussed various proportions that would form the basis of our understanding of overtones, musical intervals, and musical scales. Western music eventually incorporated various tuning systems and temperaments before turning almost exclusively to equal temperament in the 18th century of the Common Era. It is a bit of an oversimplification, but the problem with the pure Pythagorean intervals based on the overtone series is that the farther that one goes away from the fundamental pitch—the one to which the string is tuned—the more out of tune harmonic intervals sound. To put it another way, one particular major 3rd from the series is not proportionally the same as another, which makes them give the effect of having differing degrees of (to coin a term) major third-ness. The equal temperament that eventually became a standard in Western music basically made every interval slightly out of tune (compared with the natural intervals that Pythagoras described) so that none would be horribly out of tune sounding. However, it must be noted that Pythagorean tuning is still occasionally used today. In fact, it is one of the options that is available on *Cleartune* and other tuning apps that are currently available on smartphones, approximately 2,500 years after Pythagoras first described the monochord proportions.

Ancient music notation systems have proven themselves to be difficult to find. Historians suppose that much music that was not entirely transitory was passed along from person to person. In other words, it was probably part of an aural and oral tradition. A widely known notation system only developed slowly over time. In Western music, symbols known as neumes were beginning in the 9th century to provide reminders of melodic motion to the members of the clergy in the Roman Catholic Church. These symbols did not clearly define pitch or exact distance of one pitch to the next, so they indeed were reminders to singers who had learned the liturgical chants with which the symbols were associated as part of the oral tradition. Eventually, more complex symbols known as heightened neumes provided singers with more information about melodic shape and relative distance from pitch to pitch. As this notation evolved, one could accurately sing the melody in this monophonic music even if one had never heard the tune before, provided that one understood the notation.

As musical notation continued to evolve, rhythm and meter were incorporated into the picture, and the development of the staff—which started as one line and evolved to the five-line staff used today—also helped to provide even more clarity to the melodic shapes and distances from note to note.

Until the late 15th century, however, music notation, like the text in a book, typically was printed by hand, by scribes. However, just as Gutenberg opened up the book to mass consumption by the development of printing using movable type, this technology came to music. Machine-printed music made musical works more readily available, such that publishing developed as an industry. In Europe, by the 19th century, it was possible to work as a freelance composer supporting oneself through publication and the sale of one's music. In the United States, the entire Tin Pan Alley era of approximately the 1880s through the 1940s was largely sheet music driven, particularly in the late 19th century and the first two decades of the 20th century. Much music making was done in the home, and mechanically printed sheet music provided amateur musicians with current hits, classics, and hymns. And although the term Tin Pan Alley is most closely associated with New York City, dozens of publishing houses flourished throughout the United States.

The mid-19th century cannot be overlooked with respect to how mechanical and acoustical advancements led to improvements in the construction, intonation, and playability of the flute and the clarinet. At the same time, Adolphe Sax invented the saxophone, an instrument that later would help to define big band and small group jazz, as well as R&B and early rock and roll. Another important mechanical development of the 19th century was Dietrich Nikolaus Winkel's invention of the metronome in 1812. Four years later, Johann Maelzel improved upon Winkel's design and began manufacturing the tempo-keeping devices.

The late 19th and early 20th centuries saw the rise of mechanical musical instruments, particularly the player piano. Although player pianos can still be found in the 21st century, today's music fans might most frequently associate them with saloon scenes in Western movies and television programs, or with old-timey retro ice-cream shops or pizza parlors. During the early 20th century, they played a significant role in the ragtime craze and in getting the popular hits of the day and older classics out to the public. The amateur home music making, which relatively inexpensive and readily available sheet music helped to make possible, and mechanical instruments such as the player piano and its coin-operated variant, the nickelodeon, brought music to classes of Americans that might otherwise not had as much access to music. Around the same time, the 1890s and the first several decades of the 20th century, music boxes were also enjoying popularity and would do so until the end of the 1920s when the recording industry and the Great Depression diminished their importance and prominence. The coin-operated multi-disc music boxes that the German company Polyphon Musikwerke made were also important as a predecessor, by not all that many years, to the jukebox.

It was in the early 20th century that commercial sound recording was making huge inroads, using technologies developed, but not necessarily widely available, in the late 19th century. Over time, shellac records and wax cylinders would give way to vinyl and plastic records in the mid-20th century. These developments in turn gave way to compact discs, which gave way to streaming audio and digital downloads.

The late 19th-century and early 20th-century technological developments in mechanical music, including player pianos and sound recordings, were not without controversy. Despite the fact that the U.S. Marine Band that he had led years earlier was one of the most frequently recorded and most commercially successful early recording artists, and despite the fact that his own professional concert band recorded commercially, the famous composer and bandleader John Philip Sousa penned an essay in 1906 warning about the dangers of "mechanical music." Sousa took a stand against audio recordings and player pianos in the article and warned that mechanical music would cause the death of amateur music making in the United States. With the abundance of amateur music making more than 100 years later, particularly evident on YouTube and via such apps as *Acapella* during the COVID-19 stay-at-home period, it seems that Sousa's fears were unfounded. And while it might have appeared that there was a major contradiction between Sousa's stance against commercial audio recording and player pianos versus his own success as a recording artist, it is generally agreed that Sousa's stance against mechanical music at least in part stemmed from turn-of-the-century copyright laws that did little to protect the interests of composers when their music was mechanically recorded by other performers. Sousa's concerns with composers' mechanical rights were addressed with input from Sousa in revisions to U.S. copyrights in 1909.

The 1920s also gave rise to radio, a technology that caught on quickly and that, like sound recording, changed and expanded over the years. The early use of AM radio for the transmission of music gave way by the late 20th century to the better-sounding FM. And as we entered the digital age, internet radio and streaming audio services continued to broaden the number of ways that individuals and businesses had access to broadcast music. However, before all these changes occurred, early radio was significant in bringing both popular hits and regional musical genres—such as country and blues—to a wider public. In fact, some of the 1920s' and early 1930s' U.S.-Mexico border stations broadcast live music by musicians such as Jimmie Rodgers and the Carter Family across a wide section of the United States and all the way into Canada. Country and blues musicians, many of whom included songs not previously copyrighted or published in their repertoire, were also among the favorites of some of the record companies of the 1920s and 1930s that were on the lookout for material for which they might be able to generate income through record sales and publishing rights.

In addition to the powerful border radio stations bringing what had been regional musical genres and styles across wide swaths of North America, the 1920s saw smaller, less powerful radio stations developing live programs such as *Louisiana Hayride*, *Midwestern Hayride*, and the *Grand Ole Opry* broadcasts to the airwaves. With the coming of radio networks in the 1920s and 1930s, these and the swing era big band ballroom performances would eventually be heard coast to coast. These coast-to-coast broadcasts helped to create a nationally shared popular culture in the United States.

Although it would be an oversimplification to attribute this completely to the radio and commercial recording industries, these industries, particularly after radio stations began the widespread broadcast of recorded (as opposed to live) music, helped to encourage the rise of the singing star. Whereas earlier in the Tin Pan Alley era, public interest tended to be more about the song and the songwriter, now more frequently a single recording of a song might dominate the charts based on the singing star who recorded it.

The shift in how the technology was used, and in the relationship between radio and the recording industry, was not, however, without controversy. The increase in the playing of phonograph records on the radio, as opposed to hiring studio musicians, led to a recording ban by the American Federation of Musicians that started in 1942 and lasted for two years.

Although the technology had been in existence for several decades, the 1950s saw the nationwide transmission of music expand out to visual media through television. During the 1950s and 1960s, television introduced the young Elvis Presley's hip-swiveling rockabilly to the United States through Jimmy and Tommy Dorsey's program *Stage Show* in January 1956. A few months later, Presley made a famous appearance on the *Milton Berle Show*; however, perhaps the most famous television performance by Presley was later that same year on *The Ed Sullivan Show*, probably the most watched and most important entertainment television program in the United States.

Beginning in the 1950s and extending over the period of 14 years, conductor-composer-educator Leonard Bernstein led the New York Philharmonic in a series of live television broadcasts of "Young People's Concerts." These 53 performances introduced the young people who attended the concerts as well as a large television audience to Latin American music; the music of past masters; the music of 20th-century composers such as Paul Hindemith, Gustav Holst, Aaron Copland,

Charles Ives, and others; and themes such as "Jazz in the Concert Hall" and "Folk Music in the Concert Hall." After the early 1970s, such a mass media approach to music education and music appreciation never reemerged on America's television networks.

For many years after the days of the Dorsey Brothers' *Stage Show* and the *Milton Berle Show*, television variety programs continued to feature the latest in popular music. During the 1960s, *Hootenanny* was the television source for folk revival music, until the British Invasion of 1964 brought rock and roll back to the forefront. In the wake of the British Invasion, relatively short-lived television programs such as *Shindig!* and *Hullabaloo* sprang up. The one constant connection between pop music and television for many years was Dick Clark's *American Bandstand*. Television also helped to provide access to country music through programs such as *Louisiana Hayride*, *Midwestern Hayride*, the *Porter Wagoner Show*, and *Hee Haw*. Beginning in 1971, soul and funk music was broadcast nationwide by the program *Soul Train*. In fact, over the course of *Soul Train*'s 35-year run, the program successfully bridged soul, funk, disco, and music well into the hip-hop era.

The television medium fundamentally changed with the coming of cable television and development of networks such as MTV and VH1 that focused on music videos. In fact, the August 1, 1981, debut of MTV marked the start of the music video age. Although promotional films for popular songs had been in existence for years, some of the music videos that were produced starting in the 1980s went much further in interpreting songs for the audience member. Although the music videos that have been produced in the years since have taken on lives of their own—sometimes being better known than the songs are on their own—this technology has also proven to be a double-edged sword. Although the video can work well commercially in conjunction with a song, it can also be so representative of only one interpretation of a song that it can diminish the viewer's use of their imagination. In any case, as Australia's Special Broadcast Services (SBS) writer Cameron Williams put it, MTV "was the little cable channel that could—within just a few years MTV went from being a scrappy upstart to defining the culture of the 1980s" (2017).

At the same time that all these changes were occurring in the various media and manners of consuming music, other technological changes were underfoot. Although for the most part today's traditional orchestral and concert band instruments had pretty much been established by the middle of the 19th century, the 1920s brought an upsurge in the development of new instruments, electronic instruments, and major modifications to some musical instruments that had been around in other forms for hundreds of years. Connected to many of these developments were improvements in electronic amplification.

Although the vaudeville theaters of the 1920s might have lent themselves to instrumental accompaniment by musical ensembles, some of the typical single accompaniment instruments, perhaps most notably the guitar, were ill-suited to larger halls. As a result, the resonator guitar was developed by instrument maker John Doperya for vaudeville performer George Beauchamp in the middle of the 1920s. This instrument could be played significantly more loudly than even a steel-string acoustic guitar. At the same time, the United States experienced a Hawai'ian music craze. As a result, instruments such as the ukulele and the Hawai'ian guitar—a guitar tuned to an open chord, played with a metal bar, and held in the player's lap—became part of the general American popular culture. Because of the popularity of the Hawai'ian style of playing guitar, the leading manufacturers of resonator guitars, National and Dobro, made both Spanish-style models, in which the guitar has a rounded neck and is played in the traditional position, and Hawai'ian-style models, which have a thicker square neck and are held with the strings on top and sometimes held in the player's lap.

The guitar underwent another significant technological change early in the next decade when by-then rivals Beauchamp, working in collaboration with Adolph Rickenbacker, and Dopyera, working in collaboration with Arthur J. Stimson, independently developed electric guitars around 1932–1933. One could reasonably argue that the electric guitar eventually virtually defined the sound of popular music from the 1950s. Its prominence in rock and roll, in particular, meant that the electric guitar is thoroughly linked with the generation gap, with the social and political protests of the 1960s as they played out in music, as well as with the punk culture in the 1970s through the Generation X angst that helped to define the grunge music and subculture of the 1990s.

At around the same time as Dopyera and Beauchamp were developing and marketing the first resonator guitars back in the 1920s, however, there was another revolution underway in higher-tech musical instruments. The Russian inventor and musician Leon Theremin developed the instrument that bears his name during the decade, finally patenting it in 1929. Using pairs of electronic oscillating circuits to control pitch and volume without the player actually touching the instrument and electronic amplification, the theremin is credited with being the first electronic musical instrument. As an instrument, its usage has been somewhat limited—perhaps primarily because it is extremely difficult to play accurately in tune with no tactile reference point. However, it was a staple of science fiction and mystery films of the 1950s and 1960s, including *The Day the Earth Stood Still*, *The Lost Weekend*, and *Spellbound*, and more recently has been used by rock musicians such as the band Smashmouth and the duo of former Led Zeppelin members Jimmy Page and Robert Plant. More important, some of the electronic principles developed by Theremin for his instrument found their way into Robert Moog's later synthesizers of the 1960s and 1970s. At approximately the same time as Theremin was developing his instrument, the French inventor and musician Maurice Martenot developed the ondes Martenot, an instrument that was capable of making the same kinds of electronically produced sounds as the theremin but that used a keyboard, thereby making the production of discrete accurate pitches significantly easier to produce. The ondes Martenot found its way into works by composers such as Edgard Varèse and Olivier Messiaen. In much more recent years, Radiohead keyboardist Jonny Greenwood has incorporated the ondes Martenot into some of the band's work, such as in the song "How to Disappear Completely." Interestingly, both Theremin and Martenot's use of electronic oscillators to produce musical tones was anticipated by American inventor Elisha

Gray's "Electric Telegraph for Transmitting Musical Tones," for which Gray received a U.S. patent in 1875 (Gray 1875).

The 1930s and 1940s saw refinements in sound recording technology. However, the next major innovation in the reproduction of recorded sound came right around the start of the 1950s with the debut of the 33–1/3 rpm vinyl LP and the 45-rpm plastic single. The vinyl/plastic material of these records produced less surface noise than the 78-rpm shellac discs that they replaced. New synthetic materials such as the plastics that came into common use after World War II also allowed record companies to cut more grooves per inch into records than had been possible with shellac. As a result, the album, which had previously consisted of several individual discs with one song per side—like a photo album containing multiple pictures—was replaced by a single long-playing 33–1/3 rpm disc. The 45-rpm single was relatively inexpensive and important for the developing youth culture of the 1950s and beyond, and, perhaps just as important, it was relatively unbreakable, especially compared with the fragile shellac discs that previously had been the standard for singles.

The 1960s might be thought of as a time of miniaturization and portability. Transistor radios could be powered by a small battery and could be easily carried around. And the transistor radios became increasingly smaller and more portable. Similarly, 8-track tapes and compact cassette tapes were ideal for music on the move. In this case, however, the main locomotion was by automobile. In fact, the 8-track tape was developed by a consortium that included the participation of the Ford Motor Company and General Motors. Because of its greater reliability, small size, and the fact that it did not require some album cuts to fade out before beginning on another track (only one of the inconveniences of the 8-track tape), the cassette proved to be the winner of the tape format wars. The portability of the cassette was exploited by the Sony Corporation, which, in 1979, introduced the Walkman, a portable cassette player that, because it was designed around the idea of using small headphones, was ideal for hearing music on the go by walkers and joggers alike. Not only did the Walkman increase the privacy and portability of music but it also helped to usher in a cultural shift in the Western world. As Du Gay et al. (2013) put it in *Doing Cultural Studies: The Story of the Sony Walkman*, "the Walkman was very much tied to notions of Japanese-ness and to hi-tech" (11). One could argue over the extent to which 21st-century interest in the Japanese language and culture—not to mention the popularity of manga and anime—is tied to the Sony Walkman; however, it clearly was part of a shift in Western thinking in which Japanese electronic products shifted from being considered "cheap" to being considered "cool" and high tech.

The 1960s and the 1970s were also the decades in which sound synthesis became mainstream. Although a number of individuals around the world were working on electronically synthesized music making at approximately the same time, it is the work of Robert Moog that remains the best remembered today. Moog's early interest in electronic music included building and then writing articles about building theremins (see, e.g., Moog 1961). Just as the use of vacuum tapes in radios and amplifiers gave way to the use of semiconductors in the form of transistors in these devices, such was the case with the transistor-based theremins that Moog and others were building in the 1960s. By the middle of the decade, Moog had developed a monophonic synthesizer. The Moog synthesizer and its successor, the Minimoog, quickly made an impact on popular culture. By the end of the decade, the Byrds, the Monkees, and the Beatles had used Moog's synthesizer on significant album cuts and singles, with the Beatles' use of the synthesizer on several tracks on their highly acclaimed *Abbey Road* (e.g., "Maxwell's Silver Hammer," "Here Comes the Sun," and others), demonstrating that this new instrument was not a toy, nor just a keyboard-based extension of the theremin suitable solely for sci-fi sound effects. Perhaps the album that even more fully brought the Moog synthesizer into popular consciousness was Wendy Carlos's *Switched-On Bach*.

Moog's development of an analog, monophonic synthesizer with its tone-color changes in large part defined by subtracting various overtones from common sound-wave forms was followed by other inventors moving into additive synthesis and digital sampling. The sampling technologies that emerged meant that the synthesizer could, for good or for ill, be used more effectively for imitating and substituting for the sound of real acoustic instruments. Although the use of sampling can be heard in the work of Stevie Wonder, Thomas Dolby, and others, sampling really came into its own in rap and hip-hop music. Although the turntable techniques of selecting and repeating sections of pre-recorded music that DJs such as DJ Kool Herc had pioneered back in the 1970s continued to be used in hip-hop, digital sampling allowed for a significantly greater number of bass lines, drum beats, instrumental figures, and so on to be incorporated into hip-hop recordings. In fact, if one does an even cursory examination of the songwriting credits on hip-hop recordings particularly from the 1990s through the present, one can get a sense of the prevalence of sampling by seeing the large number of individuals who receive songwriting credit. In many cases, significant portions of these lists are populated by writers of the songs that were sampled to create the new hip-hop songs.

In addition to digital technology becoming the dominant force in synthesizers of the 1980s through the present, digital audio recording also came into prominence during the 1980s, as did a digital medium for consumers of music: the compact disc, or CD. In addition to making new fully digital recordings, record companies were quick to remaster and reissue older recordings in CD format. In some cases, these reissues contained alternative takes of songs, songs that had only been released as singles, demo recordings of album tracks, and so on. The reason for this phenomenon? Compact discs, in addition to being more compact than the vinyl LPs that previously had been the norm, could hold more musical material. So, these enhanced reissues represented added value for the consumer.

Another late 20th-century technology that has fundamentally changed the way in which people compose, arrange, and print out music is the Musical Instrument Digital Interface, better known simply as MIDI, and the various multi-track studio and music notation software applications that are part of the MIDI world. MIDI provided a way to connect digital synthesizers to computers (beyond the computers that were built into the synthesizers themselves). This helped to allow musicians to layer track upon track, thus making it possible for people with limited keyboard skills to create complex musical works using their synthesizer keyboard. Music notation software, such as *Finale* and *Sibelius*, made it possible to arrange for full orchestra, concert band, or virtually any type of ensemble and produce legible full scores and transposed parts. And, if the composer or arranger made a mistake that was not caught before printing out the parts, a correction could easily be made and the part reprinted more easily than with handwritten, purchased, or photocopied sheet music.

The idea of a home recording studio was out of reach for most musicians in the 1970s. TASCAM released a 4-track recorder that used standard cassette tapes in 1979 and made multi-track recording more affordable for amateur and struggling professional musicians. Over the course of the next several years, other companies, such as Fostex, Yamaha, Akai, and others, produced similar cassette-based multi-track machines. These home studio units were much more than simple multi-track recorders; they offered mixing capabilities, panning, and other techniques that formerly were most closely associated with professional recording studios with expensive mixing boards. Although these machines are primarily associated with home studio work in the 1980s and 1990s, they are still part of the 21st-century lo-fi culture, which values analog do-it-yourself recording and mastering over high-tech digital.

Computer-based multi-track recording soon followed and can be readily accomplished on today's laptop and tablet computers using free and low-cost programs such as Apple's *GarageBand*. In fact, one of the most readily apparent developments that has taken place since the advent of computer-based home recording in the late 20th century and the first quarter of the 21st century is how much more affordable the technology has become. Similarly, today's music notation apps for tablets and iPads, albeit not necessarily as powerful as long-established desktop and laptop-based programs such as *Finale* and *Sibelius*, allow for the production of clean, clear scores and parts for a fraction of the cost of the desktop- and laptop-based notation programs of the past.

The 21st century has seen the ascendancy of streaming audio (as well as streaming media in general) and internet radio. Services and providers of these can bring to the consumer what is essentially a curated listening experience while also providing access to music that a listener otherwise might never have experienced, including a wide variety of folk and world music selections. Likewise, YouTube provides access to thousands of official music videos and recordings of old out-of-print folk, country, and blues recordings. In fact, 1920s' and 1930s' examples from these genres that one would have had to scour collectors' record stores in order to find even in the late 20th century can now be found easily. At the same time, YouTube and similar websites have allowed musicians, professional and amateur alike, access to an entire world of potential listeners. Justin Bieber's move from YouTube to a record contract is perhaps the best-remembered example of an artist getting their start by posting YouTube videos. In a 2009 interview with Desiree Adib of *ABC News*, Bieber said, "It had a hundred views, then a thousand views, then ten thousand views, so I just kept posting more videos and more videos.. . . Eventually, I got found by my manager who flew me to Atlanta to meet [recording artist and producer] Usher" (Adib 2009). In addition to artists who maintain a fan base through the site and artists such as Bieber who use the platform as a jumping-off point for superstardom, YouTube is also filled with instructional videos on playing various musical instruments, proper singing techniques, simplified explanations of music theory, and so on. Although some of this instruction might be dubious, some of it is pedagogically sound and truly useful.

The COVID-19 pandemic of 2020 demonstrated the importance of YouTube, Facebook Live, and collaboration smartphone and tablet applications such as *Acapella* to a far greater extent than to which they ever had been before. Musicians ranging from the unknown through the regionally known through international stars participated in Facebook Live concerts and, in some cases, enjoyed viewership that far exceeded what the artists would normally have attracted even in multiple live concerts in regular performance venues. When these Facebook Live events were shared on the artists' YouTube channels, another layer of potential virtual audience members was added.

Perhaps the most impressive use of technology during the pandemic, however, was made with collaboration software applications such as *Acapella*. These applications allowed groups of musicians to perform as part of the same video even though their individual audio and visual components were recorded separately from diverse locations. Professional and university choral groups, chamber groups, bands, rock bands, and so on put together numerous virtual performances using this technology. One such collaborative recording that received a particularly high degree of exposure was a performance of the university's fight song, "Stand up and Cheer," by members of Ohio University's Marching 110. The performance was featured as part of one of Ohio governor Mike DeWine's COVID-19 update broadcasts and subsequently has been posted to several Ohio television stations' YouTube channels (see, e.g., https://www.youtube.com/watch?v=q2-JnqxZcZE, accessed April 24, 2020). Other collaborative programs such as *Soundtrap*, *SoundStorming*, and *Trackd* also came into play although perhaps not as visibly to the general public.

The almost immediate widespread public exposure to musical technologies that had already been around for a while, such as Facebook Live and collaboration programs, demonstrated the impact that they can have when a true need arises. Other technologies that musicians used during 2020 included real-time collaboration programs such as Instagram Live and *JamKazam*. Applications such as *JamKazam* were designed with musicians in mind, so they significantly reduce the problem of latency—or delay—that are inherent in virtual meeting software such as *Zoom* and others. Perhaps more than highlighting the technologies themselves, the COVID-19 pandemic demonstrated just how creative musicians can be using what they have available in order to practice their art and make connections with audiences.

Chronology

6th Century BCE
The Greek mathematician Pythagoras is credited with developing a tuning system based on proportional divisions of a monochord.

3rd Century BCE
The Greek engineer Ctesibius of Alexandria is credited with inventing an organ. Ctesibius's hydraulis uses water pressure to send wind into a set of pipes.

9th Century CE
The predecessor of modern music notation, known as neumatic notation, is developed. The notation provided singers with general reminders of melodic direction and assumed familiarity with the Gregorian chant tunes they were singing.

ca. 1000
Music theorist Guido d'Arezzo uses a multi-line staff in music notation. Although there may have been predecessors, Guido is credited with inventing the use of staff lines, the basis for Western music notation systems that follow.

ca. 1450
Keyboard tablature systems are first used.

1476
Ulrich Han's *Missale Romanum*, arguably the first book that includes music notation with a staff that was printed using movable type, is published in Rome. The book includes 16 leaves of monophonic Roman plainchant.

ca. 1500
Lute tablature systems first come into use in Europe. These systems were the predecessors of guitar tablature systems that are still used today.

1501
Ottaviano Petrucci publishes *Harmonice Musices Odhecaton*, the first collection of polyphonic (multi-voiced) music printed with movable type. The book was a set of polyphonic chansons (songs).

ca. 1510
Organist and composer Arnolt Schlick of Heidelberg, Germany, describes an instrument that is considered the first modern organ. The instrument was operated

by a keyboard and included stops that allowed the organ to produce different tone colors.

ca. 1700
Bartolomeo Cristofori of Padua, Italy, invents the piano. From the Classical period (1750–1820) to the present, the piano has been the most prevalent keyboard instrument.

1711
English trumpet and lute player John Shore invents the tuning fork. Shore's invention became the most commonly used device to which musicians tune their instruments until the widespread use of electronic tuners in the second half of the 20th century.

1722
Johann Sebastian Bach composes his collection of keyboard pieces *Das wohltemperirte Clavier* (*The Well-Tempered Clavier*), a set of preludes and fugues in all 12 major and all 12 minor keys. The collection, and a similar collection composed by Bach in the 1730s and early 1740s, demonstrated that the equal temperament tuning system allowed music to be successfully composed and performed in any key, unlike earlier tuning systems in which some keys sounded more in tune than others.

1812
Dietrich Nikolas Winkel of Amsterdam invents the metronome.

1816
Living at the time in Paris, Johann Maelzel, originally from Regensberg, improves upon Dietrich Winkel's design for the metronome and begins manufacturing and popularizing the device.

1825
American inventor Alpheus Babcock patents the first cast-iron frame design for the piano. Babcock's development allowed pianos to produce greater volume and has influenced every acoustic piano made thereafter.

1844
Clarinetist and instrument maker Hyacinthe Klosé patents what he referred to as the "Boehm" system clarinet, naming his key and ring mechanism after flutist and instrument maker Theobald Boehm. Ironically, Klosé's patent predated Boehm's patent for the modern flute mechanism.

1846
Although his musical instrument patents span the period between the mid-1830s and 1850, the 14 patent applications that Adolphe Sax files in 1846 can reasonably be credited as the birth of the saxophone, still one of the most recognizable woodwind instruments today.

1847
Flutist and instrument maker Theobald Boehm patents the key mechanism for the flute that he had developed several years before. Boehm's developments continue to be the basis for flute fingering systems today.

1875
American inventor Elisha Gray receives a patent for his "Electric Telegraph for Transmitting Musical Tones," a device that used oscillators to generate tones. Although this device did not become a commercial success, it anticipated the use of oscillators in the 1920s' inventions, the theremin and the ondes Martenot, as well as well as in the 1960s' Moog synthesizer.

1877
Thomas Edison becomes the first person to record and play back his voice on his phonograph. With this, sound recording was born.

1887
Emile Berliner makes changes in Edison's phonograph and pioneers the use of flat discs for audio recording and playback. With this, the record that would, in one form or another, dominate the recording industry until the development of the compact disc in the 1980s, was born.

1896
Edwin S. Votey patents the Pianola, the first player piano.

1906
The first Victrola with an internal horn makes its debut.

1910
Organ builder Robert Hope-Jones enters the employment of the Rudolph Wurlitzer Company. The collaborations of Hope-Jones and Wurlitzer led to the development of the theater organ, an essential part of the silent film era.

1911
Peter Jensen and Edwin Pridham, both of whom worked for Magnavox, file a patent for a moving-coil loudspeaker. The two inventors later filed other patents that resulted several years later in the first PA (public address) system.

1919
U.S. president Woodrow Wilson makes a public speech in support of the League of Nations in front of over 50,000 people in San Diego, California, using a PA system designed by Peter Jensen and Edwin Pridham.

1920
Pittsburgh's KDKA makes the first commercial radio broadcast in the United States. Within four years, there were approximately 600 commercial radio stations in the United States.

1921
The Jack Dempsey vs. Georges Charpetier boxing match of July 2 becomes the first nationwide radio broadcast in the United States.

1925
The electrical recording process becomes part of recording industry technology. The new microphone-based process allowed for increased volume ranges on records and also made it possible to record quieter instruments and quieter musical passages.

1927

Working at the request of guitarist George Beauchamp, John Dopyera develops his first resonator guitar. By using a metal cone system, the resonator guitar could produce a significantly louder sound than a traditional acoustic guitar. The instrument was patented by Beauchamp in 1931.

1928

Leon Theremin patents the electronic musical instrument that bears his name. The theremin would help to provide an eerie atmosphere in the soundtracks of films such as *The Day the Earth Stood Still*, *Spellbound*, and *The Lost Weekend*, as well as the long-airing British television program *Midsomer Murders*. The theremin also influenced Robert Moog's work in the development of analog synthesizers.

Maurice Martenot invents the ondes Martenot, a keyboard-based electronic musical instrument. Although not as well known as the theremin, the ondes Martenot's pairing of electronic oscillators and a keyboard can be seen as a direct predecessor of the early analog synthesizers to come.

1930

Galvin Manufacturing Corporation introduces its car radio, the Motorola. The radio's name later became the company's name, and Motorola developed into one of the 20th century's major electronics firms.

1931

George D. Beauchamp receives U.S. patent 1,808,756 for the single-cone resonator guitar, a style that John Dopyera claimed to have invented, but had not patented, several years before.

Collaborating with composer Henry Cowell, Leon Theremin develops his Rhythmicon, a predecessor of the drum machine.

1932–1933

Working independently, the Dopyera Brothers and the team of George Beauchamp and Adolph Rickenbacker develop the first electric guitars.

1933

Rudolph Dopyera receives U.S. patent 1,896,484 for the Dopyera Brothers' spider-bridge resonator guitar, the resonator style perhaps most frequently used today.

1934

The Mutual Broadcasting System is established by four radio stations: Detroit's WXYZ, Chicago's WGN, New York's WOR, and Cincinnati's WLW.

A. J. Stimson receives U.S. patent 1,962,919 for an "electrophonic stringed musical instrument," an electric guitar.

Cincinnati radio station WLW becomes the first superstation in the United States, with a transmitting power of 500 kW.

1935

So-called border-blaster radio station XERA begins broadcasting from Mexico. This and other Mexican border stations played a significant role in disseminating

American country music from Mexico as far north as Canada through the rest of the 1930s.

1941
Les Paul builds one of the first solid-bodied electric guitars, known as "the log." Although Paul tried to interest Gibson in the instrument, it would not be until 1952 that Gibson would release the famous Les Paul solid-bodied electric guitar.

1947
The first transistor is developed by Bell Labs. A semiconductor, the transistor would go on to replace vacuum tubes enabling electronic devices to be smaller, more reliable, and to use less power.

1948
The first vinyl (polyvinylchloride) records are produced. Called long-playing discs (LPs), the new records featured less surface noise, improved sturdiness, and the ability to hold far more recorded sound than their shellac, 78-rpm predecessors.

1949
RCA Victor releases the first polyvinylchloride plastic 45-rpm single record. 45-rpm singles would quickly become a significant part of youth culture.

Inventor Harry Chamberlain develops his Rhythmate and his Chamberlain. These devices used tape loops on which the sounds of acoustic instruments were recorded to produce percussion patterns and pitched sounds, respectively.

1950
Leo Fender's Broadcaster solid-bodied electric guitar makes its debut. Because of trademark problems, the name was changed to the Telecaster, a model that is still manufactured today and that has been copied by numerous other guitar manufacturers.

1951
British code-breaker, computer scientist, and mathematician Alan Turing produces computer-generated music.

1954
The transistor radio makes its debut. Transistor radios used far less power and were significantly smaller and more portable, more reliable, and less expensive than their vacuum tube-based predecessors. The transistor radio became a mainstay of late-1950s' and 1960s' teenage life.

1956
Elvis Presley makes several appearances on U.S. national television. These appearances helped to propel Presley to instant stardom as a rock and roll singer, demonstrating the power of this relatively new electronic medium.

1957
The first commercially released stereo records become available.

American composer and chemist Lejaren Hiller and mathematician Leonard Isaacson collaboratively create the "Illiac Suite," a multi-movement string quartet composed by the ILLIAC I computer at the University of Illinois.

1958

The Philips Pavilion at the 1958 Brussels World's Fair features an 8-minute electronic music composition by Edgard Varèse: *Poème électronique*. This avant-garde music was heard by an estimated 2 million visitors to the multimedia pavilion.

1963

The Mellotron makes its debut. Within approximately five years, the instrument would be a staple of psychedelic rock and progressive rock.

1964

The British band the Beatles appear on the U.S. television program *The Ed Sullivan Show* on three consecutive Sundays in February. The band's first appearance on February 9 was reputed to have been watched by one of the largest U.S. television audiences ever.

1965

Ray Dolby of Dolby Laboratories rolls out the Dolby Noise Reduction System. This system reduced the tape hiss associated with compact cassette tapes.

The 8-track tape and in-car tape players make their debut and became an instant hit in the automotive industry.

1966

The Norelco 22RL962, manufactured by Philips, makes its debut. The unit is widely credited with being the first boombox.

1967

Robert Moog rolls out his first Moog synthesizer.

The American rock band the Byrds record "Goin' Back," reputed to be the first rock song to include Moog synthesizer.

The Monkees record and release "Daily Nightly," a song that featured the Moog synthesizer.

1968

Wendy Carlos records *Switched-On Bach*, a popular crossover album of the music of Baroque composer Johann Sebastian Bach recorded using a Moog monophonic synthesizer and much multi-tracking. Columbia Records released the album.

1969

The Beatles' *Abbey Road* is released. The album included several songs on which the Moog synthesizer is featured melodically, including "Maxwell's Silver Hammer," "Because," and "Here Comes the Sun." The instrument's ability to generate white noise figured prominently in the lengthy coda section of the song "I Want You (She's So Heavy)."

1971

The North American release of Emerson, Lake & Palmer's self-titled debut album includes the song "Lucky Man." This track, also released as a single, featured a Keith Emerson Moog synthesizer solo.

ca. 1973
DJ Kool Herc develops DJ turntable techniques, such as scratching and looping, that would play a central role in the development of hip-hop music and culture.

1976
Pollard Industries releases the Syndrum, the first commercially available electronic drum.

1977
Archie Shepp's *On Green Dolphin Street* becomes the first digitally recorded jazz album released in the United States.

Roland's MC-8 MicroComposer takes sequencing to a new higher level with its ability to sequence up to 5,200 notes.

1978
Telarc records the Eastman Wind Ensemble, under the baton of Frederick Fennell, performing wind band works of Gustav Holst and George Frederick Handel's *Music for the Royal Fireworks* digitally. This was the first digitally recorded commercial classical release in the United States.

1979
The Sony Corporation debuts its Walkman, a portable cassette unit designed to be listened to using headphones and suitable for walking or jogging.

The Fairlight CMI (computer musical instrument) digital sampling synthesizer makes its debut.

TASCAM releases its Portastudio, a cassette-based 4-track recorder that gave home users relatively inexpensive access to multi-track recording studio technology.

The Ry Cooder album *Bop 'Til You Drop* becomes the first digitally recorded album in a popular music genre released in the United States.

1980
The Linn LM-1 digital drum machine makes its debut.

The Roland TR-808, an analog drum machine, makes its debut.

1981
Cable television network MTV makes its debut with a video of the Buggles' recording of "Radio Killed the Radio Star."

Sequential Circuits' Dave Smith delivers the paper "Universal Synthesizer Interface" at the meeting of the Audio Engineering Society. The specifications of this interface were refined over the following year and rolled out commercially as the Musical Instrument Digital Interface (MIDI).

The Simmons Company releases its SDS-5 electronic drum set. This set would be used by numerous popular musicians throughout the 1980s.

1982
The Dutch corporation Koninklijke Philips N.V., better known simply as Philips, and the Japanese corporation Sony jointly develop and introduce the compact disc

(CD). The first CD release, on the Philips label, was of pianist Claudio Arrau's recording of Chopin waltzes.

The first MIDI-compatible synthesizer, the Sequential Prophet-600, makes its debut.

1983

Yamaha debuts its DX7 synthesizer, the first commercially successful digital synthesizer.

Roland Corporation rolls out its first MIDI-compatible synthesizer, the Jupiter-6, and the world's first MIDI-compatible drum machine, the TR-909.

1984

The Bose Corporation debuts its Acoustic Wave Music System.

Sony Corporation introduces the Discman, which gave the same kind of portability to the compact disc player that the company's Walkman did to the cassette player.

Sony Corporation introduces the first in-car CD player as an aftermarket product.

1985

The first manufacturer-installed car CD player becomes available in Mercedes-Benz automobiles.

1987

The Lincoln Town Car becomes the first U.S.-made automobile with a manufacturer-installed CD player.

Yamaha's Disklavier integrates acoustic piano and MIDI keyboard technology, which brought the concept of the player piano into the digital age. The Disklavier remains in production and continues to be used by composers worldwide today.

1988

Still one of the best-known MIDI-based music notation software products, *Finale* makes its debut.

1997

Antares Audio Technologies rolls out its controversial *Auto-Tune*.

1998

Cher's single "Believe," the first major hit to include the use of *Auto-Tune*, is released.

1999

The peer-to-peer file sharing platform Napster makes its debut and provided an easy way for users to share mp3 files over the internet.

2001

Apple Computer's Steve Jobs announces the development of *iTunes*.

XM Radio begins delivering its satellite radio service.

Peer-to-peer file sharing platform Napster, in its original form, is shut down after a considerable amount of controversy over its role in allowing the pirating of digital recordings. Napster eventually reemerged as a subscription-based streaming audio service.

2002
After launching its satellites in 2000, Sirius begins delivering its satellite radio service.

2004
Apple introduces its free virtual recording studio application *GarageBand*.

2005
The first YouTube video, an 18-second video of YouTube cofounder Jawed Karim visiting the zoo, makes its debut.

2008
Streaming audio service Spotify makes its debut.

YouTube Live is streamed worldwide from San Francisco and Tokyo.

2009
Vivendi and YouTube establish the Vevo service, which allows YouTube access to official music videos for popular artists.

2010
The 2010 Lexus SC 430 becomes the last automobile available with a factory-installed cassette player.

2016
Facebook premieres Facebook Live.

2020
The pandemic puts virtual meeting technology (e.g., *Zoom*), and livestreaming services such as Facebook Live and YouTube Live, in the spotlight. Numerous musicians made use of collaboration applications such as *Acapella*, *BandLab*, and others.

Amplification

If one visits some of the great cathedrals, theaters, and concert halls that were purpose-built prior to the 20th century, one might be surprised that church congregations, theatrical and concert audiences, and crowds assembled to hear popular speakers could actually hear the focus of their attention, given how large some of these buildings and facilities were. But until the early 20th century, electrical amplification did not exist.

The seeds of amplification were sewn in the early development of headphones, furthered by developments in speaker technology, particularly by Peter Jensen and Edwin Pridham of Magnavox in the 1910s. Jensen and Pridham also developed an amplifier for their public address (PA) system, which was famously used when U.S. president Woodrow Wilson spoke in front of at least 50,000 in San Diego, California, at what would later be known as Balboa Stadium in 1919.

Considering how common PA systems, keyboard amplifiers, guitar amplifiers, and so on are today, it is interesting to note that amplification took some time to become established. In fact, the development of the resonator guitar—which has a separate entry in this volume—came out of the need to increase the volume of the acoustic guitar so that it could be better heard in vaudeville theaters. The resonator guitar was developed in 1927 and was patented four years later, well after President Wilson's San Diego speech in a sporting arena in 1919. In terms of offering greater volume for relatively quiet acoustic instruments such as the guitar, the development of the resonator guitar was quickly followed by the invention of the electric guitar, with patents for pickups being issued just a couple of years after resonator guitar patents had been issued.

With the development of the electric guitar at the end of the 1920s and the instrument's refinement through the 1930s and 1940s, as well as for other instruments such as the accordion (which was very popular in the 1940s and 1950s), amplifiers were designed and marketed. In the United States, Leo Fender's K&F amplifiers of 1945 and the models he brought out under his own name the following year were originally intended for steel guitars (e.g., lap steel and pedal steel guitar, which were popular in country and Hawai'ian-style bands of the day). However, the pickups used in these instruments were similar to those used in electric guitars played in the Spanish position (e.g., strings in front of the player and oriented up and down), so Fender's models became the basis for electric guitar amplifiers.

> **Leo Fender**
>
> Although a number of individuals in the history of the connections between technology and music have had a name that instantly calls to mind a particular instrument, Leo Fender might be one of the most instantly recognizable. Christened Clarence Leonidas Fender, Leo Fender worked in radio repair as a young man. This background in electronics led to him building PA systems that were popular with Hawai'ian guitar, acoustic guitar, and lap steel guitar players. During World War II, Fender and Clayton "Doc" Kauffman, a former employee of Rickenbacker, formed the K&F Manufacturing Corporation and produced Hawai'ian guitars and amplifiers. After patenting a lap steel guitar and selling it with a K&F amplifier, Kauffman left the company. Fender eventually settled on the name Fender Electric Instrument Company for the enterprise. Fender developed the Esquire, which was subsequently renamed the Broadcaster, and then the Telecaster; this solid-bodied electric guitar, which is still manufactured today, is considered one of the greatest solid-bodied electric guitars to date. Fender then designed the Stratocaster, the Jazz Bass, and the Precision Bass, all still iconic instruments and designs that have been copied by numerous other manufacturers.
>
> In addition to guitars, Leo Fender was a leading designer of amplifiers, from the PA systems he built before teaming up with Doc Kaufman, to the dedicated lap steel guitar amplifier K&F produced, to the classic Fender amplifiers that followed. After he sold the Fender Electric Instrument Manufacturing Company to CBS in 1965, Fender went on to start both Music Man and G&L Musical Instruments, the latter of which found Fender teamed with George Fullerton. Because of his contributions to rock and roll as an instrument designer, engineer, and corporate leader, Leo Fender is one of the few nonmusicians elected to the Rock and Roll Hall of Fame. Fender was inducted into the Hall in 1992 by guitarist Keith Richards of the Rolling Stones.

With the first models introduced in the early 1950s, the Fender Twin series, including the Twin Reverb, became a standard for the electric guitar, arguably even more so after Fender released the first commercially successful solid-bodied electric guitar, the Telecaster, a couple of years after the first Twin amplifiers saw their debut. These amplifiers established a discernible clean tone that defined the Fender sound. The reverb that some of the Fender amplifiers offered virtually defined the late 1950s' and early 1960s' sound of surf music.

In the U.K., Vox and Marshall became the go-to guitar amplifiers of the 1960s and beyond. In large part, this was because international import and export were not nearly as common, cost-effective, or easy as they are today. Working with guitarist Dick Denney, Vox's Tom Jennings designed a 12-watt guitar amplifier called the AC-10 in 1959. The need for more power led to Vox's AC-30 amplifier later the same year. The AC-30 became one of the important defining features of British Invasion rock and roll, with its users including the Beatles, the Rolling Stones, the Shadows, the Dave Clark Five, the Kinks, and the Yardbirds. In the 1970s and beyond, the Vox AC-30 was associated with Mark Knopfler of Dire Straits, the Jam's Paul Weller, Queen's Brian May, Radiohead, and other artists. In the case of Weller and May, each was known for using a wall of linked Vox AC-30 amplifiers.

In the BBC documentary film *Vox Pop: How Dartford Powered the British Beat Boom*, Brian May explains and demonstrates one of the features of Vox amplifiers that gave them a fundamentally different sound than other guitar amplifiers: the lack of a negative feedback circuit. A negative feedback circuit commonly used in home audio systems allows an amplifier to produce a clean sound at a wide range of volume levels. The lack of such a circuit in the Vox AC-30 and its more powerful successors results in a gradual increase in volume beyond a certain point with creating increasing levels of distortion, itself a result of increasing emphasis on the overtones or harmonics above the fundamental pitch the guitarist is playing (Rogers 2011). It was this gradual increase in the harmonics, as well as Vox's use of the Celestion speakers designed and built by the British, that helped to define a characteristic Vox sound. In addition to the insight provided by May and other musicians, the BBC documentary on Vox amplifiers includes historical and technical information conveyed by several individuals who worked at Vox. The documentary may be of particular interest to readers seeking background information on the company and its role in the British Invasion. At the time of this writing, it is available on several YouTube channels (see, Rogers 2011, for example, at https://www.youtube.com/watch?v=54s3386KZVI, accessed January 14, 2021).

As Rich Maloof (2004) documents in his book *Jim Marshall—The Father of Loud*, at least some of the impetus for Marshall's amplifiers and their power from young, up-and-coming British guitarists, such as the Who's Pete Townshend, who let Marshall know what they and other rock musicians were looking for in electric guitar amplifiers: high volume and distortion capabilities.

Several American rock and roll musicians toured Britain in the late 1950s and early 1960s and had the opportunity to hear the amplifiers that were being made there, but that were not yet being exported to the United States. Roy Orbison was so impressed with the sound of the Marshall amplifiers he experienced on his 1963 concert tour of Britain that he became the first American to use Marshall amplifiers. According to Jim Marshall, British musicians' union regulation made it necessary for Orbison to hire British musicians to back him on the tour. "Someone in the band was using a Bluesbreaker [a Marshall amplifier model that was designed for Eric Clapton of John Mayall's Bluesbreakers] and he decided it was the amp he had to have" (Maloof 2004, 50). So, the American rockabilly musician Orbison joined British rock musicians such as Clapton and Pete Townshend in using Marshall amplifiers.

In the second half of the 1960s and beyond, the image of a stack or stacks of Marshall amplifiers became an iconic image in live rock and roll stage shows and bands' television appearances. The stacking of Marshall amplifiers began in 1965. The idea was to have a single head, which contained the actual amplifier electronics, and place it on top of two Marshall cabinets, each of which contained four 12-inch speakers. The idea came out of the need for more power and volume and the somewhat competing need to make the speaker cabinets small and lightweight enough that they could be moved and transported by roadies. Eventually, virtual walls of Marshall stacks were used by guitarists such as Jimi Hendrix and others in the late 1960s and in the 1970s.

This photo illustrates multiple aspects of amplification, as the sound from the Fender guitar amplifier is being picked up by a Sennheiser microphone. (Ruslan Lytvyn/ Dreamstime.com)

From the mid-1960s through the 1970s, the vacuum tubes that had played a significant role in defining the circuitry—and therefore the sound—of Fender, Vox, Marshall, Carvin, and other guitar, electric bass, and keyboard amplifiers increasingly gave way to less expensive, considerably smaller, lighter weight, more electronically stable, and more energy-efficient semiconductors in the form of the transistors that were used in solid-state amplifiers. Although some of the higher-end amplifiers continued to use vacuum tubes, the solid-state amplifier became the norm for many musicians. Musicians argued—and continue to argue—that tube amplifiers produce a more singing, warmer tone than amplifiers with circuit boards and transistors. However, it should be noted that some amplifier manufacturers, such as Fender, marketed even their early solid-state amplifiers as being the wave of the future, with technology that was intended to make vacuum tube-based amplifiers obsolete. Eventually, some manufacturers of solid-state amplifiers tried to emulate the sound of classic tube amplifiers by the well-known companies of the past. Other manufacturers offered and continue to offer hybrid amplifiers that have generally one circuit that is tube based with the rest of the electronics revolving around semiconductors.

In addition to amplifiers that can digitally emulate some of the classic tube-based amplifiers of the past, emulation apps have been published that accomplish the same thing, and some virtual studio software programs also allow musicians to emulate a variety of amplifiers. For example, Joel Kosche, lead guitarist of Collective Soul, said of his choice of apps for the iPad in the early 2010s, "Right now, I'm really digging on AmpKit. It's an app that's used with a small interface that allows you to play your guitar through a ton of different amp and effects models" (Harvell 2012, 102).

See Also: Distortion; Electric Guitars; Resonator Guitars

Further Reading
Harvell, Ben. 2012. *Make Music with Your iPad*. Indianapolis, IN: John Wiley & Sons.
Maloof, Rich. 2004. *Jim Marshall—The Father of Loud: The Story of the Man behind the World's Most Famous Guitar Amplifiers*. San Francisco, CA: Backbeat Books.
Rogers, Vince, director and producer. 2011. *Vox Pop: How Dartford Powered the British Beat Boom*. Video documentary. London: BBC Productions.
Teagle, John, and John Sprung. 1995. *Fender Amps: The First Fifty Years*. Milwaukee, WI: Hal Leonard.
White, Forrest. 1994. *Fender: The Inside Story*. San Francisco, CA: Miller Freeman Books.

Analog Synthesizers

The development of analog synthesizers owed a debt of gratitude to the work of Leon Theremin and the instrument that bears his name. In fact, analog synthesizer pioneer Robert Moog built theremins early in his career and adapted the oscillator circuitry of Theremin to the new transistors that were available in the 1960s. Moog also published how-to articles for other electronics enthusiasts interested in building transistorized theremins (see, e.g., Moog 1961).

Other oscillator-based electronic instruments that predate the analog synthesizer include the ondes Martenot, a late 1920s' invention by Maurice Martenot that used similar electronics as theremins but which used a keyboard. This made Martenot's instrument even more like the keyboard-based synthesizers that would emerge in the 1960s. Similarly, the Clavioline, which was developed in France by Constant Martin in the late 1940s, and the Univox Electronic Organ, developed by Tom Jennings in the U.K. at the start of the 1950s, were instruments that nearly bridged the gap between electronic organs and synthesizers. The Clavioline and Univox established a place in popular culture in the early 1960s: a modified version of the Clavioline was used for the synthesizer-like solo in Del Shannon's 1961 hit "Runaway," and a Clavioline, a Univox, or a combination of the two instruments was used for the principal melody in the Tornados' 1962 instrumental hit "Telstar."

Generally, analog synthesizers worked in two basic ways: through subtractive synthesis and additive synthesis. In subtractive synthesis, one starts with a basic tone, say, a sine wave in which all the overtones are present. One then uses the synthesizer to filter out various overtones—also known as harmonics—to create different wave forms and, hence, different tone colors. This is where the connection to earlier electronic instruments, such as the theremin, is easiest to see. It is almost as though the 1960s' analog synthesizer takes the basic wave form produced by a theremin-like oscillator and then manipulates it so that different tone colors can be produced. In additive synthesis, the synthesizer's programmer adds various overtones to the fundamental tone to create different composite wave forms and, hence, different tone colors, or they mix and balance a certain amount of, say, a sawtooth wave with a sine wave to create a composite tone.

Because Robert Moog's synthesizers are detailed in another entry in this volume, we will give only passing reference to them here. However, it is important to note that Moog's early synthesizers were analog, and they were monophonic. In other words, they could only play one note at a time. In fact, this was true not only of Moog's instruments but those of other manufacturers as well. The popularity of Moog synthesizers and the fact that they were for a time the most highly visible analog synthesizers in use give them special significance.

The early Moog synthesizers created a sensation in the second half of the 1960s and the early 1970s. Rock and pop groups such as the Byrds (e.g., "Goin' Back"), the Monkees (e.g., "Daily Nightly"), and the Beatles (e.g., "Maxwell's Silver Hammer," "Here Comes the Sun," and other tracks on *Abbey Road*) released songs that prominently included the Moog synthesizer between 1967 and 1969. One of the more famous monophonic analog synthesizer solos in popular music was Keith Emerson's extended melodic solo on Emerson, Lake & Palmer's 1970 recording "Lucky Man."

Despite the fact that early analog synthesizers were monophonic, perhaps the most famous late 1960s' use of a synthesizer was in the creation of Wendy Carlos's album *Switched-On Bach*. Because the Moog synthesizer that Carlos used was monophonic, the complex polyphonic (many voiced) texture of the works of Johann Sebastian Bach was created through numerous overdubs. A single analog synthesizer of the day could not play these works live. As part of a brief 1989 BBC program on Carlos's work in sound synthesis, Carlos discusses and demonstrates some of the techniques she used in creating *Switched-On Bach*. In addition, Carlos demonstrates subtractive and additive synthesis (see, https://www.youtube.com/watch?v=Z3cab5IcCy8, accessed January 14, 2021).

It should be noted that with the late 1960s' and early 1970s' recordings that used analog synthesizers, the instrument sounds like an electronic instrument. In other words, synthesizers could be programmed to approximate the wave forms of various acoustical instruments (e.g., a French horn), but only approximate. The wave forms created by woodwind and brass instruments tend not to be entirely consistent from note to note. In other words, a given note on a clarinet might have a slightly different audible tone quality and a slightly different wave form than another note on the same instrument played by the same musician. The aesthetics of the time, then, seemed to revolve around making the synthesizer sound like its own voice and not an imitation of an acoustic instrument.

A major change of approach came with the ARP and Solina (it was marketed under both names) String Ensemble, which was in production between 1974 and 1981. As implied by its name, this analog synthesizer re-created the sound of acoustic stringed instruments. This polyphonic synthesizer was heard on numerous recordings of the era, including the Buggles' "Video Killed the Radio Star," Elton John's "Someone Saved My Life Tonight," Rick James's "Mary Jane," and others. Because of the clarity of the synthesizer lines, James's "Mary Jane" is a particularly clear example of the emulation of acoustical instruments of which the String Ensemble synthesizer was capable.

Introduced in 1978, the Prophet-5, a five-voice polyphonic analog synthesizer, also represented a notable advance in synthesis. Perhaps, most important, the instrument included patch memory. This feature meant that the user could set up a particular wave form and its associated sound and store it into memory so that it could instantly be recalled. Notable users of the Prophet-5 include Depeche Mode, Kim Carnes, Thomas Dolby, Pat Metheny, Eurythmics, and others.

Although analog synthesizers were not as prevalent after the development of the first digital synthesizers in the late 1970s to the present, several of the old analog instruments were reissued over the years, including the Minimoog and others. Because some of the tones that were associated with analog synthesizers, such as the Moog, the Prophet-5, and the ARP String Ensemble, are so widely recognized, manufacturers of digital synthesizers have included patches for their instruments that emulate the sounds of the analog instruments. These include built-in emulations as well as plug-ins and downloadable code.

In addition, despite the popularity of digital synthesizers since the early 1980s, there seems to be a resurgence of interest in actual—as opposed to emulated—analog synthesizers. At the time of this writing, Moog offers several models of analog synthesizers, and Moog is joined by other companies such as Artuia, Korg, Behringer, CRAVE, Roland, and Novation. Early in the third decade of the 21st century, also on the market is the under-$80 Stylophone, a new edition of a rudimentary analog synthesizer used by David Bowie on his *Space Oddity* album. So, the analog synthesizer remains relevant even decades after the widespread use of digital sound synthesis.

See Also: Moog Synthesizers; Ondes Martenots; Theremins

Further Reading

Abernathy, David. 2015. *The Prophet from Silicon Valley: The Complete Story of Sequential Circuits*. Auckland, New Zealand: AM Publishing.

Moog, Robert A. 1961. "A Transistorized Theremin." *Electronics World* 65 (January): 29–32, 125.

Nelson, Andrew J. 2015. *The Sound of Innovation: Stanford and the Computer Music Revolution*. Cambridge, MA: MIT Press.

Pinch, Trevor, and Frank Trocco. 2002. *Analog Days: The Invention and Impact of the Moog Synthesizer*. Cambridge, MA: Harvard University Press.

Sewell, Amanda. 2020. *Wendy Carlos: A Biography*. New York: Oxford University Press.

Sweetwater Sound. 2004. "Virtual Analog Synths vs. Analog Synths." Sweetwater.com, April 22. Accessed January 18, 2021. https://www.sweetwater.com/insync/virtual-analog-synths-vs-analog-synths/.

Artificial Intelligence

Artificial intelligence plays a role in several of the technological categories covered in other entries in this volume. Here, we will take a broad snapshot of ways in which artificial intelligence, sometimes referred to as AI, has played a role and continues to play a role in the composition, recording, marketing, instruction, and other aspects of music and the music industry.

In a 2019 article on artificial intelligence and the music industry in the business magazine *Forbes*, author Bernard Marr cites a report by the McKinsey Global Institute (Bughin et al. 2018) that suggests that "70 per cent of companies will have adopted at least one AI technology by 2030" (Marr 2019). Marr mentions some of the ways in which artificial intelligence is already playing a role in the marketing and dissemination of music, such as The Orchard. This service, which is owned by Sony Corporation, advances beyond the genre-based music selection and curation systems used by some satellite radio and streaming audio services. Marr also mentions Warner Music Group's use of artificial intelligence in making artists and repertoire (A&R) decisions. Typical A&R decisions include such things as what prospective artists to sign and what repertoire to have them record for the greatest potential commercial success. Increasingly, Warner Bros. and some other companies have developed computer programs that compile and analyze data about artists' potential appeal, musical tastes of their strongest potential audience, and so on.

Using this type of computer-based analysis is a means of instantly identifying potential stars not been without controversy. In 2019, Lyor Cohen, YouTube's Global Head of Music, stated that "right now everybody's so drunk off the data that they're putting these kids in places prior to having any artist development" (Stassen 2020). As Stassen (2020) discusses, the debate between using artificial intelligence–gathered data and artificial intelligence–analyzed data versus using one's gut instincts is complex. There is, however, some evidence that using artificial intelligence in making A&R decisions can work. As Stassen writes, "If there was any doubt about [British artificial intelligence A&R company Instrumental founder, Conrad] Withey's advocacy for machine learning then, the two competitive UK Christmas No. 1s Instrumental scored last year and in 2018, via its artist services division frtyfve, has surely silenced any critics" (2020).

As given more detail in the Computer-Composed Music entry in this volume, artificial intelligence in various degrees has been part of the world of computer music composition for decades. Suffice it to say, from the 1950s into the 1980s, the introduction of aspects of artificial intelligence developed incrementally. They also were largely confined to academic composers using computer programs that learned either to replicate earlier art music styles or to compose avant-garde music. And in the early decades of this development, the work largely was confined to large academic institutions with mainframe computers or to composers with access to powerful mainframe computers through various consortia or electronic music centers. Home computers either were nonexistent or lacked the speed and power necessary for composing complex music using anything approaching artificial intelligence. There were experiments in the last several decades of the 20th century with artificial intelligence programs written in which computers would learn vernacular genres such as various styles of jazz and, through a MIDI interface with a synthesizer, be able to compose and perform music pretty much in real time; however, these experiments never made the transition into the mass consumer market at the time.

A giant in the world of electronics and popular music, Sony Corporation's CSL Research Laboratory, developed a system for composing pop songs using artificial intelligence that was rolled out in 2016. The software is capable of analyzing lead sheets (versions of the songs with the melody and chord symbols) of a certain artist or genre and then composing a new melody and chord progression in the style. The first two songs composed using Sony CSL *Flow Machines* software that have been released are "Daddy's Car" and "Mr. Shadow." Because of the machine-like vocals of "Mr. Shadow," the piece seems to fit in the concept of machine-written music.

Arguably, "Daddy's Car" is a bit more questionable in terms of how well it competes with human-composed pop songs. Human composer Benoît Carré wrote the lyrics for "Daddy's Car" and provided the vocal and instrumental arrangement. Although the Sony software analyzed lead sheets for songs by the Beatles in an attempt to produce a Beatles-like pop song, some listeners might find that the majority of the Beatlesque touches actually come from the lyrics, the vocal and instrumental arrangement, and the production, all of which were the products of Benoît Carré and not the computer software. In particular, the melody meanders more than the best efforts of Beatles songwriters John Lennon, Paul McCartney, and George Harrison. Listeners who are sensitive to phrase structure and the relationships of one melodic phrase to another might also find that cadence—or natural stopping—points generally are less fully defined in "Daddy's Car" than in the best of Lennon, McCartney, and Harrison's songs.

Google's artificial intelligence songwriting software, *Magenta*, also has been used to produce computer-composed music. Again, it can reasonably be argued that the results are workable. However, to date none of the products of artificial intelligence–composed pop songs seem likely to make it onto any of, say, *Rolling Stone* magazine's lists of the "500 Greatest Pop Songs of All Time." At least so far.

The work of OpenAI should also be noted. This project's website states, "OpenAI's mission is to ensure that artificial general intelligence (AGI)—by which we mean highly autonomous systems that outperform humans at most economically valuable work—benefits all of humanity" (https://openai.com/about/, accessed January 14, 2021). One of the group's projects is *MuseNet*, which OpenAI describes as "a deep neural network that can generate 4-minute musical compositions with 10 different instruments, and can combine styles from country to Mozart to the Beatles" (Payne 2019). Payne (2019) includes links to examples of *MuseNet*-based compositions that combine disparate styles and MIDI instrumentations. The examples were pregenerated using *MuseNet*'s "simple mode." In "advanced mode," *MuseNet* can compose an entirely new example each time the user interacts with the program (Payne 2019).

In considering artificial intelligence and its role in the composition and marketing of music, it is important to note that it was only in the second decade of the 21st century that it started to become significantly more present in the commercial music industry. So, it is a new technology that appears to offer increasing possibilities as the technology is developed further and industry sorts out what aspects can make the best use of artificial intelligence.

See Also: Computer-Composed Music

Further Reading

Bughin, Jacques, Jeongmin Seong, James Manyika, Michael Chui, and Raoul Joshi. 2018. "Notes from the AI Frontier: Modeling the Impact of AI on the World Economy." The McKinsey Global Institute, September 8. Accessed January 18, 2021. https://www.mckinsey.com/featured-insights/artificial-intelligence/notes-from-the-ai-frontier-modeling-the-impact-of-ai-on-the-world-economy.

Marr, Bernard. 2019. "The Amazing Ways Artificial Intelligence Is Transforming the Music Industry." *Forbes*, July 5. Accessed July 22, 2020. https://www.forbes.com/sites/bernardmarr/2019/07/05/the-amazing-ways-artificial-intelligence-is-transforming-the-music-industry/.

Payne, Christine. 2019. "MuseNet." *OpenAI*, April 25. Accessed August 20, 2020. https://openai.com/blog/musenet/.

Stassen, Murray. 2020. "Can AI-Driven A&R Transform the Music Business?" *Music Business Worldwide*, February 27. Accessed January 18, 2021. https://www.musicbusinessworldwide.com/can-ai-driven-ar-transform-the-music-business/.

Auto-Tune

Auto-Tune, which was released by Antares Audio Technologies in 1997, is a musical technology that quickly developed a bad reputation, or, perhaps it is more appropriate to say that some artists' and producers' use of the technology gave *Auto-Tune* a bad reputation. The technology takes music, for example, a singer's vocal track, and can adjust flatness or sharpness of pitch so that the singing sounds as though it is perfectly in tune. And it is this aspect of the *Auto-Tune* that is at the center of the technology's controversy. *Auto-Tune* can be used over a range of degrees, and when it is used to a high degree, it can make singing sound artificial. The perception is that *Auto-Tune* has made it possible for singers with very little vocal talent to become superstars.

Interestingly, speaking 20 years after *Auto-Tune* made its high-profile appearance on the recording industry scene with Cher's recording of "Believe," NPR Music's Steve Thompson stated, "There's a long history of artists using different vocal modifications, but it had always been disguised. You used it in ways that you would hope the audience wouldn't notice, just to make a vocal a little cleaner and clearer and more on pitch. This song ['Believe'] brings the *Auto-Tune* front and center so you hear those modifications" (Cornish 2018). Certainly, not all pre-"Believe" vocal modifications were intended not to be recognized. For example, the Artificial Double Tracking and the sending of John Lennon's voice through a Leslie speaker in the Beatles' "Tomorrow Never Knows" were meant to make Lennon's voice sound otherworldly. However, Thompson's overall point is that the use of *Auto-Tune* in the Cher recording makes the singer's voice intentionally artificial sounding. And, unlike effects such as Abbey Roads Studio's Artificial Double Tracking, sending a singing voice through a Leslie speaker meant for instruments, and other psychedelic-era modifications, the *Auto-Tune* on Cher's recording of "Believe" was not necessarily part of a widespread aesthetic movement.

Although *Auto-Tune* earned a bad reputation as a device that could help to make intonation-challenged (or, less generously, out-of-tune) singers sound more technically competent, it has been used in some recordings effectively. Arguably, the Black-Eyed Peas' 2009 release "Boom Boom Pow" is one of the better examples of how *Auto-Tune* can be used to create an almost robotic effect, something that is entirely appropriate for the modernistic nature of the song and its official music video (see, https://www.youtube.com/watch?v=4m48GqaOz90, accessed January 14, 2021).

An artist particularly strongly associated with *Auto-Tune*, T-Pain appeared on *Arsenio* and discussed his use of the technology to create an effect on his voice and not to correct pitch deficiencies (see, e.g., https://www.youtube.com/watch?v=Rjt7Sn7x-Y0, accessed June 25, 2020). Although one of the defining aspects of T-Pain's well-known song "Buy U a Drank (Shawty Snappin')" is the smooth tone color that *Auto-Tune* gives the voice, T-Pain has recorded performances of the song without *Auto-Tune*. Listening to him sing the song without the effect is interesting, particularly in a near-acoustic rendition that is available on YouTube (see, e.g., https://www.youtube.com/watch?v=daJxMIK4Y20, accessed January 14, 2021). Arguably, however, the smoothness that *Auto-Tune* lends to T-Pain's voice in the original studio version of the song helps to define the character—a seducer—that he portrays in "Buy U a Drank."

In her book *A Philosophy of Song and Singing: An Introduction*, Jeanette Bicknell includes an extensive discussion of *Auto-Tune* (Bicknell 2015, 72–79). One of the important points that the author raises is the issue of authenticity as it relates to intonation and vocal tone color. Bicknell points out well-known singers such as Bob Dylan, P. J. Harvey, Neil Young, and Nina Simone who would have sounded entirely different—and lost much of their vocal distinctiveness—had their pitch been corrected by technology such as *Auto-Tune*. To put it another way, glissandi (slides) between pitches is an integral part of, say, the entire Bob Dylan sound and aesthetic. To take that away would be to fundamentally change the character—and perhaps diminish the effectiveness—of Dylan's work. Bricknell also mentions blues singers' use of flatness on certain pitches as a form of inflection, an integral part of the genre's aesthetic that would be lost with *Auto-Tune*. The author also expresses concerns about the unreasonable expectations that the use of *Auto-Tune* can bring. At the same time, Bicknell acknowledges that frequently Auto-Tuned performers such as Selena Gomez and Kesha might indeed sound better, and their fans might prefer their voices with *Auto-Tune* because of the nature of their singing and the material with which they are associated.

Interestingly, the question of authenticity and appropriateness of pitch correction are not unique to *Auto-Tune*, although *Auto-Tune* continues to be at the forefront of discussions about pitch correction. It is important to note that in the nearly quarter century since Cher recorded "Believe," numerous virtual studio programs and apps have added some sort of pitch correction in the wake of the prevalence of *Auto-Tune*. For example, in Apple's *GarageBand*, the user can select a monophonic region within a track in a recording and correct the pitches therein to the nearest half-step. Similarly, an easy-to-install plug-in for *Audacity* allows for *Auto-Tune*-like pitch correction. So, *Auto-Tune* itself continues to be part of the commercial music industry, particularly in dance music, and it has established a legacy through the incorporation of its effects in the entire spectrum of recording software, from expensive professional-level applications to those that preloaded on laptops and tablets to those that are free downloads. And for the 2020 holiday shopping season, catalog and online retailer Hammacher Schlemmer released a home karaoke machine that includes *Auto-Tune* software "to correct a singer's pitch and enhance his or her overall performance," touting it as the first such karaoke machine on the market for the home (https://www.hammacher.com/product/only-auto-tune-home-karaoke-machine?promo=search, accessed November 16, 2020).

See Also: Virtual Studio Software

Further Reading

Bicknell, Jeanette. 2015. *A Philosophy of Song and Singing: An Introduction.* New York: Routledge.

Cornish, Audie. 2018. "20 Years of Cher's 'Believe' and Its Auto-Tune Legacy." NPR, October 22. Accessed January 18, 2021. https://www.npr.org/transcripts/659611154.

B

Boomboxes

The boombox became a mainstay as well as a stereotype of black urban life from the 1970s well into the hip-hop era. Not only were these portable stereo systems, which early on usually combined a radio and at least one cassette deck, an important part of how young people listened to their music but boomboxes also played an important part of how hip-hop music came into being. Although in many circles boomboxes have been superseded by portable music devices with headphones or earbuds, designed to make them more personal, in the 21st century, there are meetings of boombox owners and aficionados. Some of these have been covered in YouTube videos (see, e.g., https://www.youtube.com/watch?v=zKSI4yvg_KU and https://www.youtube.com/watch?v=0zDZeOaEO8Y, accessed January 14, 2021). The manufacturers of boomboxes have also adapted to the times and gradually evolving technologies with some including not only a radio and cassette player/recorder but also CD players, USB connectivity, and Bluetooth connectivity.

Generally acknowledged as the first boombox, the Norelco 22RL962, manufactured by Philips, made its debut in 1966. This, and other early boomboxes, included a radio and cassette player/recorder. The Norelco machine and other boomboxes had built-in speakers, and many had an earphone/headphone jack, so listening to the machine did not necessarily have to be a shared experience with those in the vicinity of the person playing the boombox. Be that as it may, the boombox age of the mid-1960s into the 21st century was largely defined by the communal listening experience—the antithesis of the personal listening world of the Walkman, Discman, mp3 players, iPods, smartphones, and so on. Particularly in the youth culture of the day, it was important to be seen and heard with one's boombox.

By the time the classic boombox of the 1970s arrived, many of the machines had two cassette players and a radio, along with stereo speakers. Although some units were completely self-contained, others had detachable speakers so that the stereophonic space could be enlarged. National Public Radio's (NPR) Frannie Kelley explained the importance of the double cassette decks in the musical experience of the late 1970s in a piece for NPR's *All Things Considered*. According to Kelley, the popular boomboxes of the second half of the 1970s "weren't just portable tape players with the speakers built in. You could record off the radio, and most had double cassette decks, so if you were walking down the street and you heard something you liked, you could go up to the kid and ask to dub a copy" (Kelley 2009).

After the introduction of the CD, some boomboxes would have multiple CD drawers as well as multiple cassette decks. The importance of multiple CD drawers was that one could take several discs with them without having to store them in a separate container. Not only does this suggest the growing prevalence of the compact disc throughout the 1980s and into the 1990s but it also illustrates just how easy the boombox made it to share not only one's cassette-based music with others (e.g., through dubs of one cassette to another) but also one's CD-based music with others. Although boomboxes did not have the capacity to record on CD-RWs, one could dub another person's CD-based music onto cassette.

It should be noted that Panasonic's magazine advertisements clearly tied the company's boomboxes to youth culture and urban black culture. One of the more memorable images from Panasonic was of members of the band Earth, Wind & Fire standing in the middle of a city street appearing to hold a giant Panasonic boombox up in the air. And, to a certain extent, the boombox became something of a stereotype of urban black culture in the last three decades of the 20th century. In fact, in the late 20th century, boomboxes were widely known as ghetto blasters because of their seeming omnipresent nature in low-income urban areas and in numerous films that depicted those areas.

Despite its derogatory implications, the term "ghetto blaster" was a point of humor in the 1987 James Bond film *The Living Daylights*, in which a boombox outfitted as a powerful weapon and demonstrated to the character Bond by Q, the character who designed the disguised weapons for Bond over the course of decades of Bond films. Q describes the boombox as a "ghetto blaster" after it sends its literal blast at a mannequin that is destroyed by the explosion.

Despite the negative connotations of the term "ghetto blaster" and the use of the boombox as a racial stereotype in movies, television programs, and comedy routines, boomboxes did play an important role in the African American community, particularly as the hip-hop genre was developing and a hip-hop culture was taking hold. The importance of the boombox might best be summed up by LL Cool J's 1985 song "I Can't Live without My Radio." With lyrics such as "terrorizing my neighbors with the heavy bass" and "stimulated by the beat, bust out the rhyme," the song not only extols the voluminous sonic virtues of the rapper's JVC boombox but it also links the use of the machine to provide beats over which LL Cool J can "rhyme" or rap. An LL Cool J appearance on the television program *Soul Train* that highlights the type of boombox about which he raps is available on YouTube at the time of this writing (see, e.g., https://www.youtube.com/watch?v=GdkamS5axHQ, accessed January 14, 2021).

The boombox certainly was not limited to use in African American urban community in the last several decades of the 20th century or the early 21st century, nor was its use by movie writers and producers limited to so-called blaxploitation films or as a stereotype. For example, one of the more famous uses of a boombox in a Hollywood motion picture is in the 1989 Cameron Crowe film *Say Anything*, in which John Cusack's character, Lloyd Dobler, holds a boombox playing Peter Gabriel's "In Your Eyes" over his head in his attempt to woo Ione Skye's character, Diane Court.

In the late 20th century and early 21st century, a lo-fi (for low fidelity) subculture emerged on the music scene. Musicians who were part of this wave generally eschewed digital recording technology and preferred to make their recordings

Radio Raheem carried this boombox in the film *Do the Right Thing*. The boombox is in the collection of the Smithsonian Institution. (Collection of the Smithsonian National Museum of African American History and Culture)

using inexpensive machines, such as boomboxes. The Mountain Goats, which began as a solo project of singer-songwriter John Darnielle, is perhaps one of the best-known examples of this lo-fi approach. In 2013, NPR's Celeste Headlee interviewed the Mountain Goats' Darnielle for NPR's *Weekend Edition Saturday* program on the occasion of a special reissue of the Mountain Goats' 2002 album *All Hail West Texas*. In the interview, Darnielle discussed recording the album using a Panasonic boombox and the attraction of a low-tech approach at the time of the album's recording NPR Staff 2013). Although Darnielle's *All Hail West Texas* (the album was really a solo project under the name of the Mountain Goats) is considered a classic lo-fi album, there were countless young musicians in the 1980s and into the 21st century who found the boombox to be useful in making home recordings, demo recordings, and even recordings that included overdubs. Interestingly, even a cursory perusal of various internet discussion lists (e.g., doing a Google search for "boombox overdubs" or "boombox overdubbing") offers anecdotal evidence that one of the ways in which home recordists used a pair of boomboxes to achieve overdubs and the basic effects of multitrack recording was to record themselves on cassette tape on one machine, play back the recording while simultaneously singing and/or playing along with the playback, and record the resulting combination on cassette on a second boombox.

More formally, the importance of the boombox as part of late 20th-century and early 21st-century American popular culture is suggested by the inclusion of an entire boombox collection at The Henry Ford, which was profiled on the museum's website blog in February 2020 (The Henry Ford 2020). Despite the iconic nature of turn-of-the-century boomboxes, the boombox has continued to evolve. In fact, the February 2021 issue of *Consumer Reports* magazine included a face-off review of the Ultimate Ears Hyperboom and the Braven BRV-XXL/2, both of which "receive the highest marks in our sound-quality testing" (The Editors of *Consumer Reports* 2021, 16). These current boomboxes offer Bluetooth connectivity to smartphones and tablet computers and rechargeable batteries, and the Braven BRV-XXL/2 is shaped like an old-school boombox and includes an over-the-shoulder strap for carrying. The Ultimate Ears Hyperboom also includes an optical input.

See Also: Cassette Tapes; iPads and Other Tablets; Smartphones; Transistor Radios

Further Reading

The Editors of *Consumer Reports*. 2021. "Face-Off: New (and Improved) Boom Boxes." *Consumer Reports* 86, no. 2 (February): 16.

The Henry Ford. 2020. "Boomboxes at The Henry Ford." *The Henry Ford*, February 25. Accessed June 26, 2020. https://www.thehenryford.org/explore/blog/boomboxes-at-the-henry-ford.

Kelley, Frannie. 2009. "A Eulogy for the Boombox." NPR, April 22. Accessed January 18, 2021. https://www.npr.org/2009/04/22/103363836/a-eulogy-for-the-boombox.

NPR Music. 2009. "The History of the Boombox." NPR Music. Accessed January 18, 2021. https://www.youtube.com/watch?v=e84hf5aUmNA.

NPR Staff. 2013. "'Something Being Born': On Making a Classic Album with a Boombox." *Weekend Edition Saturday*, August 10. Accessed January 18, 2021. https://www.npr.org/sections/therecord/2013/08/11/210524003/something-being-born-on-making-a-classic-album-with-a-boombox.

Techmoan. 2016. "Boom Box Time Capsule." Accessed January 18, 2021. https://www.youtube.com/watch?v=2QGWyxI-ZPs. A YouTube reflection back on the popular boomboxes of the 1980s and 1990s.

Bose, Amar G. and the Bose Corporation

A native of Philadelphia, Pennsylvania, Dr. Amar G. Bose (1929–2013) was an electrical engineer and longtime professor of electrical engineering at the Massachusetts Institute of Technology (MIT). His contributions to music, however, came with his development of speaker systems, stereo systems, automobile audio systems, and headphones through the Bose Corporation.

Amar Bose developed a love of music in part because of violin lessons he took as a child. After he had been working at MIT, Bose purchased what he thought would be a very good hi-fi system for his home, based on the system's specifications. However, Bose was dissatisfied with the audio quality. His dissatisfaction was the impetus for Bose studying the acoustics of concert halls, existing speaker systems, and auditory perception. When he first started his company in 1964, Bose's focus was on research; however, Bose Corporation quickly moved into manufacturing and brought its Bose 2201 to market in 1966. This audio system was based on Bose's research that demonstrated that the majority of sound that people hear in a great concert hall is reflected off the various surfaces in the hall. In contract, the basic paradigm in speaker design and construction was for sound to go straight out to the listener. As shown and described in the owner's manual for the 2201 speaker system (see, https://assets.bose.com/content/dam/Bose_DAM/Web/consumer_electronics/global/products/speakers/2201/pdf/owg_en_2201.pdf, accessed January 14, 2021), these spherical units contained an array of small speakers aimed in various directions, thus providing the reflected sound Bose desired.

The 2201 system was the first of many Bose products that would try to produce something closer to the reflected sound of the concert hall. Subsequent speakers took on different shapes, with most of the Bose designs increasingly looking more like conventional speakers externally, despite how different from conventional stereo system speakers they might be internally.

Bose's work did not entirely revolve around creating the perception of hall-like reflected sound for the home, however. The company has produced large speaker systems as well as small bookshelf speakers. One of its more notable products was the Acoustic Wave Music System (AWMS-1), which made its debut in 1984. The original Acoustic Wave and its successors are table-top hi-fi systems. When these systems were first introduced, one of the most interesting features was the depth of the bass sound for such a small stereo unit. Anecdotally, I was a music graduate student at the time of the Acoustic Wave's debut, and it was the subject of considerable discussion when the first one made its appearance in one of the professor's music classes. It should be noted that other manufacturers have also worked on achieving better frequency response, particularly on the bass end, from small units; however, they have done so using their own techniques. The Bose Corporation has kept tight control over its innovations and patents throughout its history.

As noted earlier, when the Acoustic Wave units first appeared, the bass response was highly unusual from such a small one-piece stereo. Likewise, the unique shape of the unit, somewhat reminiscent of a partial sine wave, provided stereo separation that might have seemed highly unlikely—if not impossible—to some potential purchasers. As writer Ernie Smith (2016) details, the unlikely fidelity, stereo separation, bass response, and volume of the early Acoustic Wave units prompted Bose to advertise the product using infomercials so that the unique capabilities of the units could be explained to consumers. Beginning in 1993, Bose also began opening retail stores so that consumers could directly experience the company's home audio products. Bose announced in January 2020 that the company would be closing most of its retail stores, as its products are now available from a wide range of online and big-box retailers.

In addition to hi-fi systems for the home and units such as the Acoustic Wave series that can be used in smaller rooms, Bose offers sound systems for larger venues, such as churches and auditoriums. Like the home units, these sound systems are notable for their compactness and clarity of sound. Likewise, several automobile manufacturers offer optional Bose car stereos. Interestingly, because the automotive systems are designed for the specific makes and models, they must be factory installed. Also, typically Bose's automotive systems are more energy efficient and therefore require less power than other high-end car audio systems.

Another important area in which Bose's technology has impacted the music-listening experience is in the company's pioneering work with noise-canceling headphones for the consumer market. As *Consumer Reports*' Thomas Germaine writes, "The story goes that Amar Bose . . . got the idea for noise-canceling headphones on day in 1978 when he couldn't hear his music over the roar of a plane engine" (2020). Noise-canceling technology had been used in aviation for years at that point, having been developed in the 1950s by Lawrence Jerome Fogel. Bose developed its aviation headsets in the 1980s and subsequently developed active noise-canceling headphones for the consumer market, in which it has been a leader since 2000. Bose has sold several models of noise-canceling headphones with ongoing technological advances, particularly in the QuietComfort series since the start of the 21st century. Despite the large number of companies producing headphones, and specifically noise-canceling headphones at the time of this writing, Bose continues to play a significant role as a designer and manufacturer of high-end headphones. For example, *Consumer Reports* ranked the Bose QuietComfort 20 at no. 1 in the category of Portable Noise-Canceling Headphones, the QuietControl 30 at no. 1 in the category of Wireless Portable Noise-Canceling Headphones, and the QuietComfort 35 Series II and the Bose Noise Canceling Headphones 700 at no. 1 and no. 2, respectively, in the category of Wireless Home/Studio Style Noise-Canceling Headphones (The Editors of *Consumer Reports* 2020).

Despite the acoustical developments of the Bose Corporation since 1964, Bose has received criticism over the years. For one thing, Bose products are expensive, even compared with some other high-end audio equipment. Another criticism of some of the units is that the frequency output, or equalization, is entirely preset; the units do not have graphic equalizers or treble, midrange, and bass controls. Some of the 21st-century units include the means to connect them to smartphones and iPods on which the equalization can be changed, thus changing the input signal that the Bose unit receives and consequently giving the user some control over the equalization. Playing a CD offers no such control on some of the units. Although this limits the listener's ability to control tonal aspects of their music, it does reinforce the idea of a distinctive Bose sound. It should be noted that this manufacturer-defined sound extends beyond just the Acoustic Wave units; references to a distinctive Bose sound can also be found in various reviews and commentaries on the company's automotive units and those do allow the user to control equalization.

In the final analysis, Amar Bose and the company he founded back in the 1960s established a new music-listening norm for consumers for home audio, small room audio, large-venue audio, and noise-canceling headphones. The name "Bose" remains synonymous with high-end sound reproduction even as the formats in which consumers listen to music (e.g., cassette, CD, and smartphones) continue to change, regardless of whether or not some of its units allow the consumer to make changes to the sound's equalization.

See Also: Headphones; Loudspeakers

Further Reading

DeBord, Matthew. 2016. "We Tried 4 of the Best Audio Systems in Cars—Here's How They Stacked Up." *Business Insider*, November 22. Accessed January 18, 2021. https://www.businessinsider.com/best-car-audio-systems-compared-2016-11.

The Editors of Bose.com. n.d. "Dream + Reach: The First 50 Years of Bose." Bose.com. Accessed April 29, 2020. https://www.bose.com/en_us/better_with_bose/dream_and_reach.html.

The Editors of *Consumer Reports*. 2020. "Noise-Cancelling Headphones." *Consumer Reports Buying Guide* 84, no. 13: 103.

Ferris, Robert. 2016. "How Amar Bose Used Research to Build Better Speakers." CNBC.com, March 24. Accessed January 18, 2021. https://www.cnbc.com/2016/03/24/how-amar-bose-used-research-to-build-better-speakers.html.

Furseth, Jessica. 2017. "Noise-Cancelling Headphones: The Secret Survival Tool for Modern Life." *Guardian*, March 16. Accessed January 18, 2021. https://www.theguardian.com/technology/2017/mar/16/noise-cancelling-headphones-sound-modern-life.

Germaine, Thomas. 2020. "Best Noise-Canceling Headphones of 2020." *Consumer Reports*, January 26. Accessed May 1, 2020. https://www.consumerreports.org/noise-canceling-headphones/best-noise-canceling-headphones-of-the-year/.

Lemley, Brad. 2004. "*Discover* Dialogue: Amar G. Bose." *Discover*, October 1. Accessed January 18, 2021. https://www.discovermagazine.com/technology/discover-dialogue-amar-g-bose.

Moskovciak, Matthew. 2006. "Bose Acoustic Wave Music System II Review." CNET.com, Accessed January 18, 2021. https://www.cnet.com/reviews/bose-acoustic-wave-music-system-2-review/.

Smith, Ernie. 2016. "The Story of the Bose Wave, the Stereo System Built for the Infomercial Era. *Vice*, December 23. Accessed January 18, 2021. https://www.vice.com/en_us/article/4xap3d/the-story-of-the-bose-wave-the-stereo-system-built-for-the-infomercial-era.

Brass Instruments

Various types of brass instruments are described in ancient writings, and therefore, this family has been in existence for millennia. Although brass instruments typically have been made out of metal, as the instruments' family name suggests, it is not the material out of which instruments are made that determine their family. For example, many saxophones from the 19th century through the present have been made and still are made out of brass; however, the saxophone is classified as a woodwind instrument. The determiner of an instrument's family is based at least as much on the method the player uses to set the air column vibrating as the material type out of which the instrument is constructed. To accomplish this on brass instruments, the player buzzes their lips while blowing into what is usually a cup-shaped mouthpiece. So, ancient instruments such as the shofar and the hunting horn are examples of brass instruments, despite the fact that they were not necessarily made of the metal.

For hundreds of years, brass instruments used the hunting horn and military bugle model of playing pitches. By changing the speed of lip buzzing—as well as by changing the interior shape of their mouth (think using different vowel and diphthong shapes, such as "oo" vs. "ee")—the player can play different pitches. On nonvalve instruments such as the hunting horn or bugle, however, the only pitches that can be played are part of the overtone series.

One of the main technical advances in the design of brass instruments came with the development of valves and their associated additional lengths of tubing. This additional valve-activated tubing allows pitches that are in between those of the overtone series to be played. The valves that eventually became a standard part of band and orchestral brass instruments functioned in two different ways and still have their adherents in the 21st century. In rotary valve instruments, the valve turns on an axis when the player pushes the key. When the valve turns, it allows air to flow through a length of tubing, thereby lowering the pitch. In the United States, the French horn is the most frequently seen and heard rotary valve brass instrument today. In some European countries, however, rotary valves are used on other instruments, including the trumpet. Piston valves open an additional length of tubing when the piston moves downward as the player pushes the key. In U.S. orchestras and bands, generally rotary valve tubas and trumpets are used. When the player depressed two keys, say, the first and second on a trumpet, the tubing associated with both of the keys/valves comes into play, increasing the interval of the pitch lowering. Both rotary and piston valves were invented, developed, and patented in the first half of the 19th century.

Because of the shapes of the instruments, how they are held when playing, and the volume that they can produce, the members of the brass family were particularly susceptible to development for marching, originally in military bands. As a result of the need for a marching bass instrument in the family, instrument maker Ignaz Stowasser patented his helicon in 1848. Adolphe Sax, best known for his invention and development of the saxophone, also patented several brass instruments in the mid-19th century, including the sax horn. Probably the best-known brass instrument invented in the 19th century, however, is the sousaphone, developed by J. W. Pepper for bandleader and composer John Phillip Sousa using some of Sousa's suggestions in the design. The sousaphone became a common sight in parade and football marching bands.

Other brass instruments invented in order to improve practicality for marching include the E-flat alto horn, which was an easier-to-hold marching substitute for the French horn, the mellophone, which is more trumpet-shaped than the French horn and the E-flat alto horn and that has supplanted the alto horn as the marching band instrument of choice for the French horn range.

In addition to developments in the external and internal shapes of brass instruments in order to make playing while mobile more practical and to improve projection in the outdoors, materials have changed with the invention of new construction products. Perhaps the best-known and most visible in the "brass" family—and the quotation marks are intentional—is the use of fiberglass to make marching sousaphones. Despite the fact that fiberglass had been developed in various stages between the late 19th century and the 1930s, it was not until the early

1960s when C. G. Conn introduced the fiberglass sousaphone. This was followed several years later by a fiberglass sousaphone from the King company, another leading manufacturer of band instruments.

One of the other areas in which technological changes that are not nearly as visible to the general public is in mouthpiece design. Particularly with the popularity of jazz in the 20th century, brass players, principally trumpet and trombone players, have searched for ways to play more loudly and to increase their pitch range. Manufacturers have developed different sizes and shapes to the cup of the mouthpiece and the attached tube that inserts into the instrument itself. Some of these were designed specifically to meet the needs of high note specialist trumpet players in jazz big bands. Other mouthpiece designs aim to help the player produce a particular type of tone on their instrument.

Just as electric guitar players increasingly turned to electronic effects pedals and recording studio-based effects for their instruments, some brass players associated either with avant-garde music or with jazz-rock fusion incorporated electronic effects into their repertoire of tone colors. Perhaps most famously, jazz trumpet player, bandleader, and composer Miles Davis began using various effects pedals as he turned to jazz-rock fusion in the 1960s and later funk and hip-hop. Most famously, Davis's 1969 album *Bitches Brew* virtually defined his brand of fusion and the types of effects that could be applied to the trumpet to make it a part of the electronic and amplified soundscape of electric guitars and electronic keyboards. Davis was not the only brass player active in jazz rock who used effects to turn their instrument into an electronic one. Trumpeter, composer, and arranger Bill Chase had echo effects incorporated into his trumpet sound for his extended solo on the 1966 Woody Herman band's recording of the song "Sidewinder." Notably, Teo Maceo produced both the Herman recording and Miles Davis's *Bitches Brew*. Chase later used echoplex on his band's recording of "MacArthur Park" (Rex 2014, 91) and, according to album cover credits, an "electric trumpet" on the track "Weird Song No. 1" on the 1974 album *Pure Music*.

Although the woodwind-like electronic wind instruments have dominated the analog and digital wind instrument/MIDI controller spaces since the 1970s, electronic wind instrument (EWI) pioneer Nyle Steiner, himself a professional trumpet player, developed an electronic valve instrument similar to an electronic trumpet as part of the group of instruments that Steiner introduced beginning in the 1970s. Perhaps because of certain technical aspects of brass playing, for example, the extensive use of the embouchure in producing the various pitches that are part of the overtone series, Steiner's electronic brass instrument did not achieve the widespread success of his and others' woodwind-based electronic instruments.

Arguably, one of the most consistently innovative multinational corporations with regard to adapting traditional acoustic instruments to the computer age, Yamaha Corporation has made innovations in recent decades that have increased the flexibility of brass (not to mention other) instruments. The company's SILENT series provides the consumer-musician with what might be described as an electronic mute, headphones, and an effects box. Although one can use, say, Yamaha's SILENT trumpet as essentially an electronic trumpet for performances and run the output from the hardware through any type of electronic effects, the product perhaps is best known as a way of making it possible to practice in practically any sound environment. The electronic mute effectively makes the instrument very quiet—albeit perhaps not entirely silent—to those around the musician. Using headphones and various settings that are available, however, the musician can hear themselves play with a natural sound. Although the SILENT series includes electronic string instruments and so on, because of the volume produced by brass instruments, the SILENT brass is perhaps the most useful part of Yamaha's technology.

Further Reading

Baines, Anthony. 2012. *Brass Instruments: Their History and Development.* Mineola, NY: Dover.

Benade, Arthur H. 1960. *Horns, Strings, and Harmony.* Garden City, NY: Anchor Books.

Herbert, Trevor, Arnold Myers, and John Wallace, eds. 2019. *The Cambridge Encyclopedia of Brass Instruments.* Cambridge, UK: Cambridge University Press.

Rex, Nicholas. 2014. *Close Up Tight: The Life and Music of Bill Chase.* Master's Thesis, M.A. in Jazz History & Research. Newark, NJ: Rutgers University.

Cassette Tapes

The compact cassette tape—most commonly known as the cassette tape—was developed by Philips and first released to the public in 1963. Although other music recording, storage, and playback systems have made their debuts in the nearly six decades since, the cassette tape is still available today and has outlived nearly all of its music-storage competitors. In fact, one audio tape technology that appeared a couple of years after the cassette, the 8-track tape, although developed and touted by a consortium that was made up of some of the biggest players in transportation and audio, came and went arguably before the cassette even reached its heyday.

Generally, commercial cassettes were released with the same track order, and the same tracks divided between the A and B sides as vinyl albums. And, like vinyl albums, cassettes needed to be turned over in the original players, although some later cassette players and decks included an additional set of heads that allowed the machine to reverse the direction of the tape at the end of a side and automatically play the second side. The music track order and distribution between A and B sides were actually a distinct advantage over the competing 8-track tape, which played in only one direction, required toggling between sets of tracks in order to hear an entire album, and sometimes had songs that faded out and then faded in again after the toggling, as well as different song orders from the original vinyl album. Cassette recording and playback technology was also more stable than that of the 8-track; unlike cassette players, the playback head actually shifted from side to side in 8-track machines, sometimes causing the head to go out of alignment, which could result in audio bleed from track to track.

One area in which 8-tracks initially held an advantage in the tape wars of the 1960s, however, was in audio fidelity, mostly because 8-track tape moved at twice the speed of cassette tape over the record and playback heads. Over the years, however, several notable improvements were made to commercially released and blank cassette tapes and players available to consumers. Cassettes were offered that used Dolby Laboratories' noise reduction technology. As this technology (covered in a separate entry in this volume) underwent changes and improvements (e.g., from Dolby A, to Dolby B, to Dolby C noise reduction), the technology was incorporated into cassette tapes.

By the 1980s, cassettes of three basic types of tape formulation were available: the original ferric oxide, chromium dioxide, and a formulation that was called "metal," despite the fact that all the formulations involved one sort of metal or another. The latter two types resulted in less noise and greater frequency response.

However, one of the challenges with the various tape formulations and with the various types of Dolby noise reduction was that optimal results were achievable only if the consumer had equipment that would accommodate the tape formulation and the type of noise reduction. For example, listening to a commercially released chromium dioxide tape using a tape deck that did not have a "chrome" switch did not take advantage of the technological advances. Similarly, listening to a cassette that was recorded using any of the Dolby noise reduction types without selecting the appropriate "Dolby" setting on the cassette player/deck did not take advantage of the technology.

Although chromium dioxide (generally known as "chrome" tapes) and metal tapes produced better audio quality than the older formulation, these improved tapes were more expensive, so they presented consumers with a bit of a trade-off. One of the interesting aspects of the release of blank chrome tapes to the consumer market was the controversy with which they were associated. It was suspected that the tapes could potentially lead to more bootleg recordings, particularly because it might be possible for someone to record a vinyl album on a chrome cassette and have a cassette copy that was of higher audio quality than a commercially produced and commercially released ferric oxide cassette. Similarly, one could theoretically put together a compilation cassette and reproduce it considerably more easily and more economically than bootlegging using vinyl. And, one with album-oriented FM radio stations broadcasting in high-quality stereo, one could record an album off the radio and own it without purchasing it. Although concerns within the recording industry about the new higher-quality cassette tape formulations in the 1980s did not reach the levels that they would with the advent of the digital audio tape (DAT) in the same decade, they were significant. And, at least one 1980s' band, Bow Wow Wow, referenced the ability to pirate music via cassette tape in their 1980 debut EMI single "C·30 C·60 C·90 Go!" which *AllMusic* band biographer Andy Kellman describes as "an ode to music piracy" that was "the first single released on cassette" (Kellman n.d.). And, as National Public Radio's Frannie Kelley explained in an *All Things Considered* piece on the importance of boomboxes in that same era, cassette boomboxes "weren't just portable tape players with the speakers built in. You could record off the radio, and most had double cassette decks, so if you were walking down the street and you heard something you liked, you could go up to the kid and ask to dub a copy" (Kelley 2009). So, the piracy was not necessarily for commercial gain but to build one's music collection. Not coincidentally, the title of the Bow Wow Wow song refers to the designations for 30-, 60-, and 90-minute cassette tapes.

One of the areas in which cassette tapes made at least something of an impact on the music world might not be as well known to the general public: its use in early home recording studio applications. TASCAM's Portastudio made its debut in 1979. This recorder allowed the user to have access to the same basic functions available in a 4-track recording studio. One could record four separate tracks, or could move two recorded to tracks to a single track, and so on. Although professional recording studios were using 8- and 16-track machines over a decade before the debut of TASCAM's cassette-based machine, and were beginning to use up to 32-track digital machines in the late 1970s, it is important to consider that a great deal could be accomplished with 4-track recording capability: the Beatles' *Revolver* and *Sgt. Pepper's Lonely Hearts Club Band* were both recorded on 4-track analog machines—professional studio quality machines to be sure but 4-track machines nonetheless.

Although initially its audio fidelity was not as good as that of the 8-track tape, its simpler mechanism, better portability, and noise reduction enhancements has kept the cassette tape alive into the 21st century. (Artoholics/Dreamstime.com)

Because cassette tapes are not as fully in the public eye in the 21st century as they were in the 1960s through the end of the 20th century, it is easy to forget that this music format was an iconic part of television advertising at several points in approximately the last third of the 20th century. Perhaps one of the most memorable advertising campaigns of the 1970s was the series of "Is it live, or is it Memorex?" commercials for Memorex cassettes. These featured the legendary jazz singer Ella Fitzgerald. One of the most memorable, which came from early in the decade, showed Fitzgerald breaking a glass with her amplified voice, followed by a cassette recording of the same passage breaking a glass. The implication was that the recording was so true to the original performance that it packed the same power and the necessary frequency response to replicated Fitzgerald's live performance. By 1974, Memorex was promoting one of its new tape formulations and took a different approach. This time, the television commercial showed Fitzgerald's producer listening to a passage and being asked to guess if it was Fitzgerald and her band performing live or if the passage was on tape. The breaking glass imagery of the original commercial from a couple of years before was reprised at the end of the later commercial, just enough to keep it in the viewing public's mind. Later commercials in the series, from the last few years of the decade, featured Fitzgerald performing with then-currently popular pop singers (e.g., Melissa Manchester) and jazz instrumentalists (e.g., Chuck Mangione). Memorex mounted another advertising campaign that featured Ella Fitzgerald in the early 1990s, the time period in which there was a resurgence of interest in cassettes. Many of these commercials are available on various YouTube sites at the time of this writing.

In the 1980s, one of the iconic print commercial images was of a young man sitting in an armchair with his clothing and hair being blown back presumably by a recording made on a cassette tape made by one of Memorex's competitors: Maxell. In addition to the print ad, which ran in numerous popular mass readership magazines and music magazines of the day, Maxell used the image in a television advertising campaign that touted the fact that their cassettes retained their audio fidelity "even after 500 plays" (see, e.g., https://www.youtube.com/watch?v=Zjf5pdJJ44Q, accessed January 14, 2021). As was the case with the Ella Fitzgerald/Memorex series, the "blown-away guy" commercials were updated in the 1990s (see, e.g., https://www.youtube.com/watch?v=JvByYedx_Hw, accessed January 14, 2021). The lingering iconic nature of the Maxell advertising campaign is evidenced by a reference to the ads in a 2020 profile of celebrity chef José Andrés in which Maria Bustillos wrote that the chef's "energy and charisma seem to blast you right back in your chair, like in the old Maxell cassette tape commercial" (Bustillos 2020, 54).

Although the cassette tape is very much still with us today (just see the variety of blank cassettes, cassette players, and cassette boomboxes that are available from major online companies such as Amazon), it has been most closely associated with the period of the 1960s into the early 1990s. An entertaining history of the cassette tape that puts the technology into its historical context, along with explanations of the various tape formulations and noise reduction technologies, has been produced by the YouTube channel Techmoan (see, https://www.youtube.com/watch?v=jVoSQP2yUYA, accessed January 14, 2021).

Lo-Fi Culture

In response to the emphasis that was placed on crystal-clear production, digital editing of sound, and so on, a lo-fi culture developed in the 1980s and 1990s, although there were some earlier precedents. Singers, songwriters, and bands made recordings on inexpensive equipment, sometimes put out their music on cassette tape—even as the CD and streaming digital music became more the industry standard—and did so with pride. From the late 20th century into the 21st century, artists such as the Mountain Goats, artist Billy Childish's group Thee Milkshakes, the Black Keys, and others maintained a do-it-yourself, punk-like aesthetic in some of their recorded productions, regardless of their genre. This celebration of the do-it-yourself, old-school analog sound transcends genre, with artists who take the approach coming from the Americana, acoustic singer-songwriter, hip-hop, country, and rock musical categories. One of the more unusual home studio books of the 21st century, Karl Coryat's *Guerrilla Home Recording: How to Get Great Sound from Any Studio (No Matter How Weird or Cheap Your Gear Is)* (Coryat 2005), caters to do-it-yourself audio recorders. Coryat's approach is based around maximizing the quality of "dynamics, frequency content, and pan position" (Coryat 2005, 6), regardless of what equipment the musicians are using.

See Also: Boomboxes; Dolby Noise Reduction; 8-Track Tapes; Multi-Track Recording

Further Reading

Bustillos, Maria. 2020. "José Andrés Is Feeding the World." *AARP: The Magazine* 63, no. 5B (August/September): 52–55, 82.

Coleman, Mark. 2004. *Playback: From the Victrola to MP3, 100 Years of Music, Machines, and Money*. Cambridge, MA: Da Capo Press.

Coryat, Karl. 2005. *Guerrilla Home Recording: How to Get Great Sound from Any Studio (No Matter How Weird or Cheap Your Gear Is)*. San Francisco: Backbeat Books.

Kelley, Frannie. 2009. "A Eulogy for the Boombox." NPR, April 22. Accessed January 18, 2021. https://www.npr.org/2009/04/22/103363836/a-eulogy-for-the-boombox.

Kellman, Andy. n.d. "Bow Wow Wow." *AllMusic*. Accessed January 18, 2021. https://www.allmusic.com/artist/bow-wow-wow-mn0000094126/biography.

Moore, Thurston, ed. 2005. *Mix Tape: The Art of Cassette Culture*. New York: Universe Publishing.

Thompson, Dave. 2020. "Cassette Tapes Are Making a Comeback." *Goldmine*, May 5. Accessed January 18, 2021. https://www.goldminemag.com/collector-resources/cassette-tapes-are-making-a-comeback.

Collaboration Software

One of the effects of the COVID-19 pandemic of 2019–2021 was that many people, musicians and nonmusicians alike, became aware of collaboration and virtual meeting applications such as *Zoom, Skype, Acapella, Trackd, PiBox*, and numerous others. For musical collaborations, some of the virtual meeting platforms carried inherent problems, most notably the fact that there were delays, or latency. This made real-time musical collaboration over the internet very difficult, particularly for anyone who tried to use one of the platforms that was designed for virtual meetings. So instead, some musicians turned to applications that specifically were designed for collaborative performances although not necessarily in real time. Facebook, YouTube, and other social media platforms were seemingly suddenly filled with virtual performances of small folk groups, rock bands, vocal ensembles, choirs, marching bands, and so on. So, how did these groups of individual musicians record their parts in different places and at different times?

As an example, in *Acapella*, one person, who must be a subscriber to the service in order to record a project of substantive duration, sets up the project and selects a visual model in which each of the musicians will appear. Then, each musician records the audio and video of their part, preferably using the application's metronome feature. The project is saved, and a link to the file is sent to the next collaborator. Because each collaborator uses the same metronome, because the application functions like a multi-track tape recorder on which the play and record heads are placed in such a way as to avoid delays during overdubbing, and because the file is resaved after each collaborator adds their part, it is possible to create rhythmically accurate music videos. There are a few caveats, however, such as the fact that Bluetooth headphones carry with them a built-in latency that can cause coordination problems between the various collaborators. Once the final musician finishes and sends the file back to the person who set up the collaboration, that lead person can change balances and publish the final video to *Acapella*'s site, as well as to social media sites such as Facebook and YouTube.

Another application that found users during the pandemic was *Davinci Resolve*. This video- and audio-editing program at one time was used primarily by professional editors and producers; however, free and low-cost versions have made it available on a wider basis. A November 2020 review at *TechRadar*'s website extols the virtues of having the ability to edit up to 2,000 music tracks in *Davinci Resolve*, as well as the design of the video-editing design of the program (Paris 2020). During the COVID-19 pandemic, *Davinci Resolve* has been used by some of the large ensembles that have recorded and published virtual performances. It should be noted that the increased use of programs such as *Davinci Resolve* and others has brought up the continuing issue of compatibility challenges between programs such as *Resolve* and *iMovie*, as well as between versions of applications and programs for Windows, Mac OS, and iPads and other tablets.

In the case of real-time collaboration, *JackTrip* open-source software, created at Stanford University by Chris Chafe and Juan Pablo Caceres, shows promise. According to the *JackTrip* website, the software claims a latency time of 25 milliseconds. JackTrip.org states that "research indicates at this latency or less, musicians can play together at a moderately fast tempo with no significant synchrony problems. By musical collaboration we are we are referring to when both sides can hear each other as if they are in the same physical room, and can respond instantly to musical cues such as a melodic change or a fluctuating tempo" (JackTrip 2020). According to a report by NPR, *JackTrip* should work well when the performers are within 500 miles of one another (Hammar and Marshall 2020). As is the case with any virtual meeting or other video- and audio-intensive applications, high bandwidth is required to keep the delay, or latency, time to a minimum.

Other applications that became especially important during the pandemic included programs, such as *Google Drive* and *Dropbox*, which enabled songwriters, composers, and performers to collaborate, albeit not in real time. The proverbial "holy grail" of collaboration for performing musicians is not necessarily fully realized at the time of this writing, primarily because some of the applications that minimize or virtually eliminate the latency—or, delay—that one experiences in traditional virtual meeting software do so at the expense of audio quality, particularly as musicians use devices such as smartphones to connect to one another. Be that as it may, Instagram Live provides the opportunity for livestreaming from remote locations and is used by some musicians to produce virtual concerts. *JamKazam* is one of the virtual meeting-type applications that improves upon *Zoom* and *Skype*-type applications by greatly reducing the latency that makes traditional virtual meeting applications virtually unusable in real-time musical collaboration. So, although the use of a program such as *JamKazam* and smartphones with very small microphones might not result in studio quality performances or recordings, they are useful for virtual rehearsals.

BandLab works particularly well for allowing songwriters to collaborate, again, without actually needing to be in the same room. The app allows audio recordings to be sent to collaborators without the problems associated with having to send large audio files in formats such as mp3 or mp4 and so on. Some users might find the application as being at least somewhat similar in basic approach to nonmusical collaboration programs such as *Dropbox*, *OneDrive*, and so on.

Aside from the collaboration technologies that were used during the COVID-19 pandemic, other types of music collaboration software have been and continue to be used. One of the best-known applications is *SoundCloud*. Over the years, *SoundCloud* has been used by songwriters, composers, and performing musicians to get their music out to the world. In part, the attraction of SoundCloud.com, Audiomack.com, and similar sites is that they allow musicians of a wide variety of levels to self-publish their material and in turn to receive feedback from other users/subscribers. In this sense, one could think of a technology such as this as being akin to a bring-your-song-type songwriters' workshop, except that instead of the workshop involving a relatively limited number of people gathered in one location, here users from around the world might offer criticisms, kudos, suggestions, and the like. One of the dangers of *SoundCloud* and some other similar sites is that the subscribers represent such a wide range of talent levels, so some higher-level musicians might find the experience to be somewhat frustrating as they wade through some of the postings. On the other hand, sites such as this open up streaming audio posting and potential commercial gain to the masses and not just to musicians with recording contracts.

It should be noted that at the core of many of the most powerful of the some of the applications covered in this volume is Digital Audio Workstation (DAW) technology. This technology is an important part not only of collaboration software but also of digital recording, such as in computer-based home studio applications. The strides that have been made in this and other technologies suggest that real-time collaboration between larger and larger groups of musicians might be possible in the not-too-distant future.

For many musicians, the first opportunity that they had to make use of collaboration software in creating and disseminating music was born out of necessity when the COVID-19 pandemic made concerts with live audiences in typical performance venues impossible. A variety of individuals and organizations developed YouTube videos and even extensive learning modules to assist musicians with the intricacies of many of the programs and apps described previously. For example, church musicians who had to turn to virtual services, Facebook Live services, and virtual Christmas concerts could make use of the aptly named *The Resourceful Church Musician* (theresourcefulchurchmusician.com) for in-depth tutorials on some of the most practical and accessible musical collaboration software.

See Also: Desktop and Laptop Computers; Virtual Studio Software

Further Reading

The Editors of *Reverb*. 2020. "11 Tools for Collaborating on Music Remotely." *Reverb*, March 24. Accessed May 15, 2020. https://reverb.com/news/ways-to-collaborate-on-music-remotely.

Hammar, Nickolai, and Colin Marshall. 2020. "Playing Music Together Online Is Not as Simple as It Seems." NPR.com, July 15. Accessed July 16, 2020. https://www.npr.org/2020/07/14/891091995/playing-music-together-online-is-not-as-simple-as-it-seems.

JackTrip.org. 2020. "JackTrip: A System for High Quality Audio Network Performance over the Internet." JackTrip.org, May 11. Accessed July 16, 2020. https://ccrma.stanford.edu/software/jacktrip/.

Paris, Steve. 2020. "DaVinci Resolve 17 Review." *TechRadar*, November 11. Accessed November 23, 2020. https://www.techradar.com/reviews/davinci-resolve-17.

Compact Discs (CDs)

Although the digital age in recorded music had begun with the fully digital recordings that were made in the late 1970s, the widespread commercial accessibility of digital music—recorded digitally, remastered digitally from analog recordings, or simply digitized from analog recordings—came in 1982 with the development and release of the compact disc, or CD. After the release of pianist Claudio Arrau's recording (the actual recording had been made in 1979) of Chopin waltzes on the Philips label, the CD quickly came to the world of pop music in 1982 with the CD reissues of a host of albums, including those by some of the most popular artists of the era, including Billy Joel's *52nd Street* and *The Stranger* and ABBA's *The Visitors*.

Although the CD was a disc and thereby similar in shape to the vinyl records that preceded it, the technology differed significantly. Compact discs truly lived up to their name, as the 4 3/4-inch diameter discs could hold over 70 minutes of music, far exceeding the capacity of 33–1/3-rpm two-sided, 12-inch vinyl albums. The digital information, stored in the well-known binary computer format of ones and zeros, is encoded on the disc with minute notches. The notches are read by a laser, so there is nothing in the CD player's mechanism that actually touches the encoded surface. In sharp contrast, shellac and vinyl records of all the various formats, sizes, and materials from the 1870s through the present all use some sort of stylus. The fact that CD technology does not use a stylus means that there is no physical wear to the data part of the disc caused by playing it. However, because a photosensitive dye is used in CDs, the disc will degrade over time, particularly if it is exposed to ultraviolet light. This is especially true of user-recordable CDs (e.g., CD-ROMs and CD-Rs). And like vinyl records, scratches on CDs can cause playback problems.

One of the challenges posed by vinyl albums with multiple pieces of music has been selecting individual tracks. Although some turntables included laser technology that would locate the silent grooves between tracks, in general, track selection has been one of the drawbacks of vinyl albums. The same challenge was posed by cassette tapes; however, some tape decks included technology that allowed them to locate the silent tape between tracks so that the user could jump from one track to another. The compact disc made track selection significantly easier than even the best technologies applied to record turntables or cassette decks. In fact, some CD players allow the user to program alternate orders of the tracks on a disc. Using a CD deck with that capability allows the user, for example, to listen to the songs of the Beatles' *Sgt. Pepper's Lonely Hearts Club Band* in the original intended order, or to move the song "Her Majesty" to its original intended spot on the second side of the Beatles' *Abbey Road*. Either operation could be accomplished by pushing buttons to select the track order. Some CD players can store re-orderings such as these examples so that the programming operation would only need to be performed once.

Once the CD became the norm for the way in which new music was released to the public, a number of things happened: some concert/classical pieces that previously had to be broken up between sides of a vinyl album or even spread over multiple vinyl discs could now be listened to in their entirety without having to turn over the record or replace one disc with another. For some recording artists, the world of commercial popular music and genres such as jazz also changed. Instead of being limited to 45 minutes or so of music, albums could be lengthier. This meant that albums could include more tracks and/or lengthier tracks. In the case of reissues of albums originally released on vinyl, some record companies included bonus alternative takes, demo recording versions of album cuts and songs released on singles, and so on as part of the package. For the consumer, in some cases that represented a high level of added value. For example, the 1993 Rykodisc reissue of Elvis Costello's 1977 debut album, *My Aim Is True*, included all the songs from the original U.K. and U.S. releases of the LP, plus two tracks that had been released only as singles, and seven additional tracks that were either early pre-fame professional recordings or demo versions. Columbia Legacy's 1997 reissue of Miles Davis's iconic 1959 LP *Kind of Blue* contains one bonus selection; however, it is a 9–1/2-minute long alternate take of "Flamenco Sketches." The *Kind of Blue* reissue includes a booklet with extensive liner notes—far more extensive than those on the back of the original LP—by popular music critic Robert Palmer (see, Palmer 1997). The *Kind of Blue* booklet also includes photographs from the original recording sessions not included with the original LP. Similarly, the 1999 Columbia Legacy reissue of Johnny Cash's 1968 album *At Folsom Prison* contains two bonus tracks from the performance and an extensive booklet that contains information about the performance, as well as reproductions of some of Cash's handwritten notes related to his performance at the maximum-security prison.

The automotive industry adapted to the popularity of compact discs, and an industry of aftermarket automotive CD players also entered the retail marketplace. Sony Corporation introduced aftermarket players in 1984. The first car to have a factory-installed CD player available came from the German company Mercedes-Benz in 1985. U.S. car manufacturers were a bit slower to enter the realm of factory-installed CD players; the Lincoln Town Car became the first American-made automobile to offer a CD player in 1987. Sony Corporation followed up its popular Walkman personal cassette-based music system with the Discman, with which one could listen to CDs using headphones. Germane to our consideration of CDs and automobiles, an offshoot of the Discman, known as the Car Discman, made it possible to listen to CDs in a car that was equipped with either an auxiliary input using an 1/8-inch jack (e.g., "miniplug") or through a cassette deck, with the electronic signal being sent from a cassette-shaped adapter to the play head of a car's cassette deck. Therefore, the Car Discman made it possible to enjoy

CDs in an automobile without having to have the factory-installed or aftermarket CD player. As the compact disc grew in popularity, the capabilities of listening to CDs in one's car grew, with various companies offering CD changers that allowed the user to have several discs available at their fingertips. With the establishment of downloadable music, internet radio, streaming audio, and other technologies in the 21st century, the car CD player—particularly the factory-installed player—has almost completely become a thing of the past at the time of this writing. Interestingly, at the time of this writing, an internet search for such topics as how one can listen to CDs in a new car yields results showing how to hook up an old 1980s Sony Discman or one of its competitors from the time to a 2010s or 2020s USB-based automotive sound system: use the 1/8-inch miniplug.

It is not just in the world of the automobile that CDs seem to have become passé for listening to music. For example, the standard in laptop computers is now not to include CD or DVD drives. This is because storage has moved to USB ("flash" or "thumb") drives and various cloud-based services. The linkage of CDs and computers is not dead, though. One can still purchase external CD or DVD drives, something that is necessary on today's laptops and tablets to move the digital information (music) from a commercial CD or an old user-recorded CD-R or CD-RW to a music application such as *iTunes*. Unfortunately, as the compact disc continues to play a less-and-less significant role in the distribution of music, the value-added aspects of the booklets that proved to be an enhancement over the album cover notes of the vinyl age has become lost; materials of this type are not routinely distributed to consumers with music downloads or with streaming audio.

Digital Recording Designations

After the development of digital recording and mastering technology, and the emergence of the CD as a media for the commercial release of new music, CD reissues of pre-digital analog recordings started to appear. For several years, these reissues were given designations that quickly showed the consumer whether the recording and mastering were done using analog or digital means. The designations consisted of three letters, using either "A" for analog or "D" for digital. The first of the three letters indicated how the recording itself had been made, the second letter indicated whether the mastering was digital or analog, and the third letter indicated the nature of the reissue. The designations commonly found were AAD, ADD, and DDD. For example, the CD reissue of the Contemporary Chamber Ensemble's 1972 analog recording of compositions of experimental composer Edgard Varèse, which reissued on CD as Elektra/Nonesuch 9 71269–2, bore the designation AAD, meaning that the original recording and mastering were done on analog equipment and the CD reissue represented the sole digitization of the material. Although this AAD designation was quite common during the infancy of the CD, increasingly earlier recordings would be digitally remastered from the original analog tapes, resulting in the ADD designation, and with the increasing adoption of digital recording, the DDD designation eventually became commonplace. Eventually, these designations ceased to be used on new releases and new reissues.

See Also: Boomboxes; Digital Recording

Further Reading

Coleman, Mark. 2004. *Playback: From the Victrola to MP3, 100 Years of Music, Machines, and Money.* Cambridge, MA: Da Capo Press.

Fine, Thomas. 2008. "The Dawn of Commercial Digital Recording." *ARSC Journal* 39, no. 1 (Spring). Accessed June 12, 2020. http://www.aes.org/aeshc/pdf/fine_dawn-of-digital.pdf.

Palmer, Robert. 1997. Liner notes for *Kind of Blue* by Miles Davis. CD reissue. Columbia Legacy CK 64935.

Computer-Composed Music

Using digital computers to compose music dates back to the 1950s. At that time, computers were mammoth, room-sized machines that were found in government research facilities, large research universities, and the like. As a result, early computer-composed music tended to revolve around academic institutions and composers/professors associated with those institutions. As home computers, laptops, and tablet computers came into usage and composition software for these machines became available, computer-composed music has become somewhat more mainstream.

In the United States, Lejaren Hiller, who held posts in both chemistry and music at the University of Illinois and was later a professor of music at the State University of New York at Buffalo, made history when he used the University of Illinois at Urbana–Champaign's giant ILLIAC I computer to compose the "Illiac Suite." This 1957 piece is widely acknowledged as the world's first computer-composed musical work. Hiller and mathematician Leonard Isaacson jointly programmed the computer to follow various standard musical rules of counterpoint in this four-movement string quartet. In addition to generating random choices and then checking them against the rules that were programmed into the machine, the ILLIAC I was also programmed to use Markov Chains in making compositional choices. The basic principle of the Markov process, named for the Russian mathematician Andrey Markov, the probability of each event depends only on the state attained in the previous event. Some aspects of Hiller and Isaacson's work in this landmark composition continued to influence computer-music composers throughout the next several decades.

Academic institution–based and research center–based use of computers to compose music resulted in several important publications, particularly from the late 1950s through the end of the 20th century. These publications documented the processes, and in some cases the specific coding, that were used to create several of the important computer compositions. Hiller and Isaacson, for example, authored the article "Musical Composition with a High-Speed Digital Computer" for 1958 publication in the *Journal of the Audio Engineering Society* (see, Isaacson and Hiller 1993 for information on a subsequent reprint of the article). The following year saw the publication of their book *Experimental Music:*

Composition with an Electronic Computer (see, Hiller and Isaacson 1959). Both the article and the book focused on the methodology that Hiller and Isaacson used for the "Illiac Suite." Similarly, the Greek-French computer music composer Iannis Xenakis wrote the 1963 book *Musiques formelles: nouveaux principes formels de composition musicale*, which was translated into English in 1971 as *Formalized Music* (see, Xenakis 1971). In 1966, French computer-music composer Pierre Barbaud wrote the treatise *Initiation a la composition musicale automatique* (see, Barbaud 1966). Barbaud's book explains his programming procedures as well as aesthetic choices that he used in his computer-composed works, as well as providing numerous lines of sample computer code that Barbaud used.

One of the challenges faced in using computers to compose music for many years was the fact that one needed to be able to program the rules and so forth into the machine in FORTRAN, C, or some other programming language. Some computer-music composers continued to program in these languages even as the logic behind their coding grew increasingly sophisticated and increasingly incorporated aspects of artificial intelligence (see, e.g., Ames 1989, 1990). On the other hand, as home computers became more common, some computer-music composers/programmers attempted to make computer composition more widespread by removing the need for high-level programming skills. One such composer, Morton Subotnick, created several software packages, including *Making Music* and *Creating Music*. Subotnick's music composition programs were aimed at encouraging young people, as young as age 5, to create musical compositions using techniques that would already be familiar to them through activities such as finger painting. The process and the compositions introduced children to the basic materials and concepts of music through play.

Although the use of artificial intelligence was once mostly limited to academic composers, it moved into experiments with computer-synthesizer combinations that could compose and play music of a variety of genres (e.g., be-bop jazz) by studying melodic patterns, typical harmonic procedures, rhythmic procedures, and so on of the genre and then composing a piece that would match the paradigm. There were some attempts to develop this into commercial applications in the 1980s and 1990s; however, the most notable and commercially successful uses of artificial intelligence in computer-music composition have occurred in the 21st century.

In the second half of the 2010s, several companies rolled out artificial intelligence software systems that analyzed pop song styles and then are supposed to compose pieces in the mold of what was analyzed. I intentionally use the words "supposed to" because there is some debate about how closely the results of some of these analyses and subsequent new compositions result in music that really resembles the source material. Examples include Google's *Magenta* open-source software and Sony CSL Research Laboratory's *Flow Machines* software. The 2016 release of "Daddy's Car," with music composed by Sony's *Flow Machines* and an arrangement and lyrics by French composer Benoît Carré, generated a considerable amount of attention in the technology-related online press (see, e.g.,

Vincent 2016). Although the song was based on the Sony software's analysis of lead sheets of songs by the Beatles, and despite the Beatleesque touches in Carré's vocal and instrumental arrangements, it can reasonably be argued that the computer-composed melody meanders more and a less-well-defined feeling of phrase structure flow than the best pop songs written by John Lennon, Paul McCartney, or George Harrison for the Beatles.

The other area in which artificial intelligence–based computer music composition can be found is in the video game industry. Examples are discussed in detail in the Artificial Intelligence entry in this volume. Suffice it to say that video game composers are writing code that uses artificial intelligence to coordinate computer-composed and performed music with choices made by gamers, the speed at which these choices are made, and so on.

Similarly, generative music applications such as Brian Eno and Peter Chilvers's *Bloom* app, which was introduced in 2008, create computer-composed (with interaction from the user) and computer-generated music that is consistent with Eno's ambient music style; however, Generative Music also has its own entry in this volume.

See Also: Artificial Intelligence; Generative Music; Video Games

Further Reading

Ames, Charles. 1987. "Automated Composition in Retrospect: 1956–1986." *Leonardo* 20, no. 2: 169–185.

Ames, Charles. 1989. "The Markov Process as a Compositional Model: A Survey and Tutorial." *Leonardo* 22, no. 2: 175–187.

Ames, Charles. 1990. "Statistics and Compositional Balance." *Perspectives of New Music* 28, no. 1: 80–111.

Barbaud, Pierre. 1966. *Initiation a la composition musicale automatique*. Paris: Dunod.

Gagniuc, Paul A. 2017. *Markov Chains: From Theory to Implementation and Experimentation*. Hoboken, NJ: John Wiley & Sons.

Hiller, Lejaren A., and Leonard M. Isaacson. 1959. *Experimental Music: Composition with an Electronic Computer*. New York: McGraw-Hill. Reprinted 1979. Westport, CT: Greenwood Press.

Isaacson, Leonard, and Lejaren Hiller. 1993. "Musical Composition with a High-Speed Digital Computer." In *Machine Models of Music*, 9–21. Cambridge, MA: MIT Press. This is a reprint of an article originally published in the *Journal of the Audio Engineering Society* in 1958.

Jones, Andy. 2018. "Brian Eno and Peter Chilvers Bloom: 10 Worlds Review." *MusicTech*, December 28. Accessed October 30, 2020. https://www.musictech.net/reviews/brian-eno-and-peter-chilvers-bloom-10-worlds-review/.

Vincent, James. 2016. "This AI-Written Pop Song Is Almost Certainly a Dire Warning for Humanity: Let's Not Rule It Out, Anyway." *Verge*, September 26. Accessed January 18, 2021. https://www.theverge.com/2016/9/26/13055938/ai-pop-song-daddys-car-sony.

Xenakis, Iannis. 1971. *Formalized Music: Thought and Mathematics in Composition*. Translated by Christopher Butchers, G. W. Hopkins, and Mr. and Mrs. John Challifour. Bloomington: Indiana Press.

Computer-Generated Music

For years, computer-generated music, like computer-composed music, revolved around larger academic institutions that owned powerful mainframe computers and the composers who were associated with those institutions. In addition, there were some electronic music centers and consortiums at which composers used the mainframe computing resources that might not otherwise have been available to them. With the advent of personal computers and the huge strides that have been made in personal computing speed and power, computer-generated music is now considerably more mainstream than it was back in the 1950s and 1960s. In the entry on computer-composed music in this volume, we were primarily concerned with music that was composed by computer and meant to be performed by human musicians. In this entry, we are looking at music that is composed by computer and is also performed by computer.

One of the more famous examples of early computer-generated sounds came in the form of the collaborative composition "*HPSCHD*" by John Cage and Lejaren Hiller. This piece, which was premiered in 1969, was scored for harpsichord and multi-track, computer-generated tape. The composition is an interesting mixture of the Western art music tradition with sounds and compositional procedures of the digital computing age. The harpsichord part is constructed using randomly arranged snippets of music by composers from the 18th century into the early 20th century. The computer at the University of Illinois made decisions on the ordering of materials, based on a FORTRAN program, and the computer generated the material included on the tape.

Even as home desktop computers and laptop computers became more numerous, more accessible, and more powerful, electronic music studios and large university-owned mainframe computers were still part of the world of computer-generated music, particularly for composers who were associated with universities. For example, William Albright, who taught composition at the University of Michigan, used the university's electronic music center's computer to generate material for a four-channel tape for his 1985 composition "Sphaera," a work for piano and tape. Although this is not the most famous example of computer-generated musical material, it is interesting to consider in terms of how it (like other compositions from the time period) bridges the gap between computer-composed music, sound synthesis, and computer-generated music. Many of the compositional decisions for the piece were made by Albright; however, he used the university's computer to generate random 12-tone rows as part of the material on the tape, as well as for generating the electronic sounds for the tape. So, the computer played the role of partial collaborative composer, generator of electronic sounds, and backdrop for the human composer's choices of electronic material (Perone 1988, 165–166).

As the computer age reached the late 20th century and entered the 21st century, there was a considerable amount of overlap between computer-generated music and artificial intelligence. One of the more recent interesting examples of work in artificial intelligence–based computer-generated music can be found in the work of OpenAI. OpenAI's *MuseNet* can compose, score, and generate 4-minute compositions in a variety of disparate styles. The application's attempt to analyze and make use of musical stylistic connections between composers, performers, and genres illustrated at Payne (2019) suggests at least a partial philosophical connection with Pandora's Music Genome Project. The examples that OpenAI provides on its website may or may not resonate with listeners as being similar in style to the style of the composer and genre on which they are based. The mix of composer style with a genre not necessarily associated with that composer, however, makes for interesting listening.

Other examples of the crossover between computer-composed and computer-generated music and the world of artificial intelligence include some of the work that has been done since the 1990s in the world of video game music. Although this is detailed in the entry on Video Games in this volume, some of the technological highlights must also be noted here.

One of the first notable major changes in approach to video game music came in the 1990s with LucasArts' *Interactive Music Streaming Engine* (*iMUSE*) technology. Developed by Michael Land and Peter McConnell, this technology tried to smooth out the transitions (which previously had been quite rough in some video games and nonexistent in others) between music cues associated with different scenes in the games. According to Michael Sweet, author of *Writing Interactive Music for Video Games*, *iMUSE* "allowed the music system to wait for a musical phrase to end before transitioning to a new cue.... This system ensured a seamless transition by waiting for the current musical phrase or melody to end instead of interrupting it" (Sweet 2015, 45). Sweet continues by explaining that the importance of this development was that it provided a better sound continuum for the player, thus keeping them more fully engaged in the entire game experience. However, in a 2016 interview, Peter McConnell explained that contrary to the reputation for interactivity that *iMUSE* had back in the 1990s, he and Land had actually created the perception of interactivity simply by composing and recording multiple versions of sound cues of different durations. The game would play the appropriate one based on what the gamer was doing, and the phrase-to-phrase, cue-to-cue connection described previously would make it appear that the game was composing the music based on what was happening in the game (Mackey 2016). So, *iMUSE* represented a step toward mainstream computer-generated music in video games, but just one step.

At approximately the same time as Land and McConnell were developing *iMUSE*, singer-composer and synthesist Thomas Dolby was taking a different approach to addressing the same technological issues in the form of his *Audio Virtual Reality Engine* (*AVRe*). According to Dolby, "In a computer game, the musical changes have to happen *instantaneously*—in real time, without unpleasant glitches or overlaps or gaps [italics in the original]" (2016, 162). Ultimately, the video game music system that Dolby and collaborator Steve Ellison developed (*AVRe*) reacted to the moves that the gamer made, as opposed to maintaining a static cue in, say, a particular room, as was the case with some of the games of the time. In this respect, not only did *AVRe* match visuals, situations, and music but it also had this interactive attribute that helped to bring the programming of music without video games closer to artificial intelligence, as the game and its music became more intertwined with the gamer's choices, speed of motion, and the potential situations in which their character might find themselves.

The entire issue of fully integrating music and game progression and gamer choices is still being pursued by programmers and composers in 2020. In fact, between 2016 and 2020, several articles appeared in magazines and online about new developments that were either being promised or made with regard to the development of fully real-time artificial intelligence–composed video game music systems that would be able to accomplish what early 1990s' technology such as *iMUSE* and *AVRe* began. These include *Melodrive* and *WolframTones*, both of which can be used to provide music that is composed essentially in real time and changes based on the progression of the game.

See Also: Artificial Intelligence; Computer-Composed Music; Generative Music; Video Games

Further Reading

Dolby, Thomas. 2016. *The Speed of Sound: Breaking the Barrier between Music and Technology*. New York: Flatiron Books.

France-Presse, Agence. 2016. "First Recording of Computer-Generated Music—Created by Alan Turing—Restored." *Guardian*, September 26. Accessed July 22, 2020. https://www.theguardian.com/science/2016/sep/26/first-recording-computer-generated-music-created-alan-turing-restored-enigma-code.

Mackey, Bob. 2016. "Day of the Tentacle Composer Peter McConnell on Communicating Cartooniness." *USGamer*, March 8. Accessed January 18, 2021. https://www.usgamer.net/articles/day-of-the-tentacle-composer-peter-mcconnell-on-communicating-cartooniness.

Morgan, Robert P. 1972. Liner notes for *Edgard Varèse: Offrandes, Intégrales, Octandre, Ecuatorial*. Reissued on CD as Elektra/Nonesuch 9 71269–2.

Payne, Christine. 2019. "MuseNet." *OpenAI*, April 25. Accessed August 20, 2020. https://openai.com/blog/musenet/.

Perone, James E. 1988. *Pluralistic Strategies in Musical Analysis: A Study of Selected Works of William Albright*. Ph.D. dissertation. Buffalo, NY: State University of New York at Buffalo.

Sweet, Michael. 2015. *Writing Interactive Music for Video Games*. New York: Addison-Wesley.

Desktop and Laptop Computers

The first usage of computers in composing and generating music involved large mainframe computers. As a result, the connections of music and computers generally were limited to musicians with connections to major research universities or consortiums. Home computers were introduced in the early 1970s; however, home computers were not mainstream until the 1980s. Beginning in the 1980s, home computer applications for notation, composition, sound synthesis, and so on brought computer-based music technology to a far wider number of users and consumers than what had been the case in the past. These applications continue to evolve through the present.

Alex Wiltshire, the author of *Home Computers: 100 Icons That Defined a Digital Generation* (Wiltshire 2020), told CNN in a telephone interview that early in the introduction of home computers, designers were primarily concerned that "not to be scary. . .[a]nything they could do to say 'Hey I could be in your house, I'm not obtrusive and I'm not going to take over'" was at the forefront of design aesthetics (Holland 2020). Apparently, people were not scared by computers, as they became quite commonplace in American households, especially as functionality increased and price decreased. By 1980s, Apple, IBM, Tandy, several long-gone kit computer companies, Coleco, Commodore, Atari, and others fought it out for domination of the home computer market. Although desktop computers have a strong connection with music technology, it was the coming of the laptop computer that has probably been even more important, because unlike the desktop computer, laptops were not tied to a single location.

Apple dominated the world of computer-based music applications in the 1980s and still leads today. The Apple II computer and the Apple IIe boasted more music-related applications than similar early home computers by Atari, Commodore, Coleco, Sinclair, and so on. Even the longtime mainframe computer giant IBM, with its IBM PC, did not seem to capture the attention of music-related programmers or users to the extent that Apple's machines did. Apple's focus on graphics and music applications was enhanced even further with the rollout of the Apple IIGS in 1986. Apple's evolving Macintosh computers continued to dominate the music world. Anecdotally, if one attends a music-related academic conference even into the third decade of the 21st century, one will see Apple laptops and tablets, in the form of iPads, far outnumbering any other laptop or tablet platform/operating system.

As examples of the kinds of software that helped to establish Apple as the home computer of choice for musicians, Will Harvey's 1986 music composition program

Music Construction Set was designed for the Apple II, and Activision's 1986 program *The Music Studio* was designed to use the expanded and enhanced sound capabilities of the Apple IIGS. When the Macintosh computer became Apple's primary desktop computer offering, it quickly became the computer of choice for MIDI applications, including music notation and sequencing software.

In the late 1990s and early 2000s, home computers—desktop and laptop—became famously associated with another application: ripping CDs. This process involved taking the tracks of a commercial compact disc, which were CDA files, and using software that would convert the files (the CD's tracks) into mp3 files. The mp3 files were not only significantly smaller in size—with little loss of audio quality—but they could also be shared over the internet. The smaller size of mp3 files enabled users to put multiple commercial CDs' worth of music on a single CD-RW. The controversy of the practice of ripping, however, came from the widespread sharing of mp3 files. This sharing meant that artists, songwriters, and record companies were receiving far less compensation from royalties and unit sales than might otherwise would have been the case.

The ripping of CDs and the damage to the recording industry increased with the introduction of Napster and other peer-to-peer filesharing sites and software. Although Napster and peer-to-peer filesharing are covered in another entry in this volume, it should be noted in our consideration of desktop and laptop computers that these were the machines that were at the forefront of the controversy of the use of peer-to-peer sites. For the most part, the widespread illegal use of these sites for sharing and bootlegging did not last long, as the sites were either shut down or brought into compliance with copyright laws to become streaming audio services shortly after their turn-of-the-millennium notoriety.

Speed, processing power, and storage capabilities were all issues with regard to some music applications with early laptops. Software that was functional on desktop machines might not work on the laptops. The high cost of laptop computers compared with desktop computers was also a significant issue. More recently, laptops have overcome these challenges, and music notation, MIDI sequencing, audio and video editing, music instructional, home studio, and even professional recording studio software can run on many of the laptops that are available. Although some desktop computers are more robust than laptops, the price differential between machines that are limited to a fixed location and those that can easily be carried from one location to another has disappeared and, in some cases, has been reversed.

See Also: Instructional Software; MP3s; Music Notation Software; Peer-to-Peer File Sharing

Further Reading

Holland, Oscar. 2020. "Designing the World's First Home Computers." *CNN*, May 3. Accessed May 4, 2020. https://www.cnn.com/style/article/home-computers-design-history/index.html.

King, Brad. 2002. "The Day the Napster Died." *Wired*, May 15. Accessed January 18, 2021. https://www.wired.com/2002/05/the-day-the-napster-died/.

Wiltshire, Alex. 2020. *Home Computers: 100 Icons That Defined a Digital Generation*. Cambridge, MA: The MIT Press.

Digital Audio Tape (DAT)

Digital Audio Tape, or DAT (sometimes redundantly called DAT tape), was for a short time an important technology, especially early in the digital recording age. Sony Corporation developed this technology, which was introduced in 1987. Although DAT decks were available at the consumer level, the technology was most closely associated with professional and live recording, either in the studio or in concert venues. For example, DAT recorders were used by some educational institutions to make digital recordings of music students' degree recitals, concerts by university ensembles, guest artist and faculty recitals, and so on. In performing this kind of function, digital audio tape and its associated recording machines took their place in the succession that had included reel-to-reel machines and high-frequency response professional cassette decks (e.g., the Nakamichi cassette decks that were especially highly regarded during the 1970s and 1980s).

It is important to note that DAT players and recorders functioned in a fundamentally different way than the compact cassettes that they resembled. Digital audio tapes actually functioned more like the videotapes of the era than cassettes, in that the tapes played in only one direction. Another important distinction between DAT technology and cassettes is that the digital information was impervious to the wow and flutter that is an inherent part of electromagnetic tape.

Although the information stored on DATs is indeed digital and had several advantages over electromagnetic tape, the physical medium itself shares a significant disadvantage as a storage medium with other magnetic tape media, such as cassettes, reel-to-reel tapes, and 8-track tapes: the physical tape degrades over time. However, arguably, because the information on the tape is digital, it is easier to make a nearly identical clone of the original tape, unlike the process of copying an analog cassette, reel-to-reel tape, and so on, in which the cloning process is not possible. So, while preserving the physical DAT indefinitely might not be possible, theoretically, preserving the digital information intact is.

The entire issue of the cloning of digital information created controversy during the period in which the tapes were most heavily used. The recording industry was concerned with DATs because copyrighted material could potentially be more easily cloned and bootlegged than had been the case with analog material. A similar controversy had arisen in the late 1970s and early 1980s when new types of Dolby noise reduction and new compact cassette tape formulations (e.g., chromium dioxide and so-called metal tapes) had noticeably improved frequency response and lowered noise levels on cassettes. In fact, the practice of pirating music using cassettes was the basis of the 1980 debut single from the new wave rock group Bow Wow Wow: "C·30 C·60 C·90 Go!"

The fact that digital audio tape could go further than high-quality analog cassettes in actually cloning music resulted in more intense attempts to control the technology by the recording industry. Even compared with the other consumer digital recording and playback technology of the late 1980s, the compact disc (CD) and recordable compact disc (CD-R), digital tape had at least one significant advantage that was feared would allow the technology to be readily used for music piracy: one could record up to 2 hours of music on a DAT, while CDs could hold significantly less at the same level of audio quality. Although concerns about

DATs were raised in countries around the world, as Clinton Heylin wrote in the book *Bootleg: The Secret History of the Other Recording Industry*, "it was in America that the record industry was at its most belligerent. The RIAA [Recording Industry Association of America] made it clear that any manufacturer who imported DAT machines, available in the Far East in the early months of 1987, into the US would be the subject of a lawsuit on the grounds of 'copyright infringement'" (Heylin 1994, 244). Several attempts at developing anti-copying code for DAT recorders were tried but ultimately failed.

In the final analysis, the consumer market for DATs and DAT recorders and players was insignificant compared with the analog cassette tape market, which still had not died out at the time of the DAT controversies. In fact, Sony Corporation, which had first introduced DATs and DAT recorders and players, discontinued production of its DAT hardware at the end of 2005. To put the nature of the DAT within the consumer market into perspective, at the time of this writing, over a decade and a half after Sony pulled out of the DAT market it had initiated, the company still manufactures boomboxes and other players and recorders for both CDs and cassettes, and Sony, Maxell, TDK, and other manufactures still produce blank cassettes.

In the areas in which digital audio tapes did make a stronger impact than in the home consumer market, including radio and live and studio recording, other electronic storage devices, including computer and external hard drives, SD cards, and so on, offered more storage space, easier editing, and less chance of degradation of the physical media.

See Also: Cassette Tapes; Digital Recording

Further Reading

Coleman, Mark. 2004. *Playback: From the Victrola to MP3, 100 Years of Music, Machines, and Money.* Cambridge, MA: Da Capo Press.

Fine, Thomas. 2008. "The Dawn of Commercial Digital Recording." *ARSC Journal* 39, no. 1 (Spring). Accessed June 12, 2020. http://www.aes.org/aeshc/pdf/fine_dawn-of-digital.pdf.

Heylin, Clinton. 1994. *Bootleg: The Secret History of the Other Recording Industry.* New York: St. Martin's.

Digital Recording

As early as 1971, digital recordings were made and commercially released in Japan by Denon using its digital pulse code modulation technology (Fine 2008). Other companies in Europe and in the United States were also developing digital technology throughout the 1970s, including Soundstream and 3M in the U.S. and Decca Records in Europe. The basic premise of these early attempts at digital recording were to improve upon the recording process itself. In other words, the end result, say, an album, would be released in analog form as a vinyl record or on cassette tape. Consumer digital playback was not part of the focus of these companies in the 1970s.

Although there were earlier digital recordings, one of the first major digital recordings by a big-name musician was avant-garde jazz saxophonist Archie

Shepp's 1977 Denon album *On Green Dolphin Street*. The first digitally recorded classical album recorded and released in the United States was the 1978 Telarc release of a recording of wind band works of Gustav Holst and George Frederick Handel's *Music for the Royal Fireworks* by the Eastman Wind Ensemble under the baton of Frederick Fennell. The Telarc recording, in particular, became famous as a demonstration of the wide dynamic range that digital recording made possible. The first digitally recorded album in a popular music genre was Ry Cooder's 1979 album *Bop 'Til You Drop*.

Some of the more notable recordings of this early part of the digital recording age were Telarc's classical releases. In particular, 19th-century and early 20th-century orchestral and wind band compositions allowed the clarity of sound and the wide dynamic range possible with digital recording to shine through arguably better than any other type of music. Works such as Igor Stravinsky's *Firebird*, Maurice Ravel's orchestration of Modest Mussorgsky's *Pictures at an Exhibition*, and Nicolai Rimsky-Korsakov's *A Night on Bald Mountain* were among the compositions recorded on Telarc by leading ensembles such as the Cleveland Orchestra and the Atlanta Symphony and Chorus.

Whether directly stated or not, it is interesting to note that many of the best-remembered early digitally recorded albums, certainly those in the jazz and classical genres, were based on the premise of capturing a live performance with as much aural accuracy as possible. The importance of the live ambience is confirmed by Telarc's use of concert venues, such as Cleveland's Severance Hall, for its recordings of the Eastman Wind Ensemble and the Cleveland Orchestra, as opposed to, say, large recording studios.

Although classical and jazz recordings of what were essentially live performances without an audience did not require extensive multi-track recording, many popular music genres did, what with the overdubs, separation of vocal tracks from instrumental tracks and the like. However, multi-track digital recording was quick to develop, and even at the time of the first fully digital recording of a pop album, Ry Cooder's 1979 *Bop 'Til You Drop*, 32-track digital multi-track recording and the types of mixing techniques previously associated with multi-track tape recording were available.

As was the case with the early compact discs that were released in the early 1980s, some music fans argued that digital recordings lacked the warmth of analog recordings. During the 1980s and beyond, various combinations of analog and digital recording, editing, and playback have been used in an attempt to capture the best of both worlds, and some digital studio programs can emulate analog recording. In any case, whether the actual recording is analog or digital, the editing process is considerably simplified through digital editing. Music that has been digitally recorded or converted from analog to digital format can be manipulated much more quickly, too, than by working with electromagnetic tape.

It is important to note that in addition to digital recordings made by companies such as Decca in Europe, Telarc in the United States, and Denon in Japan for commercial release in the late 1970s and early 1980s, there was another aspect of digital recording technology that was generating interest among musicians: digital synthesizers and the new sampling capabilities that they possessed. As detailed in this volume's entry, Digital Synthesizers, the Fairlight CMI (computer musical instrument) synthesizer, which was introduced at the end of the 1970s, was the first digital sampling keyboard. The user of the Fairlight CMI could digitally record sound and digitally edit it on the machine's computer-like display screen. A large number of other digital sampling synthesizers appeared on the market from the 1980s through the present and have continued to offer digital recording capabilities.

Perhaps one of the most important developments in digital audio recording came with the release of small portable digital recorders from Zoom and other companies. Today, handheld 4-track recorders with built-in stereo microphones can be purchased between $200 and $300, making high-quality digital field/on-location audio recordings possible for the general public. The audio quality of these machines is such that they are also used by reporters for national news outlets. The typical handheld multi-track recorder stores audio files on SD cards. The user's computer can read these cards, and the files can be loaded on the computer for mixing, editing, storage, and so on.

Although these are covered in detail in entries on Virtual Studios, iPads and Other Tablets, and Smartphones, a range of digital recording, editing, and studio applications ranging from free apps to expensive, professional-grade programs are available for the professional and home user. Digital audio recording can be done on a device, such as a smartphone, and edited on the phone and uploaded to a computer for further editing, the integration of studio effects, and so on, in a purchased program or even a free program such as *Audacity*. Or, a musician might use an application such as *GarageBand*—which is available in iPhone, iPad, and computer versions—to digitally record, edit, add preprogrammed drum tracks, and so on. In this way, high-quality digital audio recording—monophonic, stereophonic, or multi-track—is available to the masses to a degree and with a portability that was never the case with analog recording.

See Also: Compact Discs (CDs); Digital Audio Tape (DAT); Digital Synthesizers; iPads and Other Tablets; Smartphones

Further Reading

Fine, Thomas. 2008. "The Dawn of Commercial Digital Recording." *ARSC Journal* 39, no. 1 (Spring). Accessed June 12, 2020. http://www.aes.org/aeshc/pdf/fine_dawn-of-digital.pdf.

Huber, David M. 2018. *Modern Recording Techniques*. 9th ed. New York: Routledge.

Lendino, Jamie. 2019. "Apple *GarageBand* (for Mac) Review." *PC Magazine*, September 26. Accessed October 29, 2020. https://www.pcmag.com/reviews/apple-garageband-for-mac.

White, Marla. 2006. "Guerrilla Home Recording." *Music Educators Journal* 93, no. 1 (September): 23–24.

Digital Synthesizers

For the general public, the digital age in sound synthesis began in 1979 with the release of the Casio VL-1, also known as the VL-TONE. This small monophonic combination of synthesizer and calculator might seem like a quaint toy compared with the elaborate and expensive analog synthesizers that preceded it, not to mention the considerably more expensive and powerful digital synthesizers that joined it in the late 1970s and from the 1980s through the present.

The general public's introduction to digital sound synthesis certainly came in a small package. The Casio VL-1 sported a 2–1/2-octave miniature keyboard, although the instrument was just over 21-cm long. Although this monophonic instrument was inexpensive enough that it could be played at home, and seemed almost like an electronic toy, it found its way into some of the synth-pop and new wave rock recordings of the early 1980s. For example, the liner notes on the Human League's well-known 1981 album, *Dare* (which includes the group's most best-remembered hit, "Don't You Want Me"), list this small instrument as one of several synthesizers used on the album (Rushent and the Human League 1981).

Considerably more sophisticated and powerful than Casio's introductory offering to digital synthesizers, the Fairlight CMI (CMI stood for computer musical instrument) retained some of the analog synthesizer–like abilities to modify wave forms and create unique tone colors, not necessarily tied to the emulation or imitation of any acoustic instruments. The Fairlight CMI featured additive synthesis and used a screen to display a wave form, which the user could modify similarly to how one might draw on a present-day iPad or other tablet screen. Although out of the reach of the general public, this Australian-designed instrument became popular with recording artists and offered sampling and sequencing capabilities, in addition to the on-screen manipulation of tone colors.

The development of digital sound synthesis owed a debt of gratitude to work that composer and professor John Chowning did at Stanford University in the 1970s. Chowning developed synthesis based on frequency modulation (FM). Although Chowning attempted to interest major synthesizer companies in his technology, it was the Yamaha Corporation—well versed in acoustic pianos but not a major player in the synthesizer world at the time—that licensed the technology. Yamaha in turn licensed Chowning's technology to New England Digital Corporation, which used Chowning's work in the design of the first generation of the Synclavier of 1977. The Synclavier II, introduced in 1980, was a digital synthesizer.

Aside from the somewhat toy-like Casio VL-1, many of the digital synthesizers at the start of the 1980s were quite expensive—in many cases, far more powerful than the analog synthesizers that preceded them, but quite expensive. For example, the Synclavier II could at various times be purchased for virtually anywhere between approximately $25,000 and $200,000. However, the price of powerful digital synthesizers soon dropped dramatically, which brought digital synthesis up in functionality from instruments such as the Casio VL-1 and made technology that arguably could go head-to-head with a Synclavier accessible to more working musicians, studios, and educational institutions. For example, the Yamaha DX7, introduced in 1983, had a list price of approximately $2,000. The DX7 used the frequency modulation principles developed by John Chowning, just as the Synclavier had done, but unlike the 1977 Synclavier I, the DX7 had a fully functional keyboard. This feature was part of later generations of the Synclavier; however, the relatively high price of the various Synclavier models meant that university music departments—and especially computer music programs—continued to be the principal purchasers of those machines while performing musicians and more casual synthesizer users favored the lower-priced machines.

Although one could purchase an early Yamaha DX7 for a fraction of the price of an early Synclavier or a Fairlight CMI, after the release of the DX7, Yamaha, Casio, Korg, and other companies released digital synthesizers that significantly undercut the cost of the DX7. Although some of the machines released in the mid- to late 1980s and beyond lacked some of the features (and sometimes had smaller keyboards) of the DX7, they truly brought digital sound synthesis to working musicians. The use of MIDI to connect synthesizers/keyboards to computers, too, meant that not all the synthesized voices, storage, and so forth needed to be contained in the keyboard unit itself. Increasingly, the keyboard could be used as a trigger for digital synthesis software on the computer.

It seems that one of the main considerations of the manufacturers of digital synthesizers in the wake of the success of Yamaha's DX7 was producing an instrument that could be played pretty much out of the box. In fact, the digital synthesizers of the 1980s included numerous preset tone colors and voices and left little to no room for the user to engage in the same sort of construction of new individualistic tone color settings that was a regular and necessary part of sound synthesis during the analog days of Robert Moog's and other producers' synthesizers of the late 1960s and early 1970s.

One of the complaints about digital synthesizers, however, is that some have relied too heavily on emulating acoustic instruments and presets that had little to do with the unique, clearly electronic tone colors that had defined analog sound synthesis in the late 1960s and the 1970s. In the 21st century, a number of digital synthesizers and software companies offer patches that emulate the classic sounds of Moog and other early analog synthesizers so that those classic electronic tone colors can once again be part of the palette of possibilities for the synthesizer player.

See Also: MIDI; Sampling; Yamaha DX7 Synthesizer

Further Reading

Nelson, Andrew J. 2015. *The Sound of Innovation: Stanford and the Computer Music Revolution*. Cambridge, MA: MIT Press.

Rushent, Martin, and the Human League, producers. 1981. Liner notes for *Dare* by the Human League, LP, A&M Records SP-4892.

Shepard, Brian K. 2013. *Refining Sound: A Practical Guide to Synthesis and Synthesizers*. New York: Oxford University Press.

Distortion

The *Merriam-Webster Dictionary* defines "distortion" as "the act of twisting or altering something out its true, natural, or original state" (Merriam-Webster

Casio Computer Co., Ltd.

A major conglomerate founded just after the end of World War II, Casio today might be better recognized by consumers as a manufacturer of digital watches, cash registers, digital cameras, and calculators than for its connections to music; such is the wide diversity of electronic products that the company has produced throughout the decades.

Beginning at the end of the 1970s, Casio produced small synthesizers that were priced notably lower than many of the models produced by Korg, Yamaha, Roland, and other Japanese synthesizer manufacturers. Some of the company's products were polyphonic; however, even monophonic, diminutive instruments such as the VL-1 synthesizer, whose 2–1/2-octave keyboard is just a tad over 21-cm long, were not solely used by individuals wanting a relatively inexpensive synthesizer to play with at home; even this instrument found its way onto some of the early and highly popular synth-pop and new wave recordings of the early 1980s. In fact, the liner notes on the Human League's well-known 1981 album, *Dare* (which includes the group's most popular hit, "Don't You Want Me"), list this small instrument as one of several synthesizers used (Rushent and the Human League 1981). In fact, despite its humble size and limited capabilities, Casio's VL-1 is generally acknowledged as the first commercially available digital synthesizer. The company's synthesizers with miniature and with full-sized keys not only provided an affordable alternative to other brands during the synth-pop days of the 1980s; Casio is still making affordable portable keyboards today. At the time of this writing, several online vendors are selling the Casiotone CTS-200, 61-key, portable MIDI-compatible synthesizer on sale for approximately $120, although even the list price for the CTS-200 is under $200. Even Casio's current full-sized keyboards tend toward portability and tend to be priced lower than the full-sized keyboards of many of its competitors.

Online, 2021). From a technical standpoint, distortion in music occurs when the natural sound is changed. This distortion can occur in a variety of ways, such as when a saxophone player gets saliva buildup in the mouthpiece, which creates a buzzing sound not part of the instrument's normal tone quality. Here, we are concerned with the distortion that occurs in electronic musical instruments, generally by design but sometimes by accident.

Tonal distortion was heard in electric blues music as early as the 1940s, when sometimes guitar amplifiers would be driven to the point that the vacuum tubes could not handle the input cleanly and naturally amplified sound of the electric guitar became distorted. Basically, what occurs is that some of the overtones become more prominent, which gives the tone an edgier sound. More famously, however, the unintentional distortion of electric guitar tone played a significant role in the sound of what some writers consider to be the first rock and roll record: Jackie Brenston and His Delta Cats' (featuring Ike Turner) 1951 recording of "Rocket 88." In this case, apparently the guitar amplifier was damaged on the way to the studio. It was used nevertheless, and the tear in the speaker resulted in the buzzy distortion that is heard throughout the song in the guitar riff. Similarly, the distorted electric guitar sound on the Kinks' 1964 recording of "You Really Got Me" was the result of a slash in guitarist Dave Davies's amplifier.

The guitar tone color heard on recordings such as "Rocket 88" and "You Really Got Me" aligned perfectly with the sense of rebellion that marked the youth subculture that was developing after World War II; sociologists would later term the phenomenon of the separation in values between teens of the 1950s and 1960s and their parents as the "generation gap." In the wake of these acoustic examples of guitar distortion (the distortion on these two iconic recordings was not the result of electronics but alterations of the vibrating membrane: the speaker), manufacturers of guitar amplifiers and effects pedals began offering a variety of ways in which guitarists could deliberately add distortion to their palette of tone colors. On guitar amplifiers with built-in distortion, there are typically two volume controls: one for overdrive or gain and another to control the actual volume that the amplifier produces. Turning up the overdrive/gain will send more electrical current to the main sound-producing circuit than it can handle, which will distort the sound. Because the actual audible volume is controlled by a separate circuit, distortion can be produced even at low output volumes.

Guitar effects pedals designed for overdriven sound—essentially, lower-level distortion—work similarly. The guitar is plugged into a powered pedal (generally the power to the pedal is provided by plugging the pedal in, or by means of a 9-volt battery). The pedal provides more voltage to the amplifier than it would normally receive from the electric guitar alone and thus will distort the sound. Distortion pedals work similarly although, compared with overdrive pedals, create more distorted saturation of the sound.

As the 1960s moved into the 1970s, distortion was used by blues-based rock guitarists such as Jimi Hendrix, Eric Clapton, Carlos Santana, and others, becoming part of each guitarist's personal tonal stamp. Heavily overdriven and distorted guitar tone then became a standard part of the sound of heavy metal rock and punk rock. It continues to be used in creative ways to give extra grittiness to the electric guitar tone, to produce a more lyrical sustained sound, and to give electric guitar players one more tool with which they can create a personal and distinctive tone that defines their "sound." In fact, the ongoing nature of the popularity of distortion—one might even go so far as to say the fact that it is so fully integrated into the expected timbre of rock music—is evidenced by the fact that at the time of this writing, distortion and related effects pedals are being made by numerous manufacturers.

So, in the final analysis, distortion is an electronic effect that originally was something of a mistake or accident that, for cultural reasons, became an accepted and a celebrated part of the sonic tools that guitarists routinely include in their bags of tricks. The fact that various degrees of distortion have been used so pervasively in popular music now for over a half-century shows how the rock and hip-hop eras developed an aesthetic sense that differed fundamentally from that of mainstream popular music before the middle of the 20th century. Although distortion is primarily associated with the electric guitar, it should be noted that electronic keyboard and synthesizer players have also used the effect during this same period.

See Also: Effects Pedals; Electric Guitars

Further Reading
Hunter, Dave. 2013. *Guitar Effects Pedals: The Practical Handbook*. San Francisco, CA: Backbeat Books.

Maloof, Rich. 2004. *Jim Marshall—The Father of Loud: The Story of the Man behind the World's Most Famous Guitar Amplifiers*. San Francisco, CA: Backbeat Books.

Merriam-Webster Online. 2021. "Distortion." Accessed November 22, 2021. https://www.merriam-webster.com/dictionary/distortion.

Teagle, John, and John Sprung. 1995. *Fender Amps: The First Fifty Years*. Milwaukee, WI: Hal Leonard.

DJ-ing

Before the hip-hop age, the term "disc jockey" (DJ) probably conjured up images of someone in a radio studio selecting discs, cuing them up on the turntable, perhaps reading some copy from a local advertiser, providing some sort of snappy introduction to the upcoming record, and lowering the stylus on the turntable's arm. The DJ-ing to which we refer here goes well beyond that tradition and is largely a product of and defining component of hip-hop and disco music.

Although commercial radio—with music—dates back to the 1920s, the music that was broadcast tended to be live and, if not live, prerecorded in the radio studio for later broadcast. To put it another way, commercial radio by and large did not broadcast the latest hit singles as would become the case in the 1940s and beyond. The radio DJ was born when radio moved from employing program hosts to employing host-like individuals who spun the records.

DJs in some radio markets and some radio stations, however, did much more than play records, provide advertisements, provide snappy banter, and introduce recordings. Some were able to select the music that they played. Although there is not sufficient space to detail the practice, DJ's power to select discs led to the practice known as payola, in which record company representatives would bribe DJs to play certain records on their radio programs. Although this practice had occurred earlier, it became widely known to the general public with the payola crackdown of the late 1950s. Several prominent rock and roll DJs, including Alan Freed (who is credited with the taking the phrase "rock and roll," earlier used as a euphemism for sex in some blues and R&B songs, and using it to describe the new combination of R&B and country music that has continued to bear the name), were charged with income tax evasion; the payments they received from record companies was unreported (under the table) income.

The role of the radio DJ changed somewhat beginning in the late 1960s with the increasing popularity of FM radio, which lent itself to album-oriented rock programming. Today, some radio stations still employ DJs; however, others have increasingly focused on syndicated DJ-like programming from internationally known hosts such as John Tesh and others.

Outside of the radio industry, DJs provide music and the requisite snappy banter, for wedding receptions and other similar events as well as for karaoke. During the disco era of the 1970s, DJs in dance clubs played a particularly important role as they used multiple turntables to provide smooth, rhythmically seamless transitions between records. The techniques that disco DJs used—some of which will

Beginning in the 1970s, DJ techniques involving multiple turntables played major roles in disco music. They continue to be at the core of hip-hop music. (iStockPhoto.com)

be discussed later—became part of the hip-hop DJs' repertoire of techniques as the hip-hop genre began to take hold roughly during the disco era.

Although other individuals might have been doing similar things elsewhere, what came to be the essential DJ side of hip-hop music has been widely credited to DJ Kool Herc. Clive Campbell, better known by his professional name, DJ Kool Herc, initially used techniques of transitioning between records (e.g., vinyl discs) and isolating particular instrumental sections of the discs to create long grooves or to lengthen introductions or run-out sections that were in use in discos at the same time, this taking place in the early 1970s. In discos, DJs used these techniques, as well as making changes to the tempo of songs, in order to keep dancers moving and to maintain a desired energy level on the dance floor. By contrast, as DJ Kool Herc put it in his introduction to Jeff Chang's *Can't Stop Won't Stop: A History of the Hip-Hop Culture*, he came from "'the people's choice,' from the street" and that success only came to Kool Herc and other early hip-hop DJs if people enjoyed what they were doing (Chang 2005, xi). The difference between disco DJs and hip-hop DJs such as Kool Herc in the South Bronx and elsewhere was not entirely a distinction between a club atmosphere and the street. The degree to which hip-hop DJs had to manipulate discs, isolate break sections in songs, create loops using turntables, and so on typically was more complicated than what was done in discos. In part, this was because of the variety of the material that hip-hop DJs used. Instead of the recordings that were produced for the disco market, which tended to be roughly the same tempo and, more frequently than dancers might have realized, in the same key, DJ Kool Herc and other successful early hip-hop DJs drew from R&B, disco, rock, and funk recordings. Some of the popular recordings that provided source material included James Brown's "Give It Up, Turn It Loose" and the Incredible Bongo Band's 1973 cover of the instrumental piece "Apache," to name just two examples. The main criterion for selecting material was the beat and the groove feeling of the instrumental breaks. This material provided the basis for what would be essentially a new piece created by the DJ and the rappers with whom they collaborated.

Today, hip-hop, karaoke, and other DJs still have traditional vinyl record-based turntable technology at their disposal, with several large online audio vendors offering specialized DJ turntable setups. In addition, DJ-specific stereo headphones are available from several manufacturers. These allow the DJ to listen to two separate inputs simultaneously, one in the right ear and one in the left. And the 21st-century DJ has new computer-based tools at their disposal, including Sonic Foundry's *Acid* (which was purchased by Sony Digital Pictures in 2003), Ableton's *Ableton Live*, and other software that can be used to record loops easily and to use drag-and-drop technology to fashion what are essentially new compositions created out of preexisting musical material.

See Also: Headphones; Karaoke; Radio; Sampling

Further Reading

Chang, Jeff. 2005. *Can't Stop Won't Stop: A History of the Hip-Hop Culture*. New York: Picador.

DJ Booma. 2017. *How to Be a DJ in 10 Easy Lessons: Learn to Spin, Scratch and Produce Your Own Mixes*. London: QED Publishing.

Duke, Dan. 2015. "Jimmy Ray Dunn Starts His Own Radio Station—on the Internet." *Virginian-Pilot*, November 5. Accessed January 18, 2021. https://www.pilotonline.com/entertainment/music/article_6cef89c1-e04b-56d0-b12a-3936e362bdeb.html.

George, Nelson. 2012. "Hip-Hop's Founding Fathers Speak the Truth." In Murray Forman and Mark A. Neal, eds., *That's the Joint! The Hip-Hop Studies Reader*, 2nd ed., 44–55. New York: Routledge.

Schloss, Joseph. 2012. "Sampling Ethics." In Murray Forman and Mark A. Neal, eds., *That's the Joint! The Hip-Hop Studies Reader*, 2nd ed., 610–630. New York: Routledge.

Dolby Noise Reduction

Founded in the U.K. by Ray Dolby, Dolby Laboratories developed the first noise reduction system, Dolby A, for cassette tapes in 1965, followed by its second generation—Dolby B—three years later. Noise reduction was necessary for cassettes in part because of the slow speed of the tape, 1–7/8 ips compared with 3–3/4 ips for 8-track and even faster speeds that were used on reel-to-reel tape decks. The slow speed of the tape in cassettes led to more noise and a narrower frequency range than in other formats. Dolby's system basically compressed the dynamic range of the sound during recording and then boosted it during playback. This helped to eliminate a noticeable amount of the tape hiss that was inherent in recording at low speeds on magnetic tape. Eventually, Dolby's company moved its headquarters to the United States and became a major player in the world of film audio systems, including noise reduction and surround sound.

Dolby noise reduction became a standard part of consumer cassette tapes throughout the 1970s. In order for the system to work, however, the listener's cassette deck or portable cassette player had to include a Dolby noise reduction circuit. This is because cassettes needed to be encoded with the noise reduction and then decoded for playback. Basically, the system boosted frequencies that were not part of the normal noise spectrum of cassette background noise. On playback, those frequencies were reduced, with the net effect being a reduction in the noise. With the advent of Dolby B at the end of the 1960s, some machines included a switch so that the listener could select no noise reduction, Dolby A noise reduction, or Dolby B noise reduction. In considering the noise reduction system, it is important to note that it only worked properly when the correct setting was selected on the playback equipment.

Playing a commercially produced cassette that was encoded using Dolby A on a machine on which Dolby was not selected, for example, would produce an overly bright sound, because the high frequencies would be artificially too high. As a result, Dolby noise reduction could be something of a mixed blessing: on cassette decks and players with Dolby noise reduction, the technology noticeably improved the sound of cassettes, but when one used decks or all-in-one players that did not have the Dolby circuitry, the resulting sound might not necessarily be heard as an improvement. That being said, it should be noted that some of the all-in-one cassette players of the 1960s were not all that far removed in audio quality from small transistor radios, which also happened to have a less-than-full-range frequency response.

Although Dolby noise reduction is most closely associated with compact cassette tapes, the system found some use in other venues as well. In the early 1970s, for example, Dolby B noise reduction started to be used by some FM radio stations to further enhance their sound. As a *Popular Science* article by John Free suggests, however, this did not catch on. Free wrote in 1974, "In November 1972, this column listed 17 such stations . . . Now, according to Dolby Labs in New York, there are only six stations" (Free 1974, 50).

Much more successful than its foray into radio was Dolby's venture into the movie industry. Beginning in 1971, the company's noise reduction technology was used in films such as *A Clockwork Orange*, the 1976 remake of *A Star Is Born*, and numerous others. Dolby Laboratories also developed a digital sound system for films as well as a surround-sound system.

Dolby C noise reduction appeared in the 1980s and represented another enhancement of the technology. The combination of Dolby noise reduction and cassette tape formulations such as chrome and metal tape helped to make the audio quality of cassettes the rival of high-speed reel-to-reel tape and even—some would argue—the rival of digital formats. In fact, it could be argued that the combination of new tape formulations and the effectiveness of Dolby's new noise reduction system played significant roles in the bootlegging and sharing of music in the 1980s, particularly as boomboxes and cassette decks increasingly offered multiple players/recorders in a single unit, as these units allowed for easy copying of cassettes.

Throughout the history of music and video technology, there have been several well-known instances of competing technologies that battle it out until only one is left. For example, during the last decade of the 19th century and the early 20th century, the phonograph cylinder technology associated with Thomas Edison competed against the shellac flat records championed by Emile Berliner and favored by several companies. Eventually, the cylinders fell by the wayside. Similarly, the home video market was the scene of a battle between VHS video tape and Sony's proprietary Betamax tape; Sony lost that particular battle. In the case of noise reduction associated with cassette tapes, dbx noise reduction, invented by David E. Blackmer, competed against Dolby noise reduction during the early 1970s. Interestingly, dbx was also used for noise reduction on some vinyl albums of the era. Despite the incorporation of dbx circuits into some cassette units by leading manufacturers such as Panasonic and Sanyo, dbx did not catch on in the marketplace as Dolby had done.

At the time of this writing, several interview segments with Ray Dolby can be found on YouTube (see, e.g., https://www.youtube.com/watch?v=6_H6HmdLLCo, accessed January 14, 2021, and https://www.youtube.com/watch?v=CVP5uS3d0HU, accessed January 14, 2021). In these and other interviews with and talks by Dolby, the inventor of the noise reduction system that bears his name explains the history of noise reduction in audio applications and the particular issues that made a noise reduction system necessary for cassette tapes.

See Also: Cassette Tapes

Further Reading

Free, John R. 1974. "Look and Listen: News, Comment, and Opinion from the World of Home-Entertainment Electronics." *Popular Science* 204, no. 2 (February): 50.

Interview segments with Ray Dolby provide insight into his work and the noise reduction system that bears his name. See: https://www.youtube.com/watch?v=6_H6HmdLLCo, accessed January 14, 2021 and https://www.youtube.com/watch?v=CVP5uS3d0HU, accessed January 14, 2021.

Drum Machines

Just as the analog synthesizers that entered the music world in the 1960s had their precedents in the early electronic instruments of the 1920s (e.g., using oscillators similar to those used in the theremin) and even some experimental instruments of the 19th century (e.g., Elisha Gray's Electric Telegraph for Transmitting Musical Tones), the drum machines of the late 1970s and beyond had their precedents with other, lesser-known experiments of decades before. By the 1980s, drum machines became a regular part of various popular music genres, including hip-hop, new wave rock, and dance-oriented music.

In a brief *Smithsonian* magazine spotlight on predecessors of the electronic drum machines of the 1980s, Scheinman (2020) traces the drum machine back to Ismail Al-Jazari's water-powered mechanical bands of the 12th century. So, the concept of creating percussive sounds without a drummer has been around for nearly a millennium. Some of the large theater organs of the early 20th century included pitched percussion instruments, such as xylophone, along with snare drums, bass drums, and cymbals. Similarly, carousel organs routinely included mechanically operated percussion instruments.

The first fully electronic drum machine was developed by Leon Theremin in the early 1930s, after he had already invented and marketed the pitched electronic instrument that bears his name. Theremin's Rhythmicon was designed in collaboration with American composer Henry Cowell, who premiered the instrument. However, very few Rhythmicons were ever constructed, so the instrument remains something of a curiosity. Nonetheless, at the time of this writing, there are several YouTube videos available in which the Rhythmicon is demonstrated (see, e.g., https://www.youtube.com/watch?v=zyjOZPiW5dw, accessed January 14, 2021). The few musicians who have used this rare instrument could program elaborate rhythmic patterns, including those with complex cross rhythms that would be difficult, if not impossible, for a live human percussionist to perform. Compared with later electronic drum machines, however, the sounds produced by Theremin's Rhythmicon bear little resemblance to the sound of traditional acoustic percussion instruments. They clearly are electronic sounds produced by oscillators—similar to the theremin—however, of very short, percussion-like durations.

The 1949 Rhythmate of Harry Chamberlin used the sounds of a drummer recorded on magnetic tape to produce 14 different rhythmic patterns (Scheinman 2020). Like Theremin's Rhythmicon, the Rhythmate was not a huge commercial success. Chamberlin's device did, however, anticipate the use of tape loops that formed the basis for the inventor's Chamberlin, a keyboard-based pitched instrument, as well as for the better-known staple of 1960s' psychedelic and progressive rock, the Mellotron. In addition, the prepackaging of popular rhythm patterns anticipated the Band Box integrated into Thomas electronic organs of the 1950s. Thomas, Wurlitzer, and other manufacturers included percussion accompaniment patterns drawn from well-known dance styles, such as the rumba, tango, and waltz. Wurlitzer's Side Man was so successful that the American Federation of Musicians lobbied to have it outlawed for fear that it would put drummers out of work.

The 1960s found several companies manufacturing and marketing transistorized drum machines. Because of the small size of these devices, they found a natural home as add-ons to and as integrated components of the electronic organs of the era. Late in the decade and into the early 1970s, electronic drum machines found their way into popular music recordings. In the 1970s, programmable drum machines became more common; however, it would not be until the end of the decade that programmable analog drum machines would become more mainstream, both in terms of affordability and in terms of their usage in mainstream popular music.

One of the principal drum machines that became prominent in popular music was the Roland TR-808, which was released in 1980. Because it was an analog drum machine, the drum sounds in the TR-808 were programmed by manipulating electrical currents to simulate the sound of real drums. In addition to providing the percussion on Marvin Gaye's hit "Sexual Healing," Whitney Houston's "I Wanna Dance with Somebody," and Drake's "God's Plan," the Roland unit has continued to be used by musicians into the 21st century. As pointed out in a recent article in *Smithsonian* magazine, the irony of the popularity of the analog TR-808 is that Roland only produced approximately 12,000 of the units and only for a period of two years. By the mid-1980s, what cost approximately $1,200 a few years before could be purchased for about $100 (Abdurraqib 2020). As a result, what had been a commercial failure for Roland, one of the major names in 1980s' sound synthesis, became a staple for home recordists, hip-hop musicians, and others.

Consideration of Roland's BOSS DR-55, Dr. Rhythm, which was released in 1979, shows how early less expensive (than the more sophisticated TR-808 described previously) drum machines were designed, as well as some of the inherent challenges they posed. The analog DR-55 included bass drum, snare drum, hi-hat cymbal, and rim shot sounds. The user could program in 6 different beats based around a division of a measure into 16 parts (e.g., 4 beats of 4, 16th notes per measure) and two 12-step rhythms (4 beats with triple subdivision). The programming method, which was used by later Roland units and some other drum machines, was step by step. For example, for a quadruple-meter measure, the user would record all the events that were to occur on the first 16th note of the measure, then the second 16th note of the measure, and so on. The DR-55 can be activated with a foot switch and has a 1/4-inch output to an amplifier and can be interfaced to other synthesis devices. The degree of accent, volume, tone, and tempo all are controlled with knobs. One of the main drawbacks of this important drum machine, though, is the fact that it is battery operated, and when the batteries die, the user loses what they've programmed into the device.

The DR-55 was the first in a long series of relatively affordable drum machines that Roland released throughout the 1980s and beyond. One of the main advances during the 1980s that came to several drum machine models was the addition of rubber pads on which users could tap rhythms. Although the step-time method of entering patterns was still used and continued to be a useful method of entering events for percussion and other synthesized instruments in MIDI-based notation and composition programs, the tapping method allowed more intricate rhythms to be entered into the drum machine, including those that did not necessarily fit nicely in the four 16th notes per beat paradigm of devices such as Roland's DR-55.

Although several manufactures released numerous drum machine models during the 1980s, perhaps the most famous—because it played such a prominent role in the world of popular music of the decade—was the Linn LM-1 drum machine. Despite its prominence in recordings of the decade, the LM-1 was produced only between 1980 and 1983. Although it was more expensive than consumer-level drum machines, the LM-1 was popular among recording musicians, including the Human League (it can be heard on the group's best-known hit, "Don't You Want Me," and on other songs on the album *Dare*), Prince (who made extensive use of the Linn drum machine on demo recordings and commercially released studio recordings), Stevie Wonder, and others. Despite its presence on popular recordings of the 1980s, ironically the Linn LM-1 was produced in even smaller numbers than the Roland TR-808; only 525 LM-1 were ever made (McNamee 2009). The LM-1 was fully programmable, but its most significant feature was that it used samples of real drums. As a result, arguably it was one of, if not the, most realistic sounding drum machines of its time.

A close competitor for realism was the Oberheim DMX, which appeared a year after the Linn LM-1. Like its predecessor, the DMX's sound was built around samples of real percussion instruments. As David McNamee notes in a retrospective on these influential, albeit expensive, drum machines, the Oberheim DMX played a crucial role in the establishment of hip-hop drum machine sounds, particularly through the work of DMX Krew of Run DMC. In the genre of rock, McNamee notes the use of the DMX in New Order's new wave rock song "Blue Monday" (McNamee 2009). As was the case with Roland's move from the expensive TR-808 to more affordable consumer-level drum machines later in the 1980s, Linn and Oberheim released machines that were significantly less expensive and that sold in greater quantities than the LM-1 and DMX.

MIDI synthesizers started to include aspects of drum machines, including digital drum sounds and so forth, that could be put into a digital recording without having to use a stand-alone drum machine. However, the stand-alone machine is still alive and well today with a long list of companies associated with sound synthesis, sampling, and other electronic products for musicians, producing machines that range in price from under $100 to over $1,000. At the time of this writing, Korg, Behringer, Roland, Nord, Alesis, Singular Sound, Arturia, and others are among the brand names of drum machines that are sold by mass-market music retailers such as Sweetwater Sound and Guitar Center, as well as more generalized online retailers such as Amazon. Although the current crop of drum machines can be used for recording, just like the expensive machines of the early 1980s, many of today's drum machines are portable and adaptable enough that they are more suited to live performance than some of the drum machines of decades ago. It is interesting to note, too, that despite the fact that sound synthesis has been firmly in the digital age for decades, many of the drum machines of the 2020s use analog technology.

See Also: Digital Synthesizers; Organs

Further Reading

Abdurraqib, Hanif. 2020. "Tick, Tick, Boom: In the Whimsical World of Pop Music, Sometimes Technology Has More Impact after It's Obsolete." *Smithsonian* 51, no. 4 (July/August): 22–23.

McNamee, David. 2009. "Hey, What's That Sound: Linn LM-1 Drum Computer and the Oberheim DMX." *Guardian*, June 22. Accessed January 18, 2021. https://www.theguardian.com/music/2009/jun/22/linn-oberheim-drum-machines.

Scheinman, Ted. 2020. "Status Cymbals: A Selection of Top Answers to the Centuries-Old Musical Question, How Do You Get By without an Actual Drummer." *Smithsonian* 51, no. 4 (July/August): 23.

American Federation of Musicians and Other Musicians' Unions

The American Federation of Musicians (AFM), widely known simply as the "musicians' union," has had a love-hate relationship with various forms of music technology through the 20th century to the present. Some examples include a two-year recording ban that the union placed on its membership in the early 1940s to protest the increased broadcasting of records by radio stations. The union's concern was that radio was putting performing musicians out of work by using pre-recorded music as opposed to studio musicians. In more recent years, the AFM has led efforts in assuring that the film industry's studio musicians received royalties due to them. The technology of streaming audio created issues with royalties on several levels, and the AFM played a role in tightening up of royalties associated with digital media.

In the 1950s and 1960s, the AFM criticized the Thomas organs' Band Box, Wurlitzer's Sideman, and other early examples of drum machines because the AFM believed that such electronic devices might eliminate jobs for percussionists. The union similarly criticized the drum machines that appeared on the market in the 1980s.

The U.S. musicians' union has also protested the use of synthesizers in pit orchestras as a means of reducing the number of players of acoustic instruments (e.g., woodwinds, brass, strings, and percussion) hired for Broadway musicals. Although synthesizer players were first recognized in the New York musicians' union contract in 1987, synthesizers in the late 1980s and 1990s largely were used to supplement a traditional pit orchestra. The challenges undertaken by the AFM began occurring in the 21st century with the advent of the Virtual Pit Orchestra in which a synthesizer could be programmed with all the instrumental accompaniment for an entire show and could be played by a single person. It should be noted that musicians' unions in other countries, too, expressed concern about what extensive use of synthesizers might mean for acoustic musicians. For example, in his memoir, Thomas Dolby writes that in the late 1970s, he used to play a Micromoog and a Solina String Synthesizer, the latter of "which, two years later, the Musicians Union voted to ban because it was taking jobs away from 'real' string players" (Dolby 2016, 21).

Further Reading

Dolby, Thomas. 2016. *The Speed of Sound: Breaking the Barrier Between Music and Technology.* New York: Flatiron Books.

Edison Phonograph

American inventor Thomas Edison is well known for his work on the electric light bulb, motion picture camera, rechargeable batteries, his advocacy of direct current (DC) over alternating current (AC), and for several other influential inventions and improvements to existing technology. Edison's 1877 invention of the phonograph proved that the human voice could be recorded and reproduced mechanically. Previously, Scotsman Alexander Graham Bell had developed and patented the telephone; however, although Bell's telephone allowed the human voice to be reproduced over distance, it did not record sound. Eventually, Edison's phonograph played sound—including music—that was recorded on wax cylinders. The machines had external horns, which amplified the vibrations that a stylus picked up from the cylinders. At the time of this writing, there are several YouTube videos that show exactly how the Edison phonograph works (see, e.g., https://www.youtube.com/watch?v=fWLlbk_bI7E, accessed January 14, 2021). These videos are indispensable for showing the mechanics of the machine, particularly because 21st-century readers are not apt to encounter an Edison phonograph that they can actually see and hear in action.

In the next decade, Edison's cylinder-based machines found themselves competing with Emile Berliner's disc-based recordings. Interestingly, both formats possessed distinct advantages and disadvantages. As metal discs for music boxes proved, discs were easier to store than other shapes. The shellac out of which various companies made disc-based records was more stable and less susceptible to becoming misshapen or melted than the wax cylinders that Edison's machines used; however, arguably, the audio fidelity of Edison's cylinders and phonographs, limited though it was, was in many cases superior to that of the flat shellac discs and their players. Another plus for Edison, the sapphire stylus with which Edison's phonographs were fitted was virtually impervious to wear. In fact, some Edison machines from the late 19th century and early 20th century still are fitted with their original styluses, which function well, now over a century after the machines were built. By comparison, the steel styluses—generally referred to as needles—on the Victor Talking Machine Company's Victrolas and other companies' similar disc-playing machines wore out quickly, so users had to keep a ready supply on hand so as not to damage the records.

Any advantages that a virtually indestructible stylus playing relatively soft wax cylinders seem to have been negated at least in part by Edison's personality, as well as by the fact that Edison had numerous competitors that adopted the disc format. Biographers generally agree that Thomas Edison was a far better inventor than marketer. Edison had very little interest in the popular music of the day, as well as very little interest in jazz. As a result, his disc-based competitors were able to contract with major popular singers and band leaders of the day, leaving Edison with recording artists who were not as immediately popular with the record—or cylinder—buying public.

When considering this challenge that Edison faced, it is important to keep in mind that the record industry of the late 19th and early 20th centuries was quite different than that of the second half of the 20th century and the 21st century. Readers may or may not be familiar with the expression, "it's the singer, not the song," either from the Rolling Stones' "Singer Not the Song" or for the phrase's general use in the popular lexicon. Contrary to that old saw, in the early 20th century, more often than not, the public's attention was more on the song than on the particular singer or instrumental recording artist. So, songwriters, such as Irving Berlin, were stars of the musical world to a greater extent than songwriters would be after World War II. In the first half of the 20th century, various different recordings of the same song—to a greater extent than what typically would be the case after the popularity achieved in the 1940s and 1950s by former big band singers such as Bing Crosby, Frank Sinatra, Ella Fitzgerald, and others—fought it out on the charts. This is not to say that the performer's name on a record made no difference to the record-buying public; it did make a difference. The point is that there were significantly more versions of songs appearing on the record charts than what would later be the case or that is the case in the 21st century. So, this was the world in which Edison found himself, and generally his roster of recording artists could not compete with the artists signed to his competitors' companies.

Interestingly, although Thomas Edison and his company continued to shy away from some of the genres that were increasingly making an impact in the recording industry, Edison shifted from a focus on wax cylinders to manufacturing discs.

Thomas Edison's wax cylinder-based format was popular at the end of the 19th century and early 20th century but was soon supplanted by the more familiar flat discs that became widely known as 78-rpm records. (Library of Congress)

As is discussed more fully in the entries on Gramophones and Victrolas and on 78-rpm Records, the end of the 19th century and the first couple of decades of the 20th century found several standards for shellac discs competing against each other. This included different ways in which the grooves were cut and shaped in the records, different sizes, different thicknesses, different standard speeds, and so on. Manufacturers created their own standards in order to coordinate record and player sales. To put it another way, some records could only safely be played on certain machines, lest the discs become damaged, sound hopelessly too fast and high pitched or too slow and low pitched, and so on. When Edison began making discs, he too followed suit. Edison's thicker records and his record players never caught on to the extent that the Victor Talking Machine Company's discs, disc-cutting format, and machines did, particularly given the head start that Victor had over Edison.

Today, Edison cylinders can sometimes be found in garage sales, swap meets, and so on; however, working models are housed in some museums that are devoted to technology or to providing a snapshot of life at the start of the 20th century. There continues to be a collectors' base for Edison phonographs, although the direct connection of these machines to later sound recording and reproduction technologies is more limited than that of the competing flat discs.

See Also: Gramophones and Victrolas; 78-rpm Records

Further Reading

Steffen, David J. 2005. *From Edison to Marconi: The First Thirty Years of Recorded Music.* Jefferson, NC: McFarland.

Stross, Randall. 2007. *The Wizard of Menlo Park: How Thomas Alva Edison Invented the Modern World.* New York: Three Rivers Press.

Stross, Randall. 2010. "The Incredible Talking Machine." *Time*, June 23. Accessed January 18, 2021. http://content.time.com/time/specials/packages/article/0,28804,1999143_1999210,00.html.

Effects Pedals

Although some early electric guitars and guitar amplifiers had effects built into them (e.g., some of Adolph Rickenbacker's early guitars), the external effects pedal became the norm. Effects pedals are connected between the instrument and the amplifier and modify the electric signal that is sent to the amplifier to create different kinds of sonic effects. The story of the creation of external effects devices for electric guitars goes back to the 1940s and DeArmond's Trem Trol 800. This device used mercury, a conductive metal that is normally liquid at room temperature, which moved back and forth alternatively making and breaking electrical contact. This is what allowed the DeArmond device to make its pulsating tremolo effect. But, as was the case with many electronic devices in the 20th century, it was the transistor (as opposed to a dangerous liquid such as mercury) that facilitated the ubiquitous effects pedals of the 1960s through the present. Some amplifiers had featured, reverb, echo, tremolo, vibrato, and other effects, but the effects pedals allowed for even more variety of tone. With on/off switches located on the top of the device, the player could activate the device with their feet.

Some early pedals did not necessarily change the tone color of the guitar and included the volume pedal. Two of the Beatles' recordings from 1965 found George Harrison using a volume pedal: "I Need You" and "Yes It Is." As an example of the technological advance that even a simple device such as a volume pedal made, in a 1987 interview published in *Guitar Player* magazine, Harrison stated that in 1964 when he had difficulty coordinating the volume pedal for the studio recording of the song "Baby's in Black," "I played the part, and John [Lennon] would kneel down in front of me and turn my guitar's volume control" (Babiuk 2002, 134). That was the kind of situation that necessitated the development of greater ease of use for volume pedals. Other pedals that appeared at approximately the same time played a significantly different role: altering the guitar's tone.

The first transistorized effects pedal was the Maestro FZ-1 fuzz box, which was released in 1962. This was the source of the tone that Keith Richards achieved on the Rolling Stones' classic song "(I Can't Get No) Satisfaction," one of the Rolling Stones' biggest hits and one of the defining recordings of the British Invasion of 1964 and 1965. Although guitarists speculate whether it was through the use of the Maestro FZ-1 or a similar device made by Vox, a similar guitar tone was used by the American studio guitarist Tommy Tedesco for the instrumental opening of the theme song for the television series *Green Acres*. The fuzz tone effect on the Rolling Stones' recording adds a level of grittiness to the song, whereas in the *Green Acres* theme, Tedesco's fuzz tone—being juxtaposed as it is against the bass harmonica—adds a degree of humor to the song. One could also argue that Tedesco's tone aligns with the grittiness of the dilapidated house and nonproductive farm that is at the center of the rural television comedy.

One of the other iconic effects pedals of the 1960s and early 1970s that is particularly easy to identify by ear is the wah-wah pedal. The first commercially available wah-wah pedal made its debut in 1966. Perhaps the best-known such pedal from the device's heyday was the pedal manufactured by Vox, although companies such as King, Hammond, DeArmond, and others produced wah-wah devices for electric guitars and electronic organs. Mechanically, the wah-wah pedal functions similarly to a guitar volume pedal. Electronically, the pedal alters the electric guitar's sound by changing the tone and frequencies that go into the amplifier. Much like jazz trumpet players who used a plunger mute to change back and forth between the instrument's open sound and muted sound, the wah-wah pedal typically is toggled back and forth to create a sort of pulsating effect to the tone color.

Electric guitarists such as Jimi Hendrix, Eric Clapton, Jimmy Page, Jeff Beck, Frank Zappa, George Harrison, and others utilized the wah-wah pedal in the late 1960s when it became one of the standard parts of rock guitar tonal vocabulary. Arguably, however, the wah-wah effect came into its own in the 1970s in the funk genre, in which the device was frequently applied to rhythm guitar parts. In 1972, Musitronics released its Mu-Tron III, which produced a wah-wah pedal–like pulsation but automatically when the device was turned on. This meant that a performer could simply push the switch once and the device would create the tone-color pulsation without the player having to constantly pedal up and down. This made the wah-wah effect more easily accessible not only for electric guitarists but also for keyboard players. For example, one of the more memorable early 1970s' use of the Mu-Tron III was by Stevie Wonder on his clavinet riff throughout the song "Higher Ground."

Among other popular effects pedals that can be used with electric guitars and keyboards are digital delay pedals, which basically record and then play back any signals (e.g., music) that go into them; overdrive and distortion pedals; phase shifters, which produce a shimmering effect on the sound; octave pedals, which allow the instrument to produce sounds an octave away from where the guitarist is actually playing; compressors, which can help to produce a more uniform sustain to the guitar's sound by compressing the louder volume that is produced when a string is initially struck; and many more.

Some of the classic 1960s' and 1970s' effects pedals made by Electro Harmonix, Vox, Boss, and other companies are no longer produced, so some of today's pedals replicate their sound. And with the continuing evolution of the effects pedal in the digital age, several companies, including Zoom, produce multi-effects processors. For example, Zoom's G1X FOUR processor includes over 70 built-in effects, can model several classic vintage tube guitar amplifiers, includes a built-in drum machine/rhythm section and circuitry for producing and playing back loops, has a built-in volume pedal, and so on. The price of 70 separate effects pedals would easily run into the thousands of dollars; however, at the time of this writing, the G1X FOUR processor can be purchased from several online retailers for approximately $110.00.

Despite the availability of multi-effects pedals such as the Zoom G1X FOUR, even a cursory search of the offerings of large online retailers such as Amazon, Sweetwater Sound, Guitar Center, and others shows that there is still a plethora of single effects pedals available. Some of these focus on one particular kind of effect, say, fuzz tone, but can emulate several of the classic fuzz tone effects pedals of the past. Several manufacturers currently offer fuzz, distortion, overdrive, digital delay, noise suppression, octave and other pitch-shifting, amplifier emulator, echo, looping, analog tape emulation, and other pedals. One of the more intriguing offerings, the Electro-Harmonix SYNTH9 pedal, allows electric guitarists and electronic keyboardists to emulate the classic synthesizers of the 1970s and 1980s.

As suggested by the Electro-Harmonix unit mentioned earlier, since the 1980s, the guitar has indeed become part of the world of sound synthesis. The continuing evolution of the electric guitar as synthesizer is suggested by a July 2020 profile in the American Federation of Musicians' *International Musician* of the Boss SY-1000 guitar synthesizer, which "brings numerous musical advantages to players including ultra-articulate tracking, lightning-fast response, instantly variable tuning, sound panning/layering, and more" (The Editors of *International Musician* 2020, 24). As a fully MIDI-compatible instrument, a guitar synthesizer can play multiple roles in a rock band, with all the capabilities of an electric guitar plus all the capabilities of a synthesizer.

The effects pedal has also entered the age of home assistants, such as Amazon's Alexa. Thingamagig offers an Alexa-ready multi-effects processor that connects to the Amazon Echo Show. This effects unit includes over 300 custom tones, 180 free backing tracks, and 17 amplifier simulations. Interestingly, because the

Thingamagig is internet connected, it can receive updates and enhancements. At the time of this writing, the unit is available for purchase at Amazon for $149.99.

See Also: Distortion; Electric Guitars; Loops; MIDI

Further Reading
Babiuk, Andy. 2002. *Beatles Gear: All the Fab Four's Instruments, from Stage to Studio.* Revised ed. San Francisco, CA: Backbeat Books.
Hunter, Dave. 2013. *Guitar Effects Pedals: The Practical Handbook.* San Francisco, CA: Backbeat Books.

8-Track Tapes

In many cases, the technical innovations that helped to shape the way that people created and listened to music were made by individuals. The development of the 8-track tape cartridge in the early 1960s is a notable exception and an example of multiple industries coming together to develop something that potentially could be mutually beneficially to all of them. The initial consortium included Bill Lear of the Lear Jet Corporation and the Ford Motor Company, General Motors, Ampex, RCA Victor Records, and Motorola.

At the time, there were precedents for cartridges that contained music encoded on electromagnetic tape; however, the commercial applications of these were limited. The primary predecessors of the mature 8-track tape were tape cartridges marketed by Earl Muntz for the automotive industry. After Bill Lear was dissatisfied with the quality of a 4-track tape player that Muntz gave him to use in his aircraft, Lear became the driving force in the efforts to improve upon the earlier cartridge system. As might have been expected based on the principals involved in the project, the 8-track tape cartridge as it came to be standardized initially was squarely aimed at the transportation industry. Previous attempts at developing portable, nonradio, sound for automobiles, such as Chrysler Corporation's offering of cars with built-in record players, had not proven commercially successful, nor were they practical.

Bill Lear's work as a driving force in the development of the 8-track tape is not as strange as it might appear to be at first glance. Not only was he active in aircraft design and the design and manufacture of autopilot and other electronic devices and systems for aircraft, years earlier he had also been one the principals in the early development of car radios and one of the principals in the company that would become to be known as Motorola.

The basic premise of the 8-track tape, and one of the primary ways in which it differed from the reel-to-reel tape and the cassette tape (which eventually superseded it, although it was developed before the 8-track), was that the tape itself was recorded and played as a continuous loop. To put it another way, one could not rewind it back to the beginning, nor did one have to flip it over for the second side, as one had to do with cassette tapes. And because it used eight tracks across the width of electromagnetic tape that had no set physical start or finish, the 8-track tape presented one challenge that its tape competitors or physical media such as LP albums did not face: sometimes songs had to be broken up between the physical tracks. A song might begin, using two of the tracks (one track for each part of the stereo signal), but continue on two other tracks. One might think of this as being analogous to the two-sided singles that used to be released on 45-rpm records. When songs were too long to fit on one side of the record, they would fade out at the end of the A side and resume on the B side. In addition, because the user toggled between tracks as dictated by the tape loop, 8-track releases of albums originally designed for primary release on vinyl did not have the same sense of an "A" and a "B" side as the original album (or, most cassette releases) did.

Where the 8-track tape did have an advantage over reel-to-reel tape was in its portability. And where it had an advantage over the compact cassette tape was in audio fidelity. The tape in a cassette moved past the record and play heads at the speed of 1–7/8 inches per second (ips), whereas the 8-track tape moved at the speed of 3–3/4 ips. The greater speed led to a better frequency range and less noise than on cassettes, at least until noise reduction systems, such as Dolby, were used on cassettes. It should be noted, however, that because the physical tape was divided into eight sections—or tracks—some 8-track tapes suffered from a degree of bleed through, particularly if the heads on the playback machine were out of alignment. The alignment issue was exacerbated by the fact that because the tape essentially had four separate stereo programs all going in the same direction, the playback heads in the machine moved as the user toggled from program to program.

As noted earlier, General Motors and the Ford Motor Company were among the entities involved in the development of the 8-track tape. As might be expected, the new medium was highly touted by auto manufacturers. And with demand that was generated for in-car and in-truck 8-track players came demand for record companies to release 8-track versions of albums. The demand is suggested by a 1965 article in *Billboard* magazine, in which "Larry Finley, head of International Tape Cartridge Corp., told *Billboard* that the Ford showings have created a growing tide of orders from Ford dealers for racks containing packages which would fit the Lear-designed machine" (The Editors of *Billboard* 1965a, 1, 10). The article reported this demand for 8-track tapes in conjunction with the popularity of the forthcoming 1966 Ford Mustangs that would be fitted with 8-track tape players.

A later 1969 article in *Popular Science* magazine compared the mechanics and positive and negative attributes of the then three leading magnetic tape formats: reel-to-reel, cassette, and cartridge, noting that the 8-track cartridge was one of three different cartridge formats. Incidentally, according to the article's author, Sam Shatavsky, the 4-track cartridge was on its way out at the time and the Playtape cartridge essentially was only used for portable players marketed for use by children (Shatavsky 1969). According to the author, "Tape cartridge, ideal for automobile music players, is the most popular of all tape systems. Starting as a Lear, RCA, and Motorola venture, eight-track has also proved its worth in home players" (Shatavsky 1969, 127). The *Popular Science* test results for the various formats showed that reel-to-reel tape had the best audio fidelity and frequency response. However, compared with the compact cassette tape, which ran at a speed of 1–7/8 inches per second, the magazine's tests suggested, "Eight-track cartridge tape, with 3 3/4-i.p.s. speed, is a happy compromise for good frequency response with reasonably long recording time" (Shatavsky 1969, 129).

Perhaps because it was most closely associated with use in boats, cars, trucks, Bill Lear's aircraft, and other modes of transportation, the 8-track tape achieved a degree of notoriety as a medium for bootleggers. In fact, some truck stops and gas stations in the 1960s and 1970s had racks of bootleg versions of currently popular albums, tape albums that were claimed to be greatest-hits collections from artists who had not necessarily released legitimate greatest-hits collections, and unauthorized genre-specific collections.

Despite the challenges posed by bootlegging and the changes to album structure that were necessitated by the arrangement of the tracks on an endless tape loop, perhaps the biggest challenge faced by this technology was that the internal mechanism was not entirely reliable. Although this was also the cause of failure in cassette tapes, the 8-track tape developed an especially bad reputation. Back during the technology's heyday, it was not unusual to see discarded tape cartridges—with tape hanging out—littering the sides of roadways. As mentioned earlier, the other problem area for the 8-track tape, and one in which the compact cassette enjoyed an advantage, was in bleed-through between tracks. In a cassette, the user had to turn the tape over in order to hear what was analogous to the B side of an album. Because the 8-track tape was a continuous loop, it was the play head in the tape deck that moved in order to change from a set of tracks to another. This lack of stability in the machines' mechanisms sometimes would cause the head to go out of alignment, which would lead to bleed-through from an adjacent track.

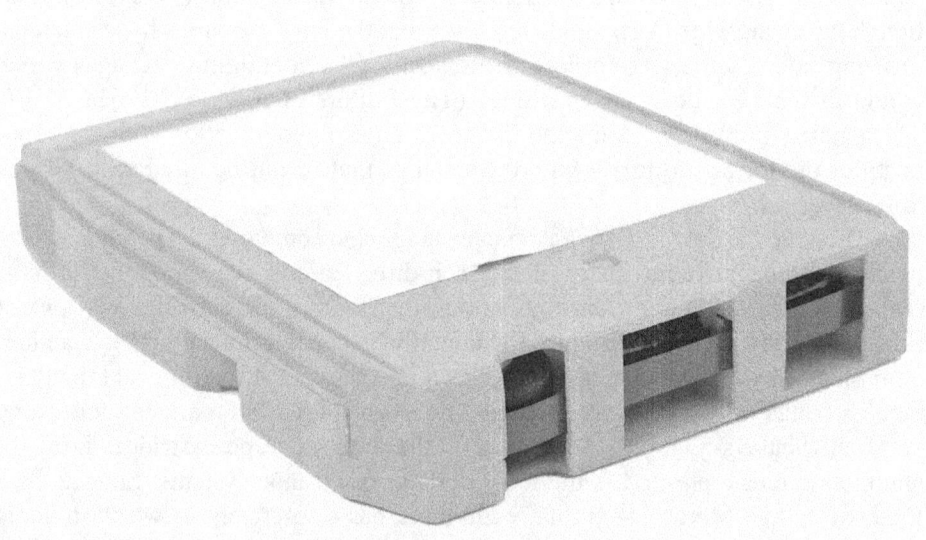

Most closely associated with cars, trucks, and pleasure boats, the 8-track reached its peak of popularity in the late 1960s and early 1970s. By the end of the 1980s, it had become a rarity. (Peanutroaster/Dreamstime.com)

By the end of the 1980s, 8-track tapes had largely fallen out of favor. The demise of interest in the format can be seen in the description of a rare 1987 release of Billy Idol's album *Whiplash Smile* by Chrysalis through the Columbia House Record Club. As *Record Collector* magazine's Tim Naylor wrote, "By the late 80s, the Columbia House and RCA record clubs only had a handful of members choosing the 8-track format and by the end of 1988, both clubs ceased to offer them to their customers" (Naylor 2015).

Despite the strong association of the 8-track tape with personal, professional, and pleasure transport (e.g., automobiles, commercial trucks, and boats), and despite the fact that the format was replaced in large part by the cassette, compact disc, and digital media, the 8-track continues to have a niche group of collectors of players and the cartridges. Naylor's article in *Record Collector* magazine shows that, at the time of publication (2015), rare 8-track releases that were quickly withdrawn by the record labels for various reasons, including a 1969 duet album featuring American singer Frank Sinatra and Brazilian bossa nova legend Antonio Carlos Jobim, and a 1982 Beatles greatest-hits collection, were estimated to be commanding prices of between $750 and $5,000.

See Also: Cassette Tapes; Reel-to-Reel Tape

Further Reading

Becker, Stephen. 2011. "8-Track Tapes Belong in a Museum." *All Things Considered*, February 16. Accessed January 18, 2021. https://www.npr.org/sections/thereccord/2011/02/17/133692586/8-track-tapes-belong-in-a-museum.

The Editors of *Billboard*. 1965a. "Lear Cartridge-Equipped Fords Getting a Fast Start." *Billboard* 77, no. 39 (September 25): 1, 10.

The Editors of *Billboard*. 1965b. "RCA Fires 175-Title Burst with Release of Stereo 8 Cartridges." *Billboard* 77, no. 39 (September 25): 3.

Finley, Larry. 1965. "Tape Cartridge Tips." *Billboard* 77, no. 39 (September 25): 10.

Naylor, Tim. 2015. "The *Record Collector* Guide to 8 Track Cartridge Stereo." *Record Collector*, February 23. Accessed January 18, 2021. https://recordcollectormag.com/articles/8-track-cartridge-stereo.

Shatavsky, Sam. 1969. "The Best Tape System for You: Reel, Cassette, or Cartridge." *Popular Science* 194, no. 2 (February): 126–129.

Electric Guitars

As noted in this volume's entry on Resonator Guitars, one of the challenges that musicians faced in the vaudeville theaters of the 1920s was with projecting the sound of guitars in large acoustical environments. Interestingly, at approximately the same time as the resonator guitar was under development and then just gaining ground in the marketplace, another more far-reaching development was also underway: the invention of the electric guitar.

As was the case with the resonator guitar, the historical timeline of development, bringing to market and patenting the electric guitar and its component parts, is convoluted. George Beauchamp, who also played a role in the development of the resonator guitar, is credited with developing the "frying pan" electric lap steel

Hawai'ian guitar, which instrument manufacturer Adolph Rickenbacker began producing in 1932. However, Beauchamp did not receive a patent for his electrical pickup technology for another five years (see, Beauchamp [1934] for information on the patent application).

At approximately the same time, John Dopyera, who also played a significant role in the development of the resonator guitar, and Arthur J. Stimson produced an electric Spanish (held against the body of the player in the conventional way and not in the lap) guitar. Like Rickenbacker and Beauchamp's electric, Dopyera and Stimson's instrument appeared in 1932. The Dobro Corporation filed a patent application in 1933 for Dopyera and Stimson's design for electric pickups (see, Stimson 1933), and the patent was granted in 1934.

The basic principle of the electric guitar that was central to both of these early designs was that a coil of wire was wrapped around a magnet (or a number of small magnets located under each of the instrument's strings). The vibration of the string(s) creates a weak electrical current in the wire, which is then fed to the amplifier. In an interesting and detailed article in *Guitar World*, electric guitar pickup designer Scott Lawing explains the physics of pickups and the various design characteristics (see, Lawing 2018).

One of the challenges posed by early amplification systems for PA systems and instrument amplifiers was the electrical hum that is inherent in some wiring systems. It is interesting to note that even in early electric guitar pickup designs, technology was being developed to minimize or eliminate this problem. Known as humbucking pickups, or humbuckers, because they buck or eliminate the electrical hum, these pickups had double coils and were available in the 1930s. Although pickup technology continues to change and advance, many of the principles still used today can be found in the instruments produced during the electric guitar's first decade of life.

It should be noted that the early Spanish-style electric guitars (held in the conventional way with the strings facing out) were hollow-bodied or semi-hollow-bodied instruments. To put it another way, they could be understood as modifications or extensions of already-existing acoustic guitar design. Like the timeline and succession of patents for the electric guitar, the timeline for the development of the solid-body electric guitar is also somewhat complex and convoluted.

Jazz guitarist Les Paul, who also made important contributions in using multi-track recording technology and tape speed manipulation in some of his recordings of the 1950s, designed a solid-body electric guitar in 1941. Paul's invention is widely known as the Log because of its slab-of-wood appearance. Although the Log ultimately led to Gibson's development and release of the Les Paul solid-body electric guitar, the Les Paul model was not patented and released until the mid-1950s. By that time, Leo Fender had already produced his Broadcaster, later renamed the Telecaster because of trademark issues with the name Broadcaster, making Fender's development the first commercially successful solid-body electric guitar.

Fender's Telecaster, his slightly later Stratocaster, and Gibson's Les Paul guitar—as well as other subsequent solid-body guitars—did away with the resonating chamber of acoustic guitars and hollow-body electric guitars. Because these instruments relied solely on their pickups to produce sound, the feedback that was associated with hollow-body and semi-hollow-body electric guitars was minimized. In addition, because the physical shape of the solid-body guitar's body played little or no role in the tone of the instrument, all sorts of intriguing shapes were released in the late 1950s and early 1960s. These included the Gibson Flying V guitar that was played famously by Bo Diddley and later by Dave Davies of the Kinks, the teardrop-shaped Vox guitar played for a brief period of time by Brian Jones of the Rolling Stones, and so on.

The impact of the electric guitar on popular culture has been enormous. At the time when electric guitars first started to make a significant impact on popular musical genres, the piano, woodwind, acoustic string, and brass instruments were the instruments most associated with genres such as jazz, blues, country, pop, and so on. In the world of jazz, however, Charlie Christian helped to establish the electric guitar as an important solo voice with his work at the end of the 1930s and start of the 1940s. In fact, Christian is often cited as one of the important transitional figures between the swing and bebop eras in jazz. At the same time, and in a career that lasted years longer, T-Bone Walker's brilliant technique, musicality, and showmanship helped to establish the electric guitar as perhaps the most important instrumental solo voice in blues music. As a guitarist and showman, Walker influenced virtually all subsequent blues electric guitarists, as well as rock guitarists such as Chuck Berry.

In turn, Berry and other electric guitar soloists of the second half of the 1950s, including Scotty Moore, Buddy Holly, and others, helped to make the instrument the dominant voice in rock and roll. Electric guitarists such as Dick Dale, members of the Ventures, and members of the Shadows also showed that electric guitar–based bands went beyond the stylistic realm of Berry and rockabilly guitarists. By the time of the 1964–1965 British Invasion, the prototypical rock band consisted of lead electric guitar, rhythm guitar, electric bass, and drums. Some groups included a keyboardist, but with few exceptions (e.g., the Doors, the Zombies, and bands of the progressive rock era) the most prominent solo instrumental voice of the rock era remained the electric guitar.

In some genres of popular music (e.g., disco) and throughout the hip-hop era, the electric guitar generally has not played quite the dominant role that it did during the 1960s and 1970s; however, it continues to make important popular cultural impacts. For example, the *Guitar Hero* series of video games demonstrates the instrument's continuing relevance. This series of games has brought the look and sound of instruments and musicians from earlier eras of rock and roll into the everyday lives of young people who in some cases were born decades after the music in some of the music in *Guitar Hero* games was first popular.

In considering the electric guitar as a culturally important example of musical technology, one must also consider its lower-pitched relative, the electric bass. The Precision Bass that Fender introduced in the early 1950s revolutionized the bass. As a solid-bodied instrument, Fender's Precision Bass and the later Jazz Bass (introduced in 1960) allowed bass players to reach high volume levels without unintended feedback. The Fender Basses also required a shorter reach than that of amplified acoustic double basses. The Fender Bass became such an iconic part of rock and roll, jazz, pop, and other musical genres during the 1950s and the early 1960s that it was not uncommon to see musicians' credits on some album covers to list generic names for examples such as "lead guitar," "piano," and so forth but used the term "Fender Bass" to refer to the electric bass. In this way, the electric bass was differentiated from the orchestral double bass with the words "Fender Bass" becoming almost like Band-Aid, Kleenex, and other brand names that became generic terms for products. This was particularly true in the United States—at least until the British Invasion—so ubiquitous were Fender's electric basses. Other manufacturers around the world introduced electric basses during the 1950s, including the German manufacturer Höfner. Höfner electric basses, such as the models used by Paul McCartney during the Beatles' first several years of worldwide popularity, featured a semi-hollow body and a shorter scale than Fender Basses. As was the case with the electric guitar, the number of companies offering electric basses expanded throughout the rock era as the electric bass completely or nearly completely supplanted the double bass in several musical genres. In addition to Fender and Höfner, Rickenbacker, Danelectro, Gibson, Carvin, and other companies produced iconic electric basses.

See Also: Amplification; Distortion; Effects Pedals

Further Reading

Beauchamp, George. 1934. Patent granted 1937. Patent application for "Electrical Stringed Musical Instrument." Accessed January 18, 2021. https://patents.google.com/patent/US2089171?oq=patent:2089171.

Bonds, Ray. 2003. *The Illustrated Directory of Guitars*. St. Paul, MN: Voyageur Press.

Lawing, Scott. 2018. "How Does a Guitar Pickup Really Work?" *Guitar World*, February 23. Accessed September 21, 2020. https://www.guitarworld.com/gear/how-does-a-guitar-pickup-really-work.

Port, Ian. 2019. *The Birth of Loud: Leo Fender, Les Paul, and the Guitar-Pioneering Rivalry That Shaped Rock 'n' Roll*. New York: Scribner.

Stimson, Arthur J. 1933. Patent granted 1934. Patent application for "Electrophonic Stringed Musical Instrument." Accessed January 18, 2021. https://patents.google.com/patent/US1962919?oq=Us1962919.

White, Forrest. 1994. *Fender: The Inside Story*. San Francisco, CA: Miller Freeman Books.

Electrical Recording

As detailed in the entries for Gramophones and Victrolas and the Edison Phonograph, early sound recordings were made essentially using the playback equipment in reverse. In other words, musicians had to perform into large horns that transferred the sound vibrations into movements of a cutting stylus. This method presented some limitations, including the need for sufficient volume to produce soundwaves large enough to result in enough stylus motion to cut the grooves in the master disc. Loud instruments were necessary, performers had to be situated around the horn in what might not have been ideal performing relationships, and so on; however, some particularly loud instruments, such as the bass drum, produced too much sound. Likewise, some of the early acoustic recording equipment was bulky and heavy. As a result, making field recordings—or any recordings outside the confines of the studio—was difficult. Examples of some of the early recording equipment can be seen in the 2017 Bernard MacMahon film *American Epic: When America First Heard Itself,* which has been shown on PBS and is available at the time of this writing through some streaming video services (MacMahon 2017).

Throughout the 1910s, microphone technology had improved to the point that commercial broadcast radio began to take off in earnest at the end of the decade when in-home radio receivers were marketed. Likewise, microphone technology would lead to PA systems coming into play at around the same time. An electrical method of recording was introduced in 1924; however, it had been under development by Joseph P. Maxfield and Henry C. Harrison of Bell Telephone Laboratories, using a microphone developed by Edward C. Wente. And during the 1910s and throughout the 1920s, other engineers, working independently and under the auspices of companies such as Edison, Columbia Records, Victor, Western Electric, and others, had been working on competing electrical recording systems. One of the improvements was that the microphone-based system could handle a broader range of dynamic (volume) levels. Not only was a wider range possible but some loud instruments—such as the bass drum—that caused problems in the acoustic recording process could now be recorded without having to use substitute instruments in their place. Likewise, relatively quiet instruments, such as the double bass, which previously were difficult to record, could now be used. As a result, the use of tuba in jazz bands decreased, and the double bass took its place as the foundation of jazz groups.

The electrical recording method made an immediate impact even outside the recording studio. Ralph Peer, who produced early recordings of country and blues musicians for companies such as Okeh, Victor, and Columbia, "first started making annual trips to Atlanta in 1923, and the next year added New Orleans, and he also did some sessions in St. Louis. But he was never happy with the quality of recording he could get on the road using the old acoustic process.... When electrical recording came in, they could get vastly better-quality recordings, and it made the process more portable" (MacMahon et al. 2017, 15). In so far as Peer would discover and make some of the first recordings of influential artists such as the Carter Family, Jimmie Rodgers, and the Memphis Jug Band in the diverse regions in which these musicians lived and worked, the electrical recording system helped to make it possible for Americans across the country to hear regional musical genres, such as blues and country, that previously had only been heard within the regions of their genesis. In fact, the recordings that Peer and others made of country musicians using the new electrical process led to country music soon becoming the best-selling music in the record industry. One of the attractions of country music, as well as rural blues, to Peer and other recording industry executives who made recordings outside of major urban areas was that the regional musicians tended to include folk songs and other noncopyrighted material in their repertoire; their music was not as highly influenced by Tin Pan Alley popular songs as was the work of urban musicians. This increased the potential for income for the record companies.

> *Half-Speed-Mastered Recordings*
>
> From a technological standpoint, the half-speed mastered LPs that enjoyed a period of popularity among audiophiles beginning in the late 1970s might seem relatively minor. The basic concept of half-speed mastering was not particularly complex: classic albums from the pop, classical, rock, country, jazz, and R&B genres were reissued in versions in which the master discs were cut at half the playback speed. That process alone led to greater audio fidelity, less noise and distortion, and greater dynamic (volume) range—in short, an audiophile experience that brought the listener closer to the sound of the original master tapes.
>
> The company that pioneered these audiophile discs, Mobile Fidelity Sound Lab, was led by Brad Hill, whose first commercial venture in the recording industry came in the late 1950s with recordings of steam locomotives. Hill had also made a major impact on the easy-listening music world in the second half of the 1960s with his Mystic Moods Orchestra recordings. These recordings mixed easy-listening pop music, largely string orchestra driven, with environmental sounds (e.g., rain and thunder from a storm) and sounds from easily identifiable machinery (e.g., a train passing from one stereo channel to another).
>
> Mobile Fidelity's Original Master Recordings not only used the half-speed mastering technique but also pressed the records on special high-quality vinyl. The recordings were also mastered using Dolby noise reduction. This combination of technique and materials produced audiophile versions of albums such as the Beatles' *Magical Mystery Tour*, a recording of major Tchaikovsky orchestral concert overtures and symphonic poems performed by the London Symphony Orchestra under the baton of André Previn, Pablo Cruise's album *A Place in the Sun*, Pink Floyd's *The Dark Side of the Moon*, Crystal Gayle's *We Must Believe in Magic*, and George Benson's *Breezin'*, just to name several of widely diverse styles and genres.

Although some may have argued that the new process decreased the feeling of warmth in the sound of records—similar to the charges that would be launched when comparing early CD recordings to vinyl albums in the early 1980s—the new electrical process not only allowed record companies to more easily make field recordings, as described earlier, but the use of microphones and the electrical process also made it possible to record orchestras in their native concert halls and soloists and chamber groups in recital halls. This is because the instrumentalists and singers did not need to be directly in front of the horn on an acoustic recording machine. And because the process allowed for the recording of quieter sounds than had been possible using the old acoustical process, the instrumentation changes that had been made for earlier recordings no longer were necessary. Music originally composed for stringed instruments but reorchestrated for woodwinds and, (especially) directional brass instruments could now be recorded in their original instrumentations.

The electrical process for recording is what really paved the way for our modern understanding of the work of recording technicians and producers. Certainly, there were no such thing as overdubs, as even with microphones and electrical translation of the sound waves into the cutting of grooves into the master disc, all recording in the late 1920s and through the 1930s was direct to disc. However, the introduction of an electrical current to the process meant that volume levels could be adjusted so that the master disc, from which negatives for stamping would be made, sounded as good and clear as possible.

The electrical process was also used as something of a marketing device by record companies, as a way of distinguishing their recordings from those of other companies that perhaps had not made the jump to electrical recording. For example, on the paper sleeves for its electrically recorded discs, Victor described the merits of its new Orthophonic recordings and the new brilliance, clarity, and volume that they made possible on older Victrolas. At the same time, the company also touted its new Orthophonic Victrolas, which the company claimed produced even better sound than older Victrola models. Other companies included phrases such as "Electrically Recorded" on their record labels, particularly after the process gained widespread acceptance in the industry and with the public.

See Also: Gramophones and Victrolas; 78-rpm Records

Further Reading

Daniel, Eric D., C. Denis Mee, and Mark H. Clark, eds. 1999. *Magnetic Recording: The First 100 Years*. New York: John Wiley & Sons/IEEE Press.

MacMahon, Bernard, and Allison McGourty, with Elijah Wald. 2017. *American Epic: The Companion Book to the PBS Series*. New York: Touchstone.

MacMahon, Bernard, director. 2017. *American Epic: The First Time America Heard Itself*. New York: Touchstone. Documentary film.

Electronic Keyboard Instruments

Keyboard-based instruments have played multiple significant roles in the development of music technology for centuries. Because the various types of keyboard-based synthesizers and electronic organs are detailed in other entries in this volume, here we will consider electronic keyboard instruments that are not synthesizers or electronic organs per se. This includes several kinds of electric pianos and several electronic keyboard instruments that bridge the gaps between pianos, organs, and synthesizers.

Some readers will undoubtedly be familiar with some of the electric pianos that made impacts on the world of popular music in the 1950s, 1960s, and 1970s, through the work of musicians such as Ray Charles, the Doors' Ray Manzarek, the Zombies' Rod Argent, and others and their use of instruments made by Wurlitzer, Hohner, Rhodes, and Fender-Rhodes, not to mention various jazz keyboardists (e.g., Chick Corea and Herbie Hancock) who favored the electric piano during the same period. The history of the electric piano actually goes back before these well-known examples from the third quarter of the 20th century.

Kurzweil

Futurist, author, and inventor Ray Kurzweil might be better known to some readers as a TED Talk presenter, for his novels, and because of his work in Optical Character

> Recognition, particularly as used in the Kurzweil Reading Machine; however, he has also played a significant role in the development of synthesizers and digital keyboards. Kurzweil and his Kurzweil Music Systems made a strong entry into the world of digital music in collaboration with singer-songwriter-keyboardist Stevie Wonder beginning in the early 1980s. The company's digital pianos and synthesizers of the 1980s were considered to be some of the best instruments of their type, particularly since many of the less expensive instruments of the time did not feature the keyboard touch or solidness of Kurzweil instruments. In 1990, Kurzweil Music Systems was purchased by South Korean piano manufacturer Young Chang, which in turn was purchased by the conglomerate Hyundai several years later. At present, Kurzweil Music Systems manufactures a wide variety of professional performance keyboards, digital pianos for the home, speakers, electronic arranging units, and other gear (see, https://kurzweil.com/products/, for detailed information on the company's current products, accessed May 28, 2020).

In the late 1920s and early 1930s, Neo-Bechstein developed an electric piano that worked on similar principals to the electric guitars that were being introduced at the same time. The Neo-Bechstein instrument used electromagnetic pickups that created a weak electrical signal based on the vibrations of the piano strings. As was the case with electric guitars, this signal was sent to an amplifier. Other than the use of pickups, the physical mechanism of the Neo-Bechstein instrument was that of a standard acoustic piano.

Baldwin created its version of a similar instrument approximately four decades later. The Baldwin Electro-Piano had hammers and strings like an acoustic piano but, like its Neo-Bechstein predecessor, had electromagnetic pickups, the electric signal from which was amplified. As a product of an era in which musical genres such as rock and jazz were increasingly amplified, an instrument that sounded like an acoustic piano, but that could be amplified without the feedback sometimes associated with microphones, offered possibilities. And Baldwin's Electro-Piano gained at least a modest following among such musicians; however, compared with the non-string-based electric pianos of the era, it did not sell particularly well. Perhaps the Electro-Piano's biggest claim to fame was jazz pianist/composer Dick Hayman's *Concerto Electro*, specifically written and recorded by Hayman for the Baldwin instrument.

The most successful electric pianos, however, were those that used alternatives to metal strings. The instrument with perhaps the most interesting history is the Rhodes electric piano. Harold Burroughs Rhodes developed a small, school-desk-sized instrument after World War II as part of his attempts to offer group piano lessons to returning service personnel. Rhodes's instrument used small metal tines instead of strings; however, similar to instruments such as the earlier Neo-Bechstein and the later Baldwin Electro-Piano, electromagnetic pickups generated weak electrical signals from the vibrations of the tines, which were then amplified. The advantage of the tines was that they would hold their tuning far better than strings, which were subject to changes of tension. Eventually, Rhodes collaborated with Leo Fender to produce a larger instrument that worked on the same principals. After CBS bought Fender in 1965, the 73-key Fender Rhodes electric piano appeared on the scene. Rhodes and Fender Rhodes instruments have been heard on countless popular music recordings from the 1960s to the present. One of the most notable is the Doors' "Riders on the Storm," in which keyboardist Ray Manzarek makes use of the instrument's tremolo setting (labeled as "Vibrato" on the instrument) to imitate the shimmering of rain.

The Rhodes and Fender Rhodes instruments were not, however, the only non-string-based electric pianos at the time. Both Hohner and Wurlitzer produced similar instruments. Like the Rhodes instruments, these electric pianos also played significant roles in the popular music of the rock era. Most famously, Ray Charles toured with a Wurlitzer electric piano in the 1950s and used the instrument on one of his most famous recordings: the 1959 song "What'd I Say." Because the studio version of the song opens with Charles playing the opening riff on the electric piano, it is easy to hear the tone color of the Wurlitzer. The Zombies' Rod Argent played the Hohner Pianet electric piano on the group's 1964 British Invasion hit "She's Not There." This also is a particularly notable track, as Argent's extended electric piano solo makes it easy to hear the tone color of the Hohner instrument. Although not as closely associated with American rock musicians of the early to mid-1960s, the Zombies and other British bands of the period used Hohner's electric piano models fairly extensively.

Hohner was also associated with the Clavinet, an electronic keyboard that was string based. Like string-based electric pianos, electromagnetic pickups generated current from the vibration of the strings, with the current then being amplified. However, unlike the more piano-like electronic keyboard, the unusual striking point of the hammer against the strings on the Clavinet gave it a unique and highly percussive tone color on its attacks. This made it particularly suitable to funk music, with Stevie Wonder being perhaps the keyboardist most closely associated with the instrument.

Some of the electronic keyboard instruments of the pre-synthesizer era anticipated the types of clearly electronic sounds later associated with the synthesizer and might be thought of as instruments that bridge the gaps between electronic organs, electric pianos, and the analog synthesizers that would emerge in the last few years of the 1960s. The two best known of these were the Univox, made by the Unicord Corporation, and the Clavioline, made and marketed in Europe by Selmer and in the United States by Gibson. Arguably, the Clavioline was the most synthesizer-like of these transitional instruments, despite the fact that it had been invented by Constant Martin in the late 1940s, two decades before Robert Moog debuted his synthesizer. The connection between the Clavioline and early analog synthesizers was that instruments such as the Moog synthesizer and the Clavioline used electronic oscillators to generate sound. Perhaps the most famous use of the Clavioline was in Del Shannon's 1961 No. 1 hit single "Runaway." In the 1970s, Unicord became better known for its Univox Compac-Piano, an electric piano that performers such as Edgar Winter and Billy Preston played while carrying the instrument around their necks. Doing an internet search for terms such as "Bill Preston Univox" or "Edgar Winter Univox" can yield photographs of Unicord's print advertising from the mid-1970s, at which time Preston and Winter used and endorsed the portable keyboard.

One of the advantages of many electronic keyboard instruments and digital synthesizers is portability, making them popular for use in performance venues and at home. (Anikasalsera/Dreamstime.com)

Roland Corporation

Founded in Osaka, Japan, by Ikutaro Kakehashi in 1972, Roland Corporation played a major role in the development of synthesizers and drum machines in the 1970s, 1980s, and 1990s. Kakehashi and the company he founded are also credited with being one of the central moving forces in the establishment of MIDI as a computer-synthesizer interface standard. Its Boss Corporation unit pioneered drum machines and advanced-feature metronomes with units such as the Dr. Rhythm and Dr. Beat, respectively. It should be noted that the Dr. Rhythm and Dr. Beat units available today are considerably more sophisticated and powerful than their earlier namesakes, as the Boss Corporation products have evolved with the technology over time. Roland's 1977 MC-8 MicroComposer was one of the most powerful sequencers of its day, and like the company's Boss products, today's Roland sequencers are dramatically more advanced than their predecessors. Based on the number of Roland keyboards and keyboard amplifiers that one sees at concerts, in churches, and in studios, these are still considered leading performance units. The company is also known for its Cube amplifiers for guitar, bass guitar, and keyboard. The Cube amps are notable for the amount of power and volume they produce, considering their compact size. Some of the Cube amplifiers are battery powered and are popular with street performers and musicians who perform in small clubs.

See Also: Organs

Further Reading
Lenhoff, Alan, and David Robertson. 2019. *Classic Keys: Keyboard Sounds That Launched Rock Music*. Denton: University of North Texas Press.

Pareles, Jon. 2001. "Harold Rhodes, 89, Inventor of an Electronic Piano." *New York Times*, January 4. Accessed June 14, 2021. https://www.nytimes.com/2001/01/04/arts/harold-rhodes-89-inventor-of-an-electronic-piano.html.

Rogers, Vince, director and producer. 2011. *Vox Pop: How Dartford Powered the British Beat Boom*. Video documentary. London: BBC Productions.

Electronic Tuners

Since the English trumpet and lute player John Shore invented the tuning fork in 1711, this device was the standard by which instrumentalists tuned their instruments up to the widespread use of electronic tuners in the second half of the 20th century. Tuning forks are still available today and might most frequently be seen used by timpani players to tune their drums. However, with the miniaturization, ability to power devices with batteries, and affordable pricing that came with the use of transiters in electronic devices beginning in the 1960s, the electronic tuner is now the standard for string, woodwind, and brass players.

Although the well-known Stroboconn, by Conn, dates back to 1936, these early electronic tuners were not made for the consumer market of individual musicians and music students. Early tuners were far too expensive to establish a consumer market for electronic tuners. However, over the years, the Stroboconn was found in educational institutions, including college and university music departments and high schools. This and subsequent strobe-based tuners by Conn, Peterson, and other manufacturers were also important for instrument designers, manufacturers, and repair technicians, as well as makers of instrument components such as clarinet and saxophone mouthpieces. Because of their design and the information conveyed by the rotating wheels of strobe tuners, these machines allow the user not just to check the tuning of the fundamental pitch being played to see if it is flat, sharp, or in tune but also to enable musicians, instrument makers, technicians, and so on to see the tuning characteristics of the instrument's overtones. The overtones in large part define an instrument's tone color, and adjusting the tuning of overtones is one of the ongoing concerns of instrument makers, component makers, and technicians.

Over the years, the C. G. Conn corporation has been known for its woodwind (particularly, its saxophones) and brass instruments. Between the debut of the Stroboconn in 1936 and the late 1960s, it was also the principal source for a succession of strobe-based tuners. However, Peterson strobe tuners made their debut in 1967. Throughout the late 1960s and the 1970s, Conn and Peterson tuners were increasingly smaller in size and less expensive, which brought the market for electronic tuners within the reach of some professional musicians.

The basic principle of stroboscopes, whether they be tuners, the strobes that were used on some turntables in the 1970s and 1980s, or in other applications, is that periodic flashing of light hits a rotating object and when the aspect being measured (e.g., the frequency of a particular musical pitch or the speed at which a record player/turntable is revolving). In the case of a strobe tuner, the rotating disc will not appear stationary in the display until the player raises a flat note or lowers a sharp note until the frequency is where it should be, giving the tuning standard being used. Peterson Tuners has posted a short video on YouTube in which the mechanics of strobe tuners are explained and demonstrated (see, https://www.youtube.com/watch?v=U8KHeyT6xB0, accessed January 14, 2021).

The next major development in electronic tuners can be in the form of smaller battery-powered devices. Although Korg, which is also well known for its relatively inexpensive digital synthesizers, was the leader, other companies followed suit. The first of Korg's offerings came in 1975 with the WT-10, a tuner that used a needle meter. Although specifics have changed in subsequent years, variations in the basic concept of the needle meter, with flat to the left of center and sharp to the right of center, remain on many electronic tuners today.

The classic Korg, Seiko, and other similar tuners of the late 20th century and early 21st century were relatively small in size, even compared to the tuners of the late 1970s and 1980s (e.g., the Korg DT-2 measures approximately 11-cm wide, 7-cm deep, and 3.5-cm tall at its highest point; the Seiko ST777 measures approximately 9.5-cm wide, 8.2-cm deep, and 2-cm tall), and were affordable enough that they brought access to electronic tuners even more into the mainstream for musicians and music students.

These devices typically either could be set to particular tuning standard presets (e.g., A=438, 440, 442, or 444 on the Korg DT-2) or could be set to various whole numbers within a certain range (e.g., between A=410 and A=450 on the Seiko ST777). Some of the electronic tuners of this type included a speaker so that the user could hear the correct tuning note and try to match it on their instrument, while others did not produce an audible tone. In addition to a small built-in microphone, many of these tuners (e.g., both the Korg and Seiko tuners cited previously) included an input and output jack for connection via 1/4-inch plugs, such as what one would find on an electric guitar cable. Typically, this input and output would mute the output while the device was turned on and create a bypass when the device was turned off. This way, a guitarist could tune their instrument without listeners hearing the tuning process. An offshoot of the miniaturization of electronic tuners has been the built-in tuners that are offered on some acoustic electric guitars that have powered output to the amplifier. In some of these instruments, like with the plug-in tuners described earlier, the player can hit a button to select the tuner, which effectively mutes output to amplifier. This allows the player to tune in relative silence.

Even as dedicated tuners were being miniaturized from the 1970s to the end of the 20th century, several companies began offering small multifunction devices. Roland Corporation's BOSS DB-11 Music Conductor, for example, which made its debut in the mid-1980s, was a metronome that included a tap mode (the user could tap a pulse and the read-out registered that tempo in beats per minute) as well as a chromatic tuner.

In the first decade of the 21st century, even more miniaturized clip-on tuners entered the marketplace. Some of these used piezo technology in which the device would pick up the instrument solely through the vibrations in the body of the instrument to which it was attached. Some of the clip-on tuners (e.g., Korg's AW-1) could be toggled between piezo and microphone mode, while other tuners used only microphone technology.

As tuner technology entered the smartphone age and the proliferations of relatively inexpensive and free apps, at the time of this writing, there are numerous such apps available for iOS and Android phones. Some of the available apps are simply tuners, while others are multifunction suites that include tuners, metronomes, and so on. Although clip-on tuners or tuners tethered to an instrument through a clip-on piezo microphone are smaller than smartphones, some of the tuner apps offer features not available in the small stand-alone tuners, such as the ability to use different tuning systems—for example, the *Cleartune* app to use equal temperament, stretch-tuned guitar temperament, Pythagorean, Pythagorean just intonation, mean tone temperament, and several others that might be of particular interest to performers of Baroque-era and pre-Baroque music.

The continuing miniaturization of electronic tuners, as well as the integration of various functions into multifunction devices can also be seen in recent smartwatches for musicians, such as the Soundbrenner Core. This watch includes a vibrating metronome feature so that the wearer can feel the pulse without the metronome producing sound, a contact tuner that picks up the frequency of an instrument from its vibrations, and a decibel meter function, and the device also functions as a wristwatch.

Korg, Inc.

Founded in 1963 as Keio Gijutsu Kenkyujo, Ltd. (Keio Electronic Laboratories), Korg remains a prolific and significant player in the world of music-related electronics. The company offers small electronic keyboards that are used primarily as MIDI controllers, full-size synthesizers, digital pianos, electronic tuners, electronic metronomes, recording consoles, and other digital music-related products. Through its acquisition of VOX, Korg also manufactures guitar amplifiers, electric guitars, and electric guitar effects pedals under the VOX name.

Korg's first release in 1963 was the Doncamatic, which the company's website describes as "Japan's first disc-based rhythm machine" (Korg n.d.). The company also made electronic organs in the pre-synthesizer part of its life. Among Korg's most notable innovations, the company claims Japan's first synthesizer, the MiniKORG, which was released in 1973, and the world's first needle-based electronic tuner. Anecdotally, when these tuners were introduced to musicians who had to rely on expensive Conn Stroboconn and Peterson strobe-based tuners, Korg's tuners made an immediate favorable impact on professional musicians and music students alike. Although numerous companies offer electronic tuners today, Korg is still recognized as one of the leading brands for quality and reliability.

Further Reading

Bickerton, R. C., and G. S. Barr. 1987. "The Origin of the Tuning Fork." *Journal of the Royal Society of Medicine* 80 (December): 771–773.

Korg, Inc. n.d. "About KORG." Korg.com. Accessed May 27, 2020. https://www.korg.com/us/corporate/.

Electronic Wind Instruments (EWI)

The Electronic Wind Instrument, or EWI, is one of the technologies of the 20th century—still used and evolving today—that can create some confusion. For one thing, the name is sometimes used generically, but the term can also refer to the "Akai EWI," a specific product name. Similarly, some prominent electronic wind instruments, such as the Lyricon, were so frequently heard in recordings of the 1970s and beyond that their names, too, are sometimes used generically for woodwind-like electronic instruments. And finally, although electronic wind instruments seemed to appear out of nowhere in the 1970s on pop recordings, and although the instruments have been associated with MIDI technology for decades now, the development of these instruments actually has a longer and quite interesting history.

We might reasonably trace the development of electronic wind instruments back to the beginning of the 1940s. Leo F. J. Arnold applied for a patent for his Electrical Clarinet in 1941 (see, Arnold 1941), with the patent being granted the following year. According to Arnold's patent application, the Electrical Clarinet had more even response from note to note than a traditional acoustic clarinet. Arnold also claimed that his invention had a wider dynamic and pitch range than the standard clarinet and that extremely high notes were easier to control. Articulation and embouchure (the lip/mouth formation, placement, and pressure) development were also said to be easier, and Arnold's fingering system, although based on that of the standard French system, or Boehm, clarinet, was simplified. Although the Arnold Electric Clarinet did not catch on, the simplified embouchure, simplified fingering system, and the ability to more easily play notes that would be difficult or impossible on an acoustic woodwind instrument all played important roles in the development of the commercially successful electronic wind instruments that would appear in the 1970s.

Of all the modern acoustic woodwind instruments, the saxophone arguably has the easiest fingering system. It is absent some of the awkward cross fingerings of the oboe, it overblows an octave so that many notes an octave apart can be played with the simple depressing of the octave key (unlike the clarinet), and so on. Similarly, the modern Boehm system flute avoids most awkward cross fingerings, and there are relatively few fingering changes that need to be made from the lowest octave to the second octave. So, it perhaps is not great surprise that the electronic wind instruments that were ancestors of Leo Arnold's Electrical Clarinet generally have been designed around the basic principles of the flute and saxophone's key mechanisms or have been based on hybrid wind instruments such as the Melodica.

In fact, perhaps the best-known electronic wind instrument development between the work of Arnold in the 1940s and the start of the heyday of electronic wind instruments in the 1970s was a wind-operated keyboard-based instrument. Like the company's standard Melodica, Hohner's Electra Melodica relied on the player to blow into a mouthpiece/wind pipe and play notes using the keyboard. Although the instrument was meant to be amplified, the intensity of the player's airstream determined volume, just like on the acoustic version of the instrument. By varying the intensity of the airstream, then, the player could create a vibrato effect.

The age of the modern electronic wind instrument came with the pioneering work of orchestral trumpet player Nyle Steiner beginning in the 1960s. Steiner's experimentation with electronic instruments culminated in the 1970s and 1980s with his brass-like Electronic Valve Instrument (EVI) and his saxophone-like EWI. Arguably, Steiner's EWI was the most intuitive of the two, because of the inherent importance of embouchure and the overtone series in the production of pitches on valve-based instruments compared with the fingering-based pitch selection of an instrument such as the saxophone. Noted saxophonist Michael Brecker made several recordings that featured Steiner's EWI, and at the time of this writing, there is a video available on YouTube in which Brecker explains how the EWI works and demonstrates it to an audience (see, https://www.youtube.com/watch?v=mFDTOgf3tB4, accessed January 14, 2021).

Although Nyle Steiner was working on his instruments in the 1960s and early 1970s, the first widely used electronic wind controller was the Lyricon, an instrument co-designed by Bill Bernardi and Roger Noble. Perhaps the most important innovation of the original Lyricon and subsequent wind controllers was its ability to interface with a synthesizer. The Lyricon was sensitive to lip and air pressure, so volume, vibrato, and so on could be controlled in a manner similar to single reed instruments such as the clarinet and saxophone. The instrument became something of a staple in the recording studios of the 1970s and 1980s and was used by saxophone players such as Kenny G, Andy Mackay, and Tom Scott. In particular, Scott's work as a frequently recorded studio musician brought the Lyricon into popular songs and albums such as Steely Dan's "Peg," Michael Jackson's "Billie Jean," Blondie's *Autoamerican*, and George Harrison's *Thirty Three & ⅓*. Ironically, perhaps the best remembered of these is "Billie Jean," which was released after the instrument's maker, Computone, had gone out of business.

Although some of the wind controllers of the past, such as the Lyricon, were designed to interface with specific synthesizers, the rollout of MIDI technology in the 1980s was accompanied by the release of several electronic wind controllers that were MIDI compatible and, therefore, could be used with a wide range of MIDI synthesizers. With this development, the electronic wind instrument—or controller—became the functional equivalent of a MIDI keyboard or any other MIDI device, except that a saxophone-like electronic instrument would play one note at a time. Although perhaps not the most practical way to do so, a MIDI-based electronic wind instrument could be used to input pitches in a music notation program, or with any other MIDI-based application, with the single-voice limitation.

Although the MIDI-based electronic wind instruments of the late 20th century offered much more than the early instruments, there were still some limitations. For example, one of the persistent challenges for players who were used to instruments such as acoustic flutes, clarinets, oboes, and saxophones was that some popular electronic instruments were activated when the player touched the keys, as opposed to when they fully depressed the keys. Since the incorporation of MIDI in the 1980s, however, electronic wind instruments have continued to become more intuitive for players and have become more adaptable. For example, some modern controllers feature multiple fingering modes. These instruments can be switched to, say, oboe, flute, or saxophone mode, which makes the transition from specific acoustic woodwind instruments more seamless than with instruments such as the Lyricon or Nyle Steiner's EWI. In addition, some instruments also include a valve mode, which allows them to be played fairly seamlessly by brass players.

At the time of this writing, the electronic wind instrument sphere is dominated by Akai, which offers several models and packages, although other manufacturers (e.g., Roland) also offer instruments. Because electronic wind instruments now use the MIDI standard, an inexpensive instrument functions as a controller and relies on the synthesizer/computer for a bank of sounds, while more expensive models essentially have built-in synthesizer capabilities including a range of preset sounds.

See Also: Brass Instruments; Digital Synthesizers; MIDI; Woodwind Instruments

Further Reading

Arnold, Leo F. J. 1941. Patent granted 1942. Patent application for "Electrical Clarinet." Accessed January 18, 2021. https://patents.google.com/patent/US2301184A/en.

Guerrieri, Matthew. 2013. "With the '70s-era Lyricon, Woodwind Met Synthesizer." *Boston Globe*, July 6. Accessed January 18, 2021. https://www.bostonglobe.com/arts/music/2013/07/06/non-event-experimental-music-concert-include-lyricon-recalling-the-lyricon-era-woodwind-like-synthesizer/kYw3FB20rj7AzFGkMMPWHI/story.html.

Swallow, Matthew J. 2016. *MIDI Electronic Wind Instrument: A Study of the Instrument and Selected Works*. Doctoral dissertation. Morgantown: West Virginia University. Accessed January 18, 2021. https://researchrepository.wvu.edu/etd/6750.

F

45-rpm Records

The introduction of the plastic 45-rpm single record at the start of the 1950s roughly coincided with the establishment of a youth culture that was the result of what later would be labeled the generation gap. In fact, in the "Inventing the Teenager" episode of his 2007 program *James May's 20th Century*, sponsored by the U.K.'s Open University, British television host May argues that the development of this technology played a significant role in the development of a distinctly teenaged subculture. And, why might it have been the 45-rpm single that was part of this development?

The plastic 45-rpm single held several significant advantages over its predecessor, the shellac 78-rpm single. For one thing, it was smaller and therefore more portable. And the 45-rpm record's portability was also enhanced by the fact that the material it was made of, polyvinylchloride (PVC), was a relatively unbreakable material compared to shellac. In addition, shellac records not only were easy to shatter but were also more susceptible to dings and scratches if one hit them against a hard surface than were the new 45-rpm singles. So now teenagers—young people tended to be the largest market for singles—could now take their records to a friend's house or to a party and not worry that dropping them might destroy them. The other advantage that the 45-rpm PVC single had was that more information could fit in a smaller area on the disc, in no small part because the grooves could be physically closer together than on a shellac record; grooves placed too close together on a shellac record were prone to collapse.

One of the odd quirks about 45-rpm plastic singles is that they were manufactured with large holes in the middle for some markets and with small holes that would fit the spindle typically used for 33–1/3-rpm albums in other markets. Generally, 45-rpm singles released for the North American markets had the large holes, while those in European markets had the small holes. This meant that singles' fans in the United States, for example, would have to use an adapter that would fit over the small spindle on the turntable, with the outside of the adapter fitting the single's hole.

Although the 45-rpm single could hold more musical information in a smaller area than the shellac singles it eventually replaced, there were limits. As some songs grew longer, particularly during the 1960s, the discs' limitations were accommodated in several ways. One common method was to break a musical selection up into two pieces, with the first part playing on the A side of the disc and the second part playing on the flip side. It should be noted that this was not entirely a 1960s' phenomenon, as some earlier songs—such as Ray Charles's 1959 single "What'd I Say"—were also broken up into two parts for single release. In the 1960s, notable examples of this procedure include Eric Burden and the Animals' 1968 antiwar song "Sky Pilot," which fades out during a mock battle sequence complete with bagpipes in a hard rock setting and then reenters in an entirely different style on the B side.

The other process that the recording industry frequently used to accommodate the temporal limitations of the 45-rpm single was to edit the album version of a piece so that it could fit on one side of a single. Perhaps the best-remembered hit that received this treatment was the 1967 Doors' song "Light My Fire," in which significant parts of the electronic organ and electric guitar solos in the middle of the full-length version were exorcised for the chart-topping single release.

It was not just because of the physical limitations of 45-rpm singles that dictated that songs either be broken up between sides of a disc or truncated from their album versions; radio played a significant role. The primary medium for singles during the heyday of the 45-rpm disc was AM radio—as opposed to album-oriented FM radio—and AM radio stations were reluctant to program lengthy singles for commercial reasons. Specifically, more breaks between songs in, say, a given hour allowed more opportunities for airing advertisements.

Introduced at the beginning of the 1950s, the plastic 45-rpm single record became a staple of the growing, post–World War II youth culture. (Michel Bussieres/Dreamstime.com)

The fact that AM radio was so closely tied to 45-rpm singles, particularly in the 1950s and 1960s, and the fact that many of the inexpensive portable record players of the 1950s and early 1960s had only one speaker also meant that consumers were more likely to find that the mixes on singles were monophonic, even after stereophonic albums increasingly grew in popularity over the course of the 1960s.

The importance of the 45-rpm single as a separate entity from vinyl albums, particularly in the 1960s, is suggested by the approach to marketing taken by the Beatles' record companies—in the U.K., EMI (1962–1968) and Apple Records (1968–1970 [the band officially broke up in 1970]), and in the United States, Capitol (1962–1968) and Apple Records (1968–1970). At least after the Beatles' first album, there were few ties between the band's albums and singles. So, highly commercially successful songs such as "Hey Jude," "Strawberry Fields Forever," "Penny Lane," "Paperback Writer," "She's a Woman," "Eight Days a Week," "Revolution," "The Ballad of John and Yoko," and others were not centerpieces of albums of the same time period; they were initially released only as 45-rpm singles. Similarly, many of the B-side songs of the Beatles' singles were not released on contemporary albums (e.g., "I'm Down"). Some of the songs that were initially released as singles later appeared on albums; however, generally that did not happen—particularly in the U.K.—until months or years had gone by (e.g., "Penny Lane" and "Strawberry Fields Forever" were released as a double-A-side single in February 1967 and later on the *Magical Mystery Tour* film soundtrack album in December 1967).

Although there has been a 21st-century resurgence in interest in the PVC vinyl album, such has not necessarily been the case with the 45-rpm single. Despite the fact that today they are largely relegated to flea markets, specialty collectors' stores, and the listening rooms of collectors, 45-rpm singles played a significant role in the development of a youth culture in the 1950s and 1960s.

See Also: 78-rpm Records

Further Reading

Coleman, Mark. 2004. *Playback: From the Victrola to MP3, 100 Years of Music, Machines, and Money*. Cambridge, MA: Da Capo Press.

Thomas, Helen, Claudine King-Dabbs, and Dan Walker, producers. 2007. *James May's 20th Century*. London: BBC Productions. Issued on DVD in 2012 as Athena Learning AMP-8788.

GarageBand

In the 1990s, several virtual recording studio and sequencing programs were released, including Sonic Foundry's *Acid pH1* and *Sound Forge*. One of the features that marked Sonic Foundry's sequencing (*Acid*) and virtual recording studio (*Sound Forge*) from the beginning of their ongoing lives was the ability they gave composers, arrangers, DJs, and so on to drag and drop musical material to different points along a composition's timeline. This drag-and-drop capability, common in word processing programs, became a standard feature in virtual studio/sequencing apps and programs. In 2003, *Acid*, *Sound Forge*, and Sonic Foundry's other offerings were purchased by Sony Digital Pictures. The following year, Apple introduced *GarageBand* in 2004 after purchasing the German music software company Emagic. Although new versions of several of the early sequencing and virtual studio programs continue to be released, and although there are other new virtual studio/sequencing programs available, Apple's *GarageBand* deserves special attention because of how accessible and ubiquitous the program is; it is part of Apple's standard offerings for Macintosh computers, iPads, and even iPhones.

Like other smartphone, tablet, and computer applications and programs, *GarageBand* has not been static since its introduction in 2004. As a result, any specifics related to a particular version of *GarageBand* are apt quickly to become obsolete and easily might supersede aspects of any earlier version. As a result, let us consider some of the basic features of the program and some of the general changes that have occurred over the course of *GarageBand*'s life to date.

GarageBand truly is an inexpensive virtual multi-track recording studio. Users can control recording levels, equalizations, and various effects for each track. If one wants to record, say, on an external electronic keyboard, electric guitar, and so on, the user needs to use an interface that allows connection from usually a quarter-inch jack (from the instrument) to whatever connection the user's device requires (e.g., USB and USB-C). Once the connection is made, the user can tap into several of *GarageBand*'s options. Some users might find that plugging an electric guitar directly into the computer using an interface is actually preferable to, say, going through their own external amplifier. This is because *GarageBand* includes several virtual amplifiers/emulations. So, the user can use *GarageBand*'s version of the famous Fender Twin Reverb amplifier and record an electric guitar track that includes some of the tonal properties of a "real" Fender Twin Reverb without having to own a classic tube amplifier. The various virtual amplifiers in *GarageBand* offer similar controls (e.g., reverb, overdrive, and equalization) as the classic amplifiers that they emulate.

Like other virtual studio programs, including expensive, professional software packages, *GarageBand* provides the ability to drag and drop sections of recordings and to create loops. The looping capabilities were particularly useful for non-percussionists who created drum tracks. One could create a track or tracks for, say, an eight-measure phrase and use it as a loop, thereby eliminating the necessity to record bass drum, snare drum, ride cymbal, and other parts for every measure of a piece. Similarly, the looping capabilities work well for repetitive pitched tracks, for example, bass lines or keyboard riffs.

Recent versions of *GarageBand* include preset drum tracks that can easily be incorporated into the user's recording. Although this gives the user a convenient way to add a percussion track quickly and accurately, this feature, like the presets that came to dominate the digital sound synthesizer market after the success of Yamaha's DX7, can be viewed as a double-edged sword: by making several preset drum tracks available and so easy to incorporate into one's recording, some of the creativity of composing and recording one's own synthesized drum tracks can be lost. The user can still create their own drum tracks; however, in recent versions of *GarageBand*, the presets appear front and center.

One of the features of *GarageBand* and many other virtual studio programs is the default synchronization of tracks. Because the program emulates professional multi-track tape recorders that have playback and record heads offset, the user can record a new track in *GarageBand* while playing along with an already-existing track in synchronization. Although this might seem like a given for any virtual recording studio software, it is important to note because some computer/tablet-based audio recording programs (e.g., *Audacity*) do not necessarily default to synchronized recording.

Although *GarageBand* is available for a variety of devices—from smartphones to laptop and desktop computers—and despite the fact that the user interface for the various versions is similar, users will probably find that some versions are easier to use and more intuitive than others. Although this will probably depend on the individual user's personal experience, this author finds the iPhone version to be limited by the size of the screen and that the iPad version seems to be the most intuitive, offering the best balance of power, usability, and features.

The continuing leadership of Apple's *GarageBand* in the inexpensive (e.g., free with Apple devices) virtual studio market is confirmed by a 2019 review of the version of the program for Macintosh computers in which *PC Magazine*'s Jamie Lendino wrote, "There's been a seismic shift in how records are made. A couple of decades ago, it took a mountain of gear to make an album. Now, you can do it with the built-in software that comes with every Apple computer, thanks to the free *GarageBand*." Lendino continues by praising the 2019 edition of the program, which "features a surprisingly serious presentation that roughly mirrors the high-end Logic Pro X digital audio workstation, or DAW" (Lendino 2019). It should be noted that the audio quality that is possible with *GarageBand* falls short of what is possible with professional studio software such as *ProTools*; however, current versions of *GarageBand* are more than adequate for demo recordings and informal releases.

See Also: Desktop and Laptop Computers; iPads and Other Tablets; Smartphones; Virtual Studio Software

Further Reading

Harvell, Ben. 2012. *Make Music with Your iPad*. Indianapolis, IN: John Wiley & Sons.

Lendino, Jamie. 2019. "Apple *GarageBand* (for Mac) Review." *PC Magazine*, September 26. Accessed October 29, 2020. https://www.pcmag.com/reviews/apple-garageband-for-mac.

LeVitus, Bob, Edward C. Bair, and Bryan Chaffin. 2018. *iPad for Dummies*. 10th ed. Hoboken, NJ: John Wiley & Sons.

Generative Music

Generative Music is the name given by musician Brian Eno to pieces that are ever changing and, more specifically, that which is produced by apps created by Eno and Peter Chilvers, such as *Bloom* (introduced in 2008). Despite its recent ties to *Bloom* and other computer, tablet, and smartphone applications, Generative Music has its roots in a wide variety of musical styles and genres from the past.

In order to fully appreciate Brian Eno's work in developing computer-based Generative Music, it is necessary to look at his background. As an artist in his youth, Eno came into contact with various aspects of performance art and the avant-garde. Eno first came to the public's attention as one of the cofounders of the group Roxy Music in the early 1970s. Eno played keyboards with the band, but at their live shows, he manned the mixing board, which gave him a major role in shaping not only the mix of sounds but also the addition of effects. To put it another way, Eno played the role of a studio record producer/recording engineer but in a live performance environment. Beginning in the mid-1970s, Eno recorded solo albums as well as collaboration with other musicians, several of whom were close to the experimental edge of popular music, including guitarist Robert Fripp. Eno also began working extensively as a producer, tending to create somewhat impressionistic soundscapes within the context of rock music, for example, on David Bowie's song "Heroes."

Eno's own compositions and recordings began exploring atmospheric ambient music in the 1970s, for example, on his aptly named 1978 album *Ambient 1: Music for Airports*. Increasingly, Eno's music emphasized atmosphere and feel over established classical or pop music forms. This move culminated in Eno's work in generative music beginning in the 21st century.

Despite the fact that some readers might think of Brian Eno's Generative Music as being entirely new, it has roots in the early 20th-century French composer Erik Satie's "furniture music," which Satie composed specifically to be background music. The idea of nonobtrusive background music also helped to give rise to Muzak and other piped-in music services beginning in the mid-20th century. Eno's move away from traditional forms and incorporation of the 20th century's avant-garde branch of minimalism also give his Generative Music ties to the work of American composers such as John Cage and Morton Feldman, the Gregorian chant and organum-influenced Estonian composer Arvo Pärt, and European experimentalists such as Karlheinz Stockhausen.

The original version of Eno and Peter Chilvers's *Bloom* app allowed iPhone users to make musical changes to the program's loops by using the touchscreen on their device. The music was accompanied by visual effects of changing shapes and colors on the screen, similar in some respects to the "Visualizer" that Apple has included as part of *iTunes* (now known as *Music*). Since its introduction in 2008, *Bloom* has expanded to include compatibility with iPads and, most recently, with Android devices. The latest incarnation of the program allows users to select from 10 different musical environments, called "worlds." As the Bloom website explains the program's original function, "Requiring no musical or technical ability, the egalitarian and user-friendly Bloom app enabled anyone of any age to create music, simply by touching the screen" (http://www.generativemusic.com/bloom 10worlds.html, accessed January 14, 2021). When the user stopped making changes, the app would take over generating new loops. With the current version of *Bloom*, it is easy for the user to incorporate the loops that the app generates into the user's own compositions using a digital audio work station. Of the 10-world version of *Bloom*, *MusicTech*'s Andy Jones writes, "If you're a fan of more ambient styles of music and that of Eno himself, then *Bloom* will definitely appeal—you really do come up with some quite mesmerising patterns [British spelling in the original]. It's not just about the notes, either, as simply staring at a riff in any World can generate random patterns on the screen, too, so you're treated to some moving visual art to accompany your—and your iOS device's—compositions" (Jones 2018).

See Also: Artificial Intelligence; Computer-Composed Music; Computer-Generated Music; iPads and Other Tablets; Loops; Smartphones

Further Reading

Jones, Andy. 2018. "Brian Eno and Peter Chilvers Bloom: 10 Worlds Review." *MusicTech*, December 28. Accessed October 30, 2020. https://www.musictech.net/reviews/brian-eno-and-peter-chilvers-bloom-10-worlds-review/.

Prendergast, Mark. 2003. *The Ambient Century: From Mahler to Moby—The Evolution of Sound in the Electronic Age*. New York and London: Bloomsbury.

Gramophones and Victrolas

As the U.S. Library of Congress puts it, "Although Emile Berliner did not invent recorded sound technology, his innovations led to its mass distribution. His flat disc recordings eventually replaced the more fragile and unwieldy Edison cylinders as consumers' sound technology of choice" (The Library of Congress n.d.). Although the machine that played flat discs was known as a gramophone, over the years, terms such as "phonograph," "gramophone," "Victrola," and so on have been thrown around somewhat interchangeably for the flat disc-playing machines.

After Thomas Edison had done his first pioneering work on sound recording using wax cylinders, Emile Berliner developed the first flat-disc recording machine in the 1880s. Berliner's development set up one of the great music technology battles of the sound-recording era: the flat disc versus the cylinder. Despite some advantages possessed by Edison's system (e.g., a virtually indestructible stylus and superior audio fidelity), the flat discs won the battle, probably because of ease of storage, not being as susceptible to melting and breakage, and because several companies latched on to Berliner's idea.

Of the companies that followed Emile Berliner's lead, the most well remembered is the Victor Talking Machine Company, and its best-known record player was dubbed the Victrola. The history of the Victor Talking Machine Company, henceforth referred to as Victor, began when Emile Berliner asked Eldridge Johnson of Camden, New Jersey, to produce a reliable spring-wound motor that could be used on Berliner's phonographs. Johnson eventually joined the numerous others who were trying to enter the phonograph market and incorporated as the Consolidated Talking Machine, which became Victor in 1901. Perhaps one of Johnson's best business decisions early in his association with phonographs was that he trademarked the soon-to-be famous "His Master's Voice." This trademark, along with the image of the dog listening to the external horn of a Victrola, became an icon of the recording industry. In 1929, RCA Corporation bought the Victor Talking Machine Company to form RCA-Victor. Although Sony Corporation later bought the company, RCA Records still exists today.

In the early 20th century, the Victrola was just one of many flat-disc players available to consumers. And by "available," it should be noted that these machines could cost the equivalent of $1,000 in today's dollars and more. Despite the fact that the flat-disc player was winning the battle with Thomas Edison's cylinder machines, there were numerous attempts to create propriety systems, in particular, to try to maximize the commercial possibilities of the linkage between record players and the records themselves. For example, some companies made discs that were meant to be played far in excess of what would become the standard 78 revolutions per minute. The players also made by these companies could play these records, even if the players made by their competitors could not. Similarly, when Edison moved into the flat record space, his discs were considerably thicker than those made by other companies, and the grooves were cut differently. Playing these records on a competitor's machine could damage the disc. These are just a couple of examples of the linkage of early 20th-century companies' use of propriety equipment and technology.

Although the term "Victrola" was specific to the record players manufactured by Victor, the term became a generic term for 78-rpm record players. This is akin to how trade names such as Kleenex are used to refer to all tissues, whether or not they are actually Kimberly-Clark's Kleenex brand. In terms of brand-name machines becoming generic terms, the generic use of Victrola might be most akin to the use of the name Hoover in the U.K. to refer to vacuum cleaners and Hoovering as the process of vacuuming. In the case of the music industry, perhaps there was good reason for this. Victor's famous dog-and-horn logo remains one of the most instantly recognizable advertising symbols in history.

The large horns on gramophones and Victrolas were integral for amplification in these nonelectric record players. (Wisconsinart/Dreamstime.com)

During the era of the eventually standard 78-rpm records, Victor made several improvements to its machines and its recording processes. For example, the early music players, whether of flat discs or wax cylinders, had external horns through which the acoustically amplified sound was produced. One of the problems associated with external horns was that they could easily be bumped and could impede access to the playing arm and the media being played (e.g., the record). Some of the larger units that Victor and other companies began manufacturing had internal horns, thus eliminating some of the challenges posed by the external horns. When electronic amplification began being used in record players, the internal-horn concept was translated into stand-alone record players with an internal speaker; stereophonic sound would not make its commercial debut until the 1950s.

During the 1910s and particularly during the 1920s, engineers working independently and as employees of Victor, Western Electric, Columbia Records, Edison, and other companies worked to develop electrical recording processes, based in part on the microphone technology that was under development for radio and public address systems. Once the electrical recording process became firmly established in the 1920s, record and record player companies used it as a marketing point for new machines, continuing to link record sales and record player sales. For example, on the paper sleeves for its electrically recorded discs, Victor included text that extolled the merits of its new Orthophonic (the name for its electrical process) recordings, including references to the brilliance, clarity, and greater volume that these recordings produced on the company's standard Victrolas. At the same time, however, the sleeves also touted the company's new Orthophonic Victrolas, which were designed to bring out the full capabilities offered by the new electrically recorded discs. Some of the models of these machines and the electrical Electrolas that also Victor made were housed in beautiful wood cabinets similar to some of the more expensive radios of the period. These machines turn the record player into an important piece of furniture for the home and suggest the importance of recorded music as it increasingly took the place of home music making on instruments.

Although the formation of RCA-Victor, the advent of electrically driven and electrically amplified record players, and the later development of vinyl 33–1/3-rpm albums and plastic 45-rpm singles might seem to have rendered the Victrolas of the early 20th century obsolete, Victrolas continues to be manufactured in locations such as India and China. In fact, Victrolas and other acoustic, spring-driven record players played roles in the Indian subcontinent long after they had largely been replaced by electrical record players in the West because they provided music in remote locations that did not have access to electricity. And today, when authentic early 20th-century Victrolas and other 78-rpm record players are difficult to find and sometimes command high prices, the iconic "His Master's Voice" logo with its dog, Victor machine, and external horn is still part of popular culture.

See Also: Electrical Recording; 78-rpm Records

Further Reading

Coleman, Mark. 2004. *Playback: From the Victrola to MP3, 100 Years of Music, Machines, and Money.* Cambridge, MA: Da Capo Press.

Kittler, Friedrich. 1999. *Gramophone, Film, Typewriter.* Translated by Geoffrey Winthrop-Young and Michael Wurtz. Stanford, CA: Stanford University Press.

The Library of Congress. n.d. "Emile Berliner and the Birth of the Recording Industry." The Library of Congress. Accessed March 30, 2020. https://www.loc.gov/collections/emile-berliner/about-this-collection/.

H

Headphones

Although one might reasonably assume that they are a more recent development in music-listening technology given the rapidity with which new headphone manufacturers and models—including those endorsed by recent and current music stars—continue to appear, the history of headphones goes back to 1880s. Late in that decade, headphones were being used in London as part of a service that provided live electronic feeds of operatic and theatrical events. Although the basic concept was akin to today's livestreaming, these London feeds were wired. The actual headphones used at the time were not portable, being supported by a metal structure. A step forward was made in the early 1890s with Ernest Mercadier's patent of hands-free telephone headphones. Because Mercadier's device was made for insertion in the ear, they are perhaps more like an 1890s' version of earbuds than headphones per se.

Writing in *SVG News*, Mark Schubin traces the development of headphones back even a decade earlier, when Thomas Edison and Ezra Gilliland developed a primitive sort of headset for telephone operators to use (Schubin 2011). The device, which resembles a boxy microphone and earpiece attached to a metal rack that the operator wore around their neck, appears more to be the progenitor of the integrated microphone and headphone of the headset than what we usually think of as headphones.

Something closer to resembling modern headphones emerged in 1910 when Nathaniel Baldwin contacted the U.S. Navy to pitch radio-listening headphones that he was building on a table in his Utah home. Although Baldwin could produce only limited numbers of these homemade devices, they "were a drastic improvement over the model then being used by Naval radio operators" (Stamp 2013). Because Baldwin was unable to keep up with the navy's demand for his headphones, Wireless Specialty Apparatus Co. collaborated with Baldwin and built a factory in Utah to manufacture headphones of Baldwin's design. Still, at this time headphones were part of the world communications and not the world of music consumption. Even as commercial radio appeared in the 1920s, the earpieces that were used did not offer any of the corded portability of the headphones that were used by the military.

During World War II, headphone technology had advanced to the point that Germany's Luftwaffe pilots used stereophonic headphones to receive radar signals to allow them to bomb targets without having visual contact with their targets. A true, commercial application for stereophonic headphones designed for listening to music, however, did not develop until well after World War II, when

John Koss released the SP3 Stereophones in 1958. Koss's design quickly followed the development of stereophonic sound and stereophonic records by the British conglomerate E.M.I. in 1957. Although Koss's original design was meant to work in tandem with his own Koss portable stereophonic turntable system, it soon turned out to be something that could be used with other stereo systems. Other manufacturers soon began including headphone jacks with their stereo equipment, and other manufacturers soon began producing and marketing headphones, with the 1/4-inch headphone jack becoming a standard.

Interestingly, although it is now a standard procedure, it was not always the case that musicians used headphones in the recording studio. According to 1996 interviews conducted by Andy Babiuk for his book *Beatles Gear*, recording engineers Geoff Emerick and Ken Townsend suggested that the Beatles might have been the first rock band to use headphones in the studio when working on overdubs at Abbey Road Studios, probably in 1966. As Babiuk puts it, "At Abbey Road, as in other studios, the common practice when overdubbing a vocal, for example, was simply to play back the existing track through loudspeakers in the studio while the performer sang along to it into a microphone. Some of the backing would inevitably 'bleed through' into the vocal microphone . . . and, irritatingly, on to the vocal track on the tape" (Babiuk 2002, 184). Whether or not the Beatles were the first recording artists to use headphones in the studio, this practice became the norm by the late 1960s. The elimination of bleed-through caused by using open speakers gave engineers and producers much more control over the clarity of the final mixes of studio recordings.

It seems that with each generation of commercial headphones, improvements have been made. One of the issues with any kind of small speaker, such as those used in headphones, is bass response. Overall, frequency response has been improved in commercial headphones, with some recent models making bass response one of their selling points. In fact, overall frequency response has also been improved for small in-ear listening devices (e.g., earbuds) to the point that speaker size essentially is no longer an issue.

One of the issues that plagued headphones over the years was their inability to block out all external noises. Although some were better than others, a major step toward eliminating external noises came with the advent of noise-canceling headphones. These powered headphones (generally, from a small replaceable battery, or a built-in rechargeable battery) have a built-in microphone that picks up low-frequency sound and electronically eliminates some of it. A 2017 article on active noise-canceling headphones in the U.K. publication *Guardian* quotes Brian Brorsbøl of Sennheiser Communications, as saying that Sennheiser's active noise-canceling headphones reduce noise by approximately 30dB, with most of the reduction coming from low-frequency noise (Furseth 2017). Interestingly, just like much of the headphone technology that predated Koss's introduction of headphones for audiophiles back in the 1950s, active noise canceling came to the consumer market via aviation where it was first used.

One of the other issues with headphones that for decades tended to make them less convenient for music fans than stand-alone home speakers or music systems with built-in speakers is the fact that they required a physical connection to a stereo amplifier, a radio, a Walkman, and so on. Inconvenient cords, poor connections of 1/4-inch jacks and the 1/8-inch stereo miniplugs that became the norm, poor connections in 1/4-inch-to-miniplug adapters, cords that would develop breaks in one or more of the wires, and so on are issues still associated with corded headphones. Although Bluetooth technology was invented in 1999, it was not until five years later that wireless Bluetooth headphones became available for the consumer market. Today, numerous companies offer wireless Bluetooth headphones that produce good audio quality and feature reliable wireless connectivity, particularly to computers, smartphones, and tablets. Even with the advantages offered by Bluetooth headphones and earbuds, there are some caveats, principal of which is latency. The slight delay that is endemic with Bluetooth technology is significant enough that corded headphones and earbuds are still the preferred listening devices when using musical collaboration software and studio software.

See Also: iPods and Other Portable Digital Music Players; Radio; Sony Walkman and Discman

Further Reading

Babiuk, Andy. 2002. *Beatles Gear: All the Fab Four's Instruments, from Stage to Studio.* Revised ed. San Francisco, CA: Backbeat Books.

Furseth, Jessica. 2017. "Noise-Cancelling Headphones: The Secret Survival Tool for Modern Life." *Guardian*, March 16. Accessed January 18, 2021. https://www.theguardian.com/technology/2017/mar/16/noise-cancelling-headphones-sound-modern-life.

Germaine, Thomas. 2020. "Best Noise-Canceling Headphones of 2020." *Consumer Reports*, January 26. Accessed May 1, 2020. https://www.consumerreports.org/noise-canceling-headphones/best-noise-canceling-headphones-of-the-year/.

Schubin, Mark. 2011. "Headphones, History, & Hysteria." *SVG News*, February 11. Accessed January 18, 2021. https://www.sportsvideo.org/2011/02/11/headphones-history-hysteria/.

Stamp, Jimmy. 2013. "A Partial History of Headphones." *Smithsonian Magazine*, March 19. Accessed January 18, 2021. https://www.smithsonianmag.com/arts-culture/a-partial-history-of-headphones-4693742/.

I

Instructional Software

Since the development of personal computers, and later tablet computers and smartphones, music instructional software has tended to focus on music theory and ear training. Music theory software drills students on the recognition and analysis of and the building of intervals, scales, and chords; analysis of musical form; and so on. Likewise, ear-training software allows students to drill aural melodic and harmonic interval recognition, chord quality recognition, rhythmic and melodic dictation skills, aural recognition of phrase structure, and other associated skills. Although some of the early software was rudimentary, some software developers have incorporated proven pedagogical techniques in their programs. In particular, the concept of mastery, in which questions and examples that students misidentify come up again in the course of a unit or module until the student displays a greater mastery of the material, is an important component of the best music instructional software packages, as it is in, for example, foreign language instructional software.

Ear-training software for high school and college students has been especially important for some programmers, particularly as this is an area that demands practice, as opposed to book learning, and it is ideally suited for the use of computers. Although there have been numerous examples over the years, let us consider just one example: *MacGAMUT*, which was developed by Ohio State University music theory professor Ann Blombach. According to the product's website, the original published version of the software, which was released in 1988, drilled students on intervals, scales, and chord qualities. The product expanded into melodic and harmonic dictation and later added a rhythmic dictation component. In 2000, Blombach expanded what had started out as an Apple Macintosh–based product into a Windows-compatible release. In 2016, *MacGAMUT* stopped shipping the software as a physical package and make it purchasable only by download, and in 2020, *MacGAMUT* was purchased by Artusi, Inc. (https://www.macgamut.com/about/, accessed January 14, 2021). Artusi has since integrated *MacGAMUT* into its own *Artusi* aural skills software.

Although *MacGAMUT* is one of the longer-lasting ear-training products still on the market and still in use at colleges and universities around the United States, it is not unique. It is worth noting that historical overview presented previously reflects the general changes in music educational software over the past several decades. Many, like *MacGAMUT*, focused on helping students to develop a small set of skills but then expanded over the years. Many also moved from a single computer platform to multiple platforms and from physical software to downloadable software. Others have gone one step further and are now cloud based.

The use of technology in music appreciation courses began modestly, with books packaged with cassette tapes of sample repertoire. In time, and as one might expect, the cassettes were replaced by CDs. In the 21st century, music appreciation and music history courseware have moved to platforms such as online textbooks that use streaming audio services to provide recordings for use in the classroom and for individual use by students, as well as some online course packages that include customizable lecture PowerPoint presentations with links to YouTube videos of repertoire, as well as online quiz and testing materials.

Noted composer Morton Subotnick, perhaps best known for his computer music, has created several educational software applications. Subotnick, however, has focused on programs for young people. For example, his 1995 program *Making Music* instructs users aged 5 and up in musical form and structure, as well as other materials of music, but in a play-like environment. Subotnick published other ear-training and composition programs as well.

It has not just been world-recognized composers such as Subotnick that have produced instructional software. Appalachian State University music theorist Jennifer Sterling Snodgrass's *InForm* from the early 21st century led advanced high school students and university music majors through form and analysis, with a focus on hearing, identifying, and charting large-scale and small-scale phrase and section structure. While *InForm* was published by Electronic Courseware Systems, my attendance of a variety of sessions of the Association for Technology in Music Instruction (ATMI), which meets annually along with The College Music Society, suggests that numerous computer-savvy university music theory and ear-training professors developed what might be termed "home-grown" instructional applications for their students that were never commercially released. It should be noted that organizations such as ATMI with its *Journal of the Association for Technology in Music Instruction* and publications such as the *Journal of Music Theory Pedagogy* and the *Journal of Technology in Music Learning* have been valuable resources for developers of music theory and ear-training software in establishing best practices in electronic music instruction. In fact, it can reasonably be argued that the development of music instructional software has played a significant role in how music theory and ear-training teachers work with their students, both using computer software and in general. To put it another way, software development has helped to put an emphasis on student learning that might not otherwise have been as strong.

In addition to published software packages that could be purchased in, say, CD-ROM form, individually downloadable, or with site licenses, at the time of this writing, there are numerous free instructional smartphone, tablet, and laptop/desktop applications. To cite just one example, *EarBeater*, an ear-training program that drills basic scales, chord qualities, intervals, and chord inversions, is available for free at the time of this writing (see, https://www.earbeater.com/online-ear-training/#/, accessed January 14, 2021). It is worth noting that the home page for *EarBeater* mentions that the current version replaces an older Flash version. Indeed, the programming languages that designers and writers of music instructional software use have evolved. In fact, one of the early reviews of Electronic Courseware Systems' publication of *InForm* stated that one of the primary problems with the program was that the Javascript programming was not compatible across browsers and that different browsers required users to make various adjustments in order to play the mp3 files of the selections on which students would do aural analysis (Perry 2006).

The fact that there has been something of a proliferation of music theory and ear-training software and inexpensive smartphone and tablet applications raises several questions, including about the quality of the numerous applications and whether there is still a need for some of the tried and true software that is still marketed in some cases decades after it first appeared. As for the inexpensive applications, one is advised to pay careful attention to reviewers, and particularly to reviews in scholarly journals and in magazines aimed at amateur performers, as these tend to look critically at the helpful and not-so-helpful features of the applications.

Some of the long-standing software packages, such as Ars Nova's *Practica Musica*, have been produced for years. A program such as *Practica Musica* has at least one distinct advantage over the newer smartphone and tablet apps: it has been used by thousands of university music students and has been vetted by their professors. Because of this, a program such as *Practica Musica* has evolved over the years. It should also be noted that typically instructional apps tend to focus on small parts of the entire music theory/ear-training instructional spectrum. The well-established, albeit more expensive, software packages tend to be broader and deeper in what they offer users, something that is of particular importance, say, for university students pursuing a professional music performance or music education degree. Some, such as *Practica Musica*, also bridge the gap between what is sometimes segregated into separate ear-training and written music theory class by including both ear-based and written recognition of chords, intervals, scales, and so on.

Another facet of educational software can be found in products such as *SmartMusic* and its successor, *New SmartMusic*, programs that function as virtual accompanists and that can be used to help students develop performance-related skills such as playing and singing with good intonation, rhythm, steadiness, and so on.

See Also: iPods and Other Portable Digital Music Players; Radio; Sony Walkman and Discman

Further Reading
Perry, Peter. 2006. "*InForm*: A Music Analysis System." *Music Educators Journal* 93, no. 1 (September): 23.

Internet Radio

The internet makes it possible for worldwide radio stations to make their broadcasts available around the world. So, radio programming coming from, say, a small university's FM station in Ohio can be listened to in Egypt. Although the programming that is likely to be enjoyed over the internet might typically be live concerts and live coverage of sporting events, theoretically, everything that a station includes in its online programming could be available worldwide. In addition to local FM and AM stations that have gone worldwide through the internet, independent operators—often, it seems, radio hosts whose contracts with traditional broadcast radio stations have expired or radio hosts who have been fired by broadcast stations—have established internet-only radio stations. In addition, the world of internet radio includes the large iHeartRadio and other similar services.

When radio stations first began turning to the internet to expand their potential listening audience, the state of some of the technology was such that often the end user had to download an audio-player application or plug-in in order to listen. Having to go through multiple steps in order to gain access to the station was, in the vernacular, clunky and certainly less convenient than simply turning on a radio and tuning to the correct frequency. There were compatibility issues with some of the plug-ins based on the computer's platform (e.g., Windows versus Macintosh OS) or the web browser that the listener was using. At the time of this writing, the state of internet radio, particularly as it pertains to local radio stations that both broadcast on the airwaves and online, is quite different. Although certainly not a scientific study, anecdotally, in summer 2020, I picked three random broadcast stations with different owners in Stark County, Ohio, and tried the live broadcast links on their websites. Accessing the stations' programming took no longer than correctly tuning into the broadcast frequency on a radio dial, and the players that were built into the websites functioned without issue.

The internet spawned some radio stations that only broadcast online. Online radio could potentially be done on a budget, would not be subject to some of the same regulations as broadcast stations, and seemed to be a natural fit for popular radio personalities who either were fired by broadcast stations or whose contracts with broadcast stations simply ran out. For example, in 2012, locally popular morning show personalities Charlotte DiFranco and Pat DeLuca could not negotiate a contract with their Stark County, Ohio, radio station, so the two formed their own internet station (Pantsios 2012). Similarly, in 2015, at Virginia Beach, Virginia, country disc jockey Jimmy Ray Dunn was let go by the radio station for which he had worked for 24 years, so he started broadcasting his live show on his own internet station, JimmyRaydio.com (Duke 2015).

Although there are several companies involved in wider groupings of individual broadcast radio stations under the umbrella of an internet radio entity, the best known and most recognizable of these is iHeart Radio. Formed in 2011, iHeart Radio collects programming from what at the time of this writing the company claims are over "850 live broadcast stations in 153 markets across America" (https://www.iheartmedia.com/stations, accessed January 14, 2021). In addition to boasting a large number of well-known local broadcast stations, these stations are known for diverse programming, including talk radio, country, hip-hop, Top 40, soft rock, pop, alternative, R&B, and so on. iHeart Radio claims over 144 million registered users of its downloadable app and that the total programming of the company (on individuals' devices, in stores, in restaurants, and so on) reaches "9 out of 10 Americans every month" (https://www.iheartmedia.com/stations, accessed January 14, 2021).

See Also: Pandora and the Music Genome Project; Radio; Satellite Radio

Further Reading

Duke, Dan. 2015. "Jimmy Ray Dunn Starts His Own Radio Station—on the Internet." *Virginian-Pilot*, November 5. Accessed January 18, 2021. https://www.pilotonline.com/entertainment/music/article_6cef89c1-e04b-56d0-b12a-3936e362bdeb.html.

Jacobson, Adam. 2019. "Behind iHeart's Product Strategy Pitch." *Radio + Television Business Report*, April 3. Accessed January 18, 2021. https://www.rbr.com/behind-ihearts-product-strategy-pitch/.

Pantsios, Anastasia. 2012. "Canton Morning Team Settles Lawsuit against Former Station." *Cleveland Scene*, July 31. Accessed June 9, 2020. https://www.clevescene.com/scene-and-heard/archives/2012/07/31/canton-morning-team-settles-lawsuit-against-former-station.

iPads and Other Tablets

When Apple Computer launched its iPad in 2010, the iPad immediately started to generate interest among musicians, particularly as the list of applications for the device grew, and continues to grow. In a little more than a decade after its introduction, the iPad, to a greater extent than other tablet computers, has been a force for music notation for printing and for display on the device itself, as a virtual recording studio, as a means of control of performance venue lighting and audio systems, not to mention organizing a library of audio files using Apple's *iTunes* and other applications.

Today, it is becoming more and more common to see classical musicians—who more typically read music notation while performing than musicians who are active in other genres—using iPads and other tablets instead of sheet music. Although a limited amount of music might be published in pdf form, musicians can scan published sheet music and use applications that are specifically designed for performing musicians. Although it is not the only such application, it has been the author's experience that many musicians are currently using *forScore* in their performances of notated music.

The basic premise of applications such as *forScore* and *eStand* for the iPad is to save the sheet music as pdf files, generally one file per song or instrumental composition. The user can then swipe the screen to turn the page or purchase an optional foot pedal that is used for page turns. At the time of this writing, one sees *forScore* being used at the time of this writing in solo performances, chamber music, choral, and orchestral performances. As one might imagine, an application such as this, particularly when used with a foot pedal, can make page turns far more convenient. In addition, performance-based applications that are based on the use of common cross-platform file formats, such as pdf, also allow musicians to magnify or shrink the music notation, as best fits their requirements.

forScore and *eStand* allow the user to write music editorial marks and other annotations on the pdf files and delete pages if there are cuts in the song—a situation sometimes faced by pit orchestra musicians. In fact, the ability to write in editorial marks on the pdf files makes applications such as *forScore* and *eStand* particularly useful for pit orchestra musicians. And if, say, a show's entr'acte is used in abbreviated form for the bows at the end of the show, the pdf file of the entr'acte can be duplicated at the end of the show's set of pdf files, with the cuts marked. Perhaps less recognized, but just as important for performing musicians, is the fact that typically the battery capacity of iPads and other such devices tends to be sufficient for use during long shows, unlike some of the battery-operated music stand lights on the market. At the time of this writing, one can find several tutorials that explain and demonstrate the features of *forScore* on YouTube (see, e.g., https://www.youtube.com/watch?v=TQYaxXgnvGA, accessed January 14, 2021).

The use of tablet and iPad-based applications such as *forScore* and *eStand* also became part of suggested protocols for improving the safety of performing musicians during the COVID-19 pandemic. Specifically, in an August 2020 article in *International Musician*, the American Federation of Musicians' Theatre, Touring, and Booking Division director Tino Gagliardi suggested the following changes to the preparation and distribution of music: "a) Adjustments to rental and touring music distribution procedures to protect performers; b) PDFs of advance books; c) Touring librarian safety in handling returned parts and rental parts; [and] d) Encourage iPad or EStand technology" (Gagliardi 2020, 9).

One of the distinct advantages that the iPad and some other tablets possess over, say, laptop computers, is that some of the applications for recording, mixing, and notating music—particularly those that specific to the iPad/tablet—tend to be less expensive than those that are specific to Windows, MacOS, and other laptop and desktop computer operating systems. For example, the iPad application *Notion* is priced at $14.99 at Apple's App Store at the time of this writing. Although the program is not as powerful as high-end music notation programs for laptop and desktop computers—such as *Finale* and *Sibelius*, to name just two—it is a fraction of the cost of the high-end, computer-based notation programs, including less powerful programs such as MakeMusic's *Finale PrintMusic*.

The free download *Audacity*, which is available for iPads as well as for laptop and desktop computers, is another application that is useful for editing of sound files for length, adding effects, looping, and so on. As a user of the iPhone, iPad, and laptop version, I would suggest that the tablet version of this particular program seems to be more intuitive and generally easier to use than the Mac OS version. Similarly, the iPad version of Apple's *GarageBand* seems in many ways to be more intuitive than the Mac OS version. Changing inputs (e.g., from a USB microphone to a guitar or microphone input from a device such as a PreSonus AudioBox) seems to be more seamless.

On the flip side, it should be noted that some software that is available in tablet and laptop/desktop versions has more limited functionality on iPads and other tablets. For example, audio editing in multi-camera/multi-sound input projects in Apple's *iMovie* is more limited in the current iPad version than it is in the Windows version of the program.

Currently, iPads appear to be the tool of choice for remote control of theatrical lighting and sound systems. The fact that the devices are as powerful and portable as they are means that these systems can be controlled from within the audience space, as opposed to from a lighting and/or sound booth. As a result, the audio and visual perspectives of the sound/lighting technician(s) can be exactly or nearly exactly the same as that of an audience member.

The iPad has seen perhaps more than its fair share of publications, including such titles as *iPads for Dummies*, *My iPad for Seniors*, *Make Music with Your iPad*, and others. It is interesting to note that all these volumes address the subject of music on the iPad but in different ways depending on the books' audiences. For example, *My iPad for Seniors* (Miller 2020) includes subjects such as playing and purchasing music from the *iTunes* Store, listening using the Siri voice interface, and using specific streaming audio services such as Amazon Music Unlimited, Amazon Prime Music, Apple Music, Pandora, Spotify, Tidal, and YouTube Music. *iPads for Dummies* not only includes those same subjects but also includes more of a focus on organizing albums, playlists, libraries, and the like in *iTunes* (LeVitus et al. 2018). By contrast, publications such as *Make Music with Your iPad* focus on music notation, recording, virtual studio (e.g., *GarageBand*), virtual MIDI instrument, drum machine, amplifier emulation (e.g., *AmpKit*), and other such applications (Harvell 2012). For example, Harvell's *Make Music with Your iPad* includes an extensive section on using *GarageBand* as a virtual studio and songwriting/arranging platform, including suggestions on how to differentiate sections in a composition, bouncing tracks like one would do in an analog tape-based studio when the number of tracks was limited (e.g., on a 4-track or 8-track studio deck), and so on.

See Also: Boomboxes; Music Notation Software; Virtual Studio Software

Further Reading

Gagliardi, Tino. 2020. "From the Theatre, Touring, and Booking Division—Safety Protocols for a Return to Work for Pit Musicians." *International Musician* 118, no. 8 (August): 8–9.

Harvell, Ben. 2012. *Make Music with Your iPad*. Indianapolis, IN: John Wiley & Sons.

LeVitus, Bob, Edward C. Bair, and Bryan Chaffin. 2018. *iPad for Dummies*. 10th ed. Hoboken, NJ: John Wiley & Sons.

McDonald, Glenn, and Ken Mingis. 2019. "The Evolution of the iPad." *Computerworld*, October 1. Accessed January 18, 2021. https://www.computerworld.com/article/3269331/the-evolution-of-the-ipad.html.

Miller, Michael. 2020. *My iPad for Seniors*. 7th ed. Carmel, IN: Que Publishing.

iPods and Other Portable Digital Music Players

Although they continued to be available and marketed after the arrival of the smartphone, iPods and other portable digital music devices, sometimes generically called MP3 players, were at their peak of use and of controversy in the last decade of the 20th century and the beginning of the 21st century. These devices helped to bring portable digital music to the masses and pointed the way to a future in which digital files would continue to gain prominence in music consumption while digital physical media's role gradually decreased.

The development of the compressed music file format known as mp3 is detailed in the entry for MP3s in this volume. However, it is important to note that this compressed file format grew out of an attempt to use ISDN phone lines for the transmission of music—not for something that could be stored and played using small stand-alone devices. In 1986, several aspects of this technology-under-development were patented, and in 1987, the German institution Fraunhofer-Gesellschaft became involved in researching low bit-rate audio encoding. In 1988, the Motion Picture Experts Group was formed to try to create a standard for encoding music in a compressed format. The mp3 file format was finally recognized by the International Standards Organization in 1992. The importance of this recognition is that it gave potential manufacturers of what came to be known as mp3 players a single standard, thus avoiding the kind of VHS-versus-Betamax battles that had occurred during the development and standardization of other media.

One of the challenges faced by users of the early mp3 players is that transferring files from one's desktop or laptop computer was not always the easiest task. Similarly, organizing the music files so that they were easily accessible also was not necessarily intuitive and differed from one mp3 player to another and from one computer-based software application to another. Some of the popular early mp3 players that were available in the late 1990s had storage capacities that were not necessarily large enough to suit the needs of some users. Another challenge with some of the inexpensive mp3 players was that some offered the user no—or, perhaps only limited—ability to change the equalization of the output to control the bass, mid-range, treble, and so on.

However, mp3 players offered portability and in a considerably smaller, lighter, and less susceptible to misplaying manner than some of the portable CD players of the time. In addition, because they were headphone based with auxiliary output, they could easily be plugged into car audio systems, which increasingly were including auxiliary (stereo mini plug) inputs.

The end of the 20th century and the start of the 21st century was a particularly fast-paced time in terms of mp3-related developments. Although some of the early mp3 players suffered from the challenges mentioned previously, these players and the computers to which they needed to be attached in order to upload music onto the players were at the center of one of the major music controversies of the age of recorded music. Peer-to-peer networking services such as Napster allowed users to share music files without payment to record companies, artists, songwriters, and so on. The controversy this created and the damage to the music industry it caused are detailed in the Peer-to-Peer File Sharing entry in this volume. Suffice it to say, mp3 players and the computers that were used to load them with music were at the center of the issue.

On a more positive note for portable music players, the recorded music industry received a considerable amount of publicity with Apple Computer's Steve Jobs's October 23, 2001, announcement of the iPod. Apple's device quickly became a hit, with the iPod and Jobs making the cover of *Newsweek* magazine on July 26, 2004. The cover proclaimed that the iPod was "The Must-Have Music Player Everyone Is Talking About." The reasons for the interest—the buzz—that the iPod generated were several in number. Perhaps most important, this small device had enough memory to hold many times the number of musical selections of other mp3 players of a similar size. The iPod also featured a variety of equalization settings, most of which were named after particular musical genres. Although this might have suggested to the user who was a fan of hard rock music that they should use the "Rock" setting for the most optimal sound, one could choose other settings and perhaps find that the "Jazz" setting actually made for a more satisfying listening experience for a particular track or for a particular album. Along with genre settings, iPods over the years have also included settings that boost the bass, boost the treble, are designed for playback through small speakers, and so on.

iPod users also found that the interface with Apple's *iTunes* allowed for relative ease in building and maintaining a library of mp3 files; it certainly was a more user-friendly system than those associated with many of the earlier mp3 players. Albums that the user burned from a CD would be grouped together in the iPod as an album; however, one could search the iPod for individual song/selection titles, or by artist, or, perhaps the most important feature, by building a playlist. The user could take any of the tracks stored on the device and construct a playlist, which we might think of as a virtual album that could contain an assortment of selections. Athletes who worked out to music found the playlist feature to be especially useful, because they could essentially program the pace of the workout (e.g., warming up, working up to maximum workout speed, and cooling down).

The iPod might have generated a considerable amount of interest and might have offered features that were difficult to find in other mp3 players of the first several years of the 21st century, but it did present some challenges. Perhaps most important, the early iPods would only sync with Apple's Macintosh computers, so if one's desktop or laptop computer was not a Mac, one was out of luck as far as the iPod went. The iPod was also more expensive than many other mp3 players, although the iPod offered more features and was easier to use than some of its competitors. Eventually, Apple expanded the original proprietary nature of the iPod and *iTunes*.

One thing that is undeniable about the linked *iTunes* application and the iPod is that the pairing fundamentally changed the way in which at least a generation of computer users and music fans conceived of music. The concept of the playlist expanded the notion of the user-constructed

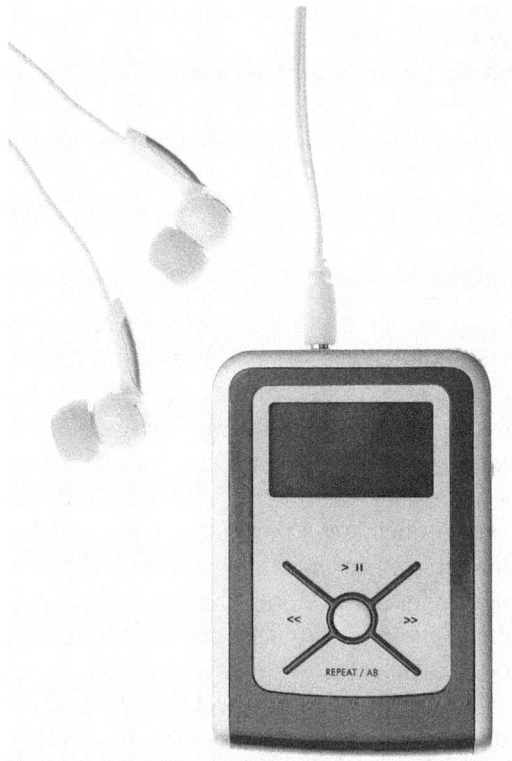

Beginning in the late 1990s, mp3 players began replacing portable cassette and portable CD players as the media of choice, particularly for joggers, runners, and walkers, because of their small size and portability. (iStockPhoto.com)

(some might go so far as to say "user-composed") album. Although music fans had been creating cassette mix tapes that might offer them an assortment of songs and/or instrumental pieces by a variety of artists on a single cassette, the iPod/*iTunes* playlist was far easier and faster to create. To users who grew up with iPods and iTunes in the early 21st century, the concept of the playlist as a coherent set of user-selected tracks became a significantly more important and commonly found way of organizing and consuming music than it had ever been in the past. Because this phenomenon has outlasted the period of the greatest popularity of the iPod and was a feature perhaps more closely associated with Apple's *iTunes* (now known as *Music*) on the various Apple devices (e.g., laptop and desktop computers, iPhones, and iPads), this is detailed in the entry in this volume on *iTunes*.

Although they continued to be manufactured, iPods and other mp3 players were rendered virtually obsolete by iPhones and other smartphones that offered many of the same music-organizing and portable music-listening features of the earlier one-function devices. Unlike mp3 players, however, smartphones also offered consumers the benefit of having small speakers. As a result, headphones or earbuds were not necessary with smartphones the way they were with mp3 players.

See Also: MP3s; Peer-to-Peer File Sharing

Further Reading
Aamoth, Doug. 2014. "Watch Steve Jobs Unveil the iPod 13 Years Ago." *Time*, October 23. Accessed January 18, 2021. https://time.com/3533908/ipod-turns-13/.
Coleman, Mark. 2004. *Playback: From the Victrola to MP3, 100 Years of Music, Machines, and Money*. Cambridge, MA: Da Capo Press.

iTunes

During the late 1990s, there were several programs that were in use for organizing one's digital music in mp3 format and for getting it onto the computer and from the computer on to an mp3 player. These had their individual quirks and style; however, the lack of uniformity between the various applications meant that there was not one clear standard and not one particular program that necessarily stood out among the rest. Apple Computer introduced both its iPod music player and *iTunes* in 2001. Over the years, *iTunes* also became a platform for the company to sell digital music and other media content, and it became a standard cross-platform program for music storage and organization on Windows and Mac OS computers alike. *iTunes* also provides the interface between personal computers and Apple's iPods (and successors), iPads, and iPhones.

The number of mp3 players on the market in the 1990s was staggering; however, many of the more affordable units offered limited storage capacity, and the software used to interface the players to users' computers differed from company to company. Apple's introduction of the iPod and *iTunes* in 2001 changed all that, although non-Apple mp3 players continued to be produced. *iTunes* offered users a more intuitive way to turn digital files from compact discs into the smaller mp3 files. In addition, it offered a more convenient way to organize these files.

Eventually, *iTunes* also offered alternative file formats that enabled users to select the audio quality of the material that they housed on their computer and/or iPod (and later iPad and iPhone). The choice of file type was important because mp3 files could be saved on a CD-ROM for playback on many of the user's CD-based devices (e.g., car CD player, laptop computer, and desktop computer). And, if one wanted to focus on audio quality, one could use *iTunes* to save digital files from a CD in a format that rivalled CD quality.

Eventually, *iTunes* included a music sales component. One could (and still can) purchase individual tracks or albums in downloadable digital format. Today, musicians can offer tracks and albums for sale as downloads on *iTunes* (and other sources) without requiring the backing of a record label. At the same time, one could argue that Apple's focus on downloading music helped to encourage the loss of the value-added material available through physical media such as a CD. To put it another way, with *iTunes*' downloads, gone were the liner notes and booklets that were published with CDs. One of Apple's download competitors, Amazon, offered what might be considered the best of both worlds: when one purchased certain CDs through Amazon, the consumer would also receive mp3 files of the tracks through Amazon Music.

Apple's marketing of music through *iTunes*—as well as the organizing protocols the program made (and still makes) available—was part of, if not at least partially responsible for, a paradigm shift in music-listening and purchasing patterns. The ability to create playlists and to purchase individual tracks that might otherwise have only been available through the purchase of an album seemed to be linked to what social observers saw as a move to a playlist mentality, particularly among college-aged users of *iTunes*, beginning in the first decade of the 21st century.

Collegiate music educators engaged in discussions in the 2000s and 2010s about how to deal with what they perceived as a fundamental change in the music-listening and music-organizing habits of their students. Particularly after *iTunes* had become firmly established and had been used along with iPods to organize music, one particular aspect of the *iTunes* structure seemed to be at the root of this fundamental music consumption change: the playlist. Anecdotally, this author was part of roundtable discussions that instructors of music in general studies courses had at several annual meetings of the College Music Society in the 2000s and 2010s, and the perception among college and university music educators that the students of this time period had developed a fundamentally different way of organizing and conceiving of music was a topic of discussion a number of times. The playlist feature of *iTunes*—and the iPod—seemed to be linked to a track or playlist mentality among students. Earlier generations, particularly those who grew up as music consumers in the 1960s through the first decade of the 21st century, certainly thought of some of their music as individual songs or individual movements of multi-movement works; however, especially after the middle of the 1960s, the album reigned supreme for many music fans. What was seen by college faculty in the 2000s and 2010s was that many students now organized their music by using playlists. They might still have albums that they had purchased through the *iTunes* store or that they had "ripped" from compact discs using *iTunes*;

however, now they were increasingly seeing music as individual, discrete chunks. As a result, a favorite playlist might include all fast movements from symphonies, with the slow movements being exorcised, or a playlist might include music by the Beatles, Green Day, Lady Gaga, and P. Diddy.

During the late 2000s and early 2010s, it was not uncommon in my own music appreciation and American music classes for general education students to have students ask to write papers on their favorite playlist rather than on their favorite album. In particular, the playlist provided a way for college students—whether they be involved in intercollegiate athletics or just using their music for casual workouts—to program a workout pacing for using lifting, jogging on a treadmill, and so on. The *iTunes* playlist mentality, then, empowered users to create their own "albums." Of course, during the heyday of the cassette tape, it was also possible to create playlist-like tapes and to mix tracks from a variety of artists or across genre boundaries. *iTunes* made the process easier and easier to modify (e.g., substituting one track for another, adding to a playlist, and deleting tracks from a playlist while keeping the file in one's library).

Another important consideration to keep in mind regarding *iTunes* is that it has not been nor it is now dependent on the consumer using it for multiple devices. So, a consumer can use *iTunes* solely on their computer without needing to own a compatible smartphone, tablet, or iPod. In fact, *iTunes* is available for Mac OS and Windows machines.

As file compression standards have changed over the years, *iTunes* has kept pace. At the time of this writing, the program's standard formats are Advanced Audio Coding (AAC) and MPEG-4, although one can still play mp3 files in the program. The AAC format produces compressed files that are close to the size of mp3 files; however, the audio quality is better. And AAC offers flexibility with an even higher bit rate for even better audio quality.

We purposely refer to "the program" in the previous paragraph because in 2020, Apple made a change to the designations of its music applications. In effect, the company retired the old multi-faceted *iTunes* and replaced it with multiple better-differentiated applications. One of the challenges with *iTunes* for the consumer was that the word itself meant different things: it was an organizing and music-playing application on computers, tablets, and smartphones; however, it was also a store for purchasing downloadable music, videos, movies, and so on. As the music industry increasingly moved to streaming music and subscription models in the 2010s, another layer of complication was added. Although *iTunes* has been replaced by separate music-playing (*Music*), music-streaming (*Apple Music*), and downloadable music (*iTunes Store*) entities, the spirit, design, and functionality of the old *iTunes* remain.

See Also: iPads and Other Tablets; Smartphones

Further Reading

Aamoth, Doug. 2014. "Watch Steve Jobs Unveil the iPod 13 Years Ago." *Time*, October 23. Accessed January 18, 2021. https://time.com/3533908/ipod-turns-13/.

LeVitus, Bob, Edward C. Bair, and Bryan Chaffin. 2018. *iPad for Dummies*. 10th ed. Hoboken, NJ: John Wiley & Sons.

J

Jukeboxes

A coin-operated record player that allows the consumer to play a recording for a small amount of money, typically in a bar, restaurant, or some similar establishment, the jukebox helped to define the way in which people obtained their music—particularly current top hits—during the devices' heyday of the 1940s, 1950s, and 1960s. Jukeboxes can still be found today, with machines over the years moving from shellac records to vinyl/plastic records, to CDs, and now to streaming technology.

Although the jukebox is defined as a coin-operated record player, its roots can be found in the coin-operated mechanical music devices—such as the well-known player piano and the advanced disc-based music boxes—that were produced and largely successful from the 1890s into the 1920s. However, it should be noted that Louis Glass and William S. Arnold invented and patented a coin-operated phonograph in 1890. The drawings and description of Glass and Arnold's device clearly show something that was quite a long distance from the glitzy jukeboxes of the mid-20th century; however, this device, as well as the player pianos and other coin-operated live music devices that soon followed, demonstrated that even in the 1890s, the concept that people might pay a small amount to hear a little music when they were perhaps at a restaurant, a saloon, and so on was on the minds of entrepreneurs.

The importance and prominence of the jukebox as a medium in the mid- to late 20th century are suggested by the fact that the music industry trade magazine *Cash Box* (sometimes published as *The Cash Box*), which was published between 1942 and 1996, included separate record charts for sales and jukebox plays up until the late 1950s. In addition, the magazine published regional reports from radio station disc jockeys that showed what records were favorites in regional markets (see, e.g., the January 5, 1957, edition of *Cash Box* at https://www.americanradiohistory.com/Archive-All-Music/Cash-Box/50s/1957/CB-1957-01-05.pdf, accessed January 15, 2021). Both *Billboard* and *Cash Box* moved to using combinations of sales, radio activity, and jukebox plays to determine chart rankings, suggesting a decrease in the prominence of jukeboxes as a commercial medium by the end of the 1950s.

It is interesting to note that back in the days of separate charts for jukebox activity and record sales, there were sometimes discrepancies between the charts. For example, in the January 5, 1957, edition of *Cash Box*, Jim Lowe's recording of "Green Door" was no. 3 in record sales and no. 2 in jukebox plays; however, significantly more dramatically, Bill Doggett's recording of "Honky Tonk" was no. 7

Jukeboxes, which were especially popular in the mid- to late 20th century, typically offered numerous choices for listeners in restaurants and bars. (Marco Clarizia/Dreamstime.com)

on the jukebox charts but no. 19 on the record charts. This suggests that consumers were considerably more likely to spend pocket change to hear "Honky Tonk," perhaps at a honky tonk (bar) than they were to purchase the record.

J. P. Seeburg, a company most closely associated with jukeboxes, extended the jukebox concept into private homes for a time. As Joseph Lanza writes in his study of easy listening and piped-in music, *Elevator Music*, "among the biggest background music suppliers aside from Muzak was J.P. Seeburg, a jukebox manufacturer in Chicago that maintained a library of up to 7,000 recordings. Seeburg had the advantage [over Muzak] of adding private houses to its customer roster: 570 new homes erected in the Detroit suburb of Westchester Village were equipped with Seeburg's Select-o-Matic units specially made for hi-fi use" (Lanza 1994, 53). Seeburg's collaboration with the developer of Westchester Village was covered by *Billboard* magazine, which included a photograph of the living room of one of the homes so that the reader could put the Select-o-Matic units in context (see, Wickman 1955, 51). This was the heyday of the jukebox, as suggested by several articles in a special issue of *Billboard*.

During the jukebox's glory years, a typical machine in a diner, bar, and so on might hold fewer than 50 up to 70 records, depending on the model. Because jukeboxes of the time were made to play singles—first shellac 78-rpm singles up into the early 1950s and then plastic 45-rpm singles—which had an A side and a B side, a typical jukebox might be capable of playing 100 or more songs. Jukebox jobbers would typically change these out at establishments with which they had contracts as songs fell out of popularity, and new popular singles were released.

The jukebox became particularly iconic within teenage culture during the 1950s and early 1960s. Singer-songwriter Chuck Berry, who had a penchant for writing songs that captured the teenaged popular culture of the late 1950s, included a reference to high school students making their way to the "juke joint" after school where they "drop a coin right into the slot" to hear their favorite music in his 1957 song "School Days." This scene of 1950s' high school life continued to represent the era for years thereafter, particularly in period-piece television programs such as *Happy Days* and period-piece films such as *American Graffiti*.

When the user followed Berry's lead and inserted a coin in the slot and used the jukebox's push buttons to select a song or instrumental piece to play, the mechanism would move that record, with the appropriate side turned toward the stylus, and the record would be played. Users might select several songs, or multiple users might select pieces; these selections would be stored, and the records would be played in the order in which they were selected.

Some diners and other restaurants would eventually have small boxes at each table that controlled a central record-playing jukebox, thus eliminating the need for the customer to actually walk over to the central device. However, an increasing orientation toward albums that began in the late 1960s, and a general decline in the importance of the jukebox in the music industry led to the machines becoming less numerous over time. With the advent of the CD, CD jukeboxes were available. Users could select an entire disc or individual tracks. Later, with the CD becoming less important, new jukebox models were released that used later technologies. For example, at the time of this writing, Rock-Ola markets its Rock-Ola Bubbler Digital Music Center 90th Anniversary model. The machine is U.S. built, features an internal hard drive that can store up to 13,000 CDs' worth of music, is compatible with smartphones and tablet computers for Bluetooth streaming, and can be linked to an establishment's integrated audio system (for full technical specifications and to see a photograph of the device, see, https://www.rock-ola.com/products/bubbler-90th-anniversary, accessed January 15, 2021).

See Also: 45-rpm Records; Music Boxes; Player Pianos; 78-rpm Records

Further Reading

Dietmeier, Bob. 1955. "Can't Hardly See No End to Artists, Song Hits, Developments: Music, Juke Box Industries Rich in All Categories, Says Operator Poll." *Billboard* 67, no. 13 (March 26): 1, 147.

Glass, Louis, and William S. Arnold. 1890. Patent for Coin-Operated Phonograph. Accessed January 18, 2021. https://patents.google.com/patent/US428750A/en.

Lanza, Joseph. 1994. *Elevator Music: A Surreal History of Muzak, Easy-Listening, and Other Moodsong*. New York: St. Martin's Press.

Lanza, Joseph. 2004. *Elevator Music: A Surreal History of Muzak, Easy-Listening, and Other Moodsong*. Revised and expanded ed. Ann Arbor: University of Michigan Press.

Wickman, Jim. 1955. "New Types of Equipment, New Types of Locations." *Billboard* 67, no. 13 (March 26): 50–51.

Karaoke

Still a major worldwide commercial sing-along enterprise today, karaoke first became available in 1971. Karaoke grew out of a request that Japanese musician Daisuke Inoue received in 1969 when he was playing keyboards in bars. A small-company executive asked Inoue to record backing tracks for songs in keys that were suitable for the executive's voice so that he could sing along with the songs in drinking establishments. According to Inoue, after the first batch of recordings was met with the executive's approval, "the idea for the Juke 8 [the first karaoke machine] dawned on me: you put money into a machine with a microphone, speaker, and amplifier, and it would play the music wanted to sing" (Inoue quoted in Madrigal 2013).

Inoue worked with electronics experts and devised the aforementioned Juke 8. The machine used 8-track tapes for the music's playback and was a self-contained unit. However, because Inoue was unaware of the commercial potential for the device, he did not apply for a patent. The first karaoke—a Japanese word that roughly translates as "empty orchestra"—machine to be patented, then, was Roberto del Rosario's 1975 Sing Along System. The word "karaoke" has been said to be based on a reaction to the use of recorded music when a musicians' strike in Japan rendered a theater pit empty of its pit orchestra.

Although the holder of the 1975 patent for the Sing Along System was from the Philippines, karaoke has continued to be recognized as a Japanese entertainment form that has gained worldwide popularity. Daisuke Inoue's original Juke 8 might have been designed around the use of 8-tracks, but over the decades since karaoke's inception, the systems have used technologies such as CDs and digital music files.

It is important to note that the phenomenon of sing-alongs was not new when karaoke came into being. For example, there had been the 1961–1964 U.S. television program *Sing Along with Mitch*, in which record producer, arranger, and conductor Mitch Miller would lead viewers in singing well-known songs, with the lyrics appearing across the bottom of the television screen. Before that, various organizations from sororities and fraternities to social clubs and civic groups used published songbooks for informal group singing. Some civic groups (e.g., Kiwanis clubs) still include an element of singing in their regular meetings.

Several things made karaoke different from those earlier examples of the sing-along. Perhaps most important, karaoke provided a way to show off one's talent (or possibly lack thereof) to one's friends in a social setting, often at a party or in a bar; it was not a sing-along in the sense of the precedents mentioned earlier. And unlike the singing in social and civic clubs, which still seems rarely to go beyond well-known patriotic songs and Tin Pan Alley hits from late 19th century and first third of the 20th century, karaoke typically includes a wide range of repertoire, including current hits. And unlike the middle-of-the-road pop arrangements featured in programs such as *Sing Along with Mitch*, karaoke became a worldwide phenomenon during the synthesizer age, when synthesized backing tracks could be created that come surprisingly close to the arrangements and production sound of the famous recordings of the songs.

A karaoke subculture developed in Japan. It can be argued that its export to other countries, particularly to those in the West, was part of a still-ongoing fascination with Japanese culture that since the last quarter of the 20th century has included extensive Japanese language study in American high schools, colleges, and universities; increases in the popularity of manga; and so on.

As karaoke moved from a primarily Japanese phenomenon to a worldwide phenomenon from the 1980s into the 1990s, the sociology of karaoke was examined extensively in scholarly papers, articles, and books in the 1990s and early 21st century, for example, in Xun and Tarocco (2007) and Lamb et al. (1994). In the introduction of their paper on karaoke culture in the United States, the latter team of researchers explained the nature of the culture and its attraction for researchers as follows:

> The karaoke bar is a culture unto itself: participatory, eclectic, convivial, habitual, and liberating. There is singing, drinking, camaraderie, and wish-fulfillment. Karaoke gives everyone a chance to be the star, if only for a night, if only for one song. Karaoke, which involves singing to a soundtrack in front of a live audience, has become a part of the culture of American bars in the 1990s, in much the same way as the "Urban Cowboy" phenomenon did in the 1980s, and thus is worthy of research. (Lamb et al. 1994)

Although karaoke primarily began and has continued to be associated purely with entertainment at parties and bars, the genre has also been used for more serious social and political commentary and protest. For example, Brenner (2018) examines the use of karaoke by young Kachins during Myanmar's Kachin rebellion as a form of participatory political rebellion that goes beyond political and social commentary.

Still, it is commercial karaoke that arguably has made the greatest impact around the world. According to a market analysis published by the Indian company QY Research, the "Global Karaoke Market is likely to touch US$ 4,693.7 million [$4.6937 billion]... by the end of 2026" (QY Research 2020). Although the number projected in the study may be optimistic, particularly given that the report was published in the middle of the global COVID-19 pandemic and before the full economic and social impact of the pandemic is understood, it is still impressive and suggests the economic impact of this still highly popular form of musical entertainment.

Today's karaoke setups cover a wide range in terms of technical capabilities and cost. Sweetwater Sound has published a "Karaoke Buying Guide" on its website that provides the company's recommendations for microphones, amplification, speaker systems, apps and software, mixtures, display screens, table stands, equipment carts, and so on (see, https://www.sweetwater.com/insync/karaoke-buying-guide/, accessed November 6, 2020). Although this list of items might be necessary for professional karaoke setups, lower-priced all-inclusive karaoke setups and machines can be found through numerous online vendors. At the time of this writing, for example, Amazon is advertising karaoke setups ranging in price from less than $100 to over $3,000. Many of these machines use karaoke computer software or tablet applications that essentially run the entire process and can provide access to a variety of backing tracks for singers. Some of the lowest-priced units are specifically designed for children and provide access to child-friendly musical content. Some of the karaoke machines currently in use today require CD-ROMs that contain both the musical tracks and the lyrics for projection. This might be interesting to note for readers who have followed the downturn in use and sales of CDs and CD-ROMs in favor of digital files that can reside on other media, particularly as computers increasingly do not contain CD/DVD drives. And, just in time for the 2020 holiday shopping season, catalog and online retailer Hammacher Schlemmer released a home karaoke machine that includes "Auto-Tune software" "to correct a singer's pitch and enhance his or her overall performance," touting it as the first such karaoke machine on the market for the home (https://www.hammacher.com/product/only-auto-tune-home-karaoke-machine, accessed November 16, 2020).

The world of karaoke apps is not limited to applications that are used with professional or lower-priced dedicated karaoke machines. At the time of this writing, there are numerous low-priced and free apps for iPhones and Android smartphones. For example, Yokee's *Karaoke* app is available as a free download (with advertising). The advertising content can be eliminated with a subscription. This app is designed for the individual user. The user selects a track—and these range from recent hits back to pop classics, such as Louis Armstrong's recording of "What a Wonderful World"—and sings along with the backing track into the phone's microphones, with the lyrics displayed on the phone's screen. The singer and backing track are recorded, and the recording can be saved and shared with friends or with a wider karaoke community. By allowing for practice and multiple recording attempts, an app such as this becomes almost a small virtual studio, limited as it might be. However, the sharing feature—the connection to the user's friends and a wider karaoke community—retains some of the social nature of classic karaoke.

See Also: Compact Discs (CDs); 8-Track Tapes; Microphones; Smartphones

Further Reading
Brenner, David. 2018. "Performing Rebellion: Karaoke as a Lens into Political Violence." *International Political Sociology* 12, no. 4 (December): 401–417.
Drew, Rob. 2004. "'Scenes' Dimension of Karaoke in the United States." In Andy Bennett and Richard A. Peterson, eds., *Music Scenes: Local, Translocal and Virtual*, 64–79. Nashville, TN: Vanderbilt University Press.
Lamb, C., J. E. Burns, J. Scaffidi, and J. Murdock. 1994. "Karaoke: Research with a Two Drink Minimum." Paper presented at the annual convention of the Association for

Education in Journalism and Mass Communication, Washington, DC. Accessed January 18, 2021. http://www2.southeastern.edu/Academics/Faculty/jeburns/karaoke.html.

Madrigal, Alexis C. 2013. "Someone Had to Invent Karaoke—This Guy Did." *Atlantic*, December 18. Accessed January 18, 2021. https://www.theatlantic.com/technology/archive/2013/12/someone-had-to-invent-karaoke-this-guy-did/282491/.

QY Research. 2020. Global Karaoke Market Research Report 2020. Accessed November 22, 2021. https://www.marketstudyreport.com/reports/global-karaoke-market-research-report-2020?gclid=CjwKCAiAnO2MBhApEiwA8q0HYaV2eCpQPr0kZBntaI62MhwQNHLMLaRDmXugShYtAyBEEA6mMPevnhoCHBkQAvD_BwE.

Ugrešić, Dubravka. 2011. *Karaoke Culture*. English translation by David Williams. Rochester, NY: Open Letter.

Xun, Zhou, and Francesca Tarocco. 2007. *Karaoke: A Global Phenomenon*. London: Reaktion Books.

L

Leslie Speaker

Perhaps many people think of speakers and amplifiers as being separate components of sound systems both in the home and in large venues such as theaters and auditoriums. The Leslie speaker, named for its inventor, Donald Leslie, has been produced since the early 1940s, and it combines amplification and a sound chamber that rotates in front of the speaker to produce a pulsating tremolo/vibrato effect.

Leslie's invention was designed to help make Hammond electronic organs produce a sound more like the popular theater organs of the 1930s, particularly with respect to their tremolo effects; however, Hammond initially was not receptive to Leslie's proposed collaboration. Eventually, however, the Leslie speaker became an integral part of the tone of Hammond electronic organs, particularly the Hammond B-3, which became an iconic part of jazz organ playing. At the time of this writing, there are several YouTube videos published that explain how the Leslie speaker works in conjunction with the Hammond organ (see, e.g., young YouTube sensation James Pavel Shawcross's video at https://www.youtube.com/watch?v=G5fI3X9BdrQ, for a demonstration that provides a good view of the internal mechanics of the Leslie speaker, accessed January 15, 2021).

It should be noted that beginning at the start of the 1940s, Donald Leslie applied for and received several patents related to his speaker. As an example of the work that Leslie was doing during the decade, let us consider his 1945 patent application (patent granted in 1949) for the "Rotatable Tremulant Sound Producer," U.S. patent 2489653. Leslie's application states that "it is an object of this invention to impose pitch tremolo or vibrato, by mechanical means, on a musical tone. It is another object of this invention to provide means for operating sound producers incorporating air actuators to secure vibrato effects in a simple and effective manner. It has been found that cyclic motion at an appropriate rate of a channel forming means utilized for transmitting a tone suffices to impart vibrato to the tone. Thus, it is another object of this invention to obtain vibrato or tremolo effects by cyclic motion . . ." (Leslie 1945). What Leslie described here and the physics behind his speaker system relate to the Doppler effect. The *Oxford Dictionary* defines the Doppler effect, named for the 19th-century Austrian mathematician and physicist Christian Doppler, as follows: "An increase (or decrease) in the frequency of sound, light, or other waves as the source and observer move towards (or away from) each other. The effect causes the sudden change in pitch noticeable in a passing siren, as well as the red shift seen by astronomers" (https://www.lexico.com/definition/doppler_effect, accessed January 15, 2021). So, the pitch change that we experience as a locomotive horn or ambulance siren that approaches and then passes us is accomplished in the Leslie speaker by means of the spinning mechanism. And it is important to note that the speaker produces a pitch vibrato and not just a feeling of pulsation based on quick volume changes.

Although numerous jazz organists made use of the expressiveness that the Leslie speaker lent to the instrument's sound, the most famous jazz organist associated with the Hammond organ and its Leslie speaker was Jimmy Smith. Most of the live videos of Smith and most of his studio recordings feature the Leslie speaker's vibrato. However, as mentioned earlier, it is what happens inside the Leslie speaker that produces the effect, so watching a demonstration that shows the inside of the speaker is highly recommended. The connection of Leslie's invention with the Hammond organ is confirmed by the fact that Mark Vail's book, *The Hammond Organ: Beauty in the B* (Vail 2002), devotes an entire chapter to the Leslie speaker and amplification system. Similarly, Dave Limina's *Hammond Organ Complete*, which focuses on jazz performance style and techniques, includes introductory material dedicated to the Leslie system as well as suggestions on how to use the Leslie system to achieve the sound of the great jazz organists of the past (Limina 2002).

Either intentionally or because guitar amplifiers had been damaged and a Leslie speaker was handy, nonorganist rock musicians—principally electric guitarists—made use of the Leslie during the 1960s. Running an electric guitar through a Leslie system was particularly common during the psychedelic era, principally 1966 through approximately 1968. The Beatles' George Harrison used a Leslie speaker for some of the group's recordings, as did Jimi Hendrix, members of Pink Floyd, the Beach Boys, and numerous others. And it was not just electronic instruments such as the organ and guitar that received the Leslie treatment. To create something of an otherworldly sound on the Beatles' song "Tomorrow Never Knows," John Lennon's voice was run through a Leslie speaker.

Although Donald Leslie's integrated amplification/speaker/tremolo system originally was intended for one particular brand of electronic organ, and although it dates from the 1940s, Leslie speakers that operate and produce sound much the same way that they have for over 70 years are still produced and widely available today through mass market music retailers such as Guitar Center, Sweetwater Sound, Musician's Friend, and others. In addition, several companies manufacture Leslie speaker emulators that enable the user to get the rotating speaker tremolo effect from a traditional amplifier, in addition to electric guitar effects pedals that produce the Leslie speaker rotating sound.

See Also: Amplification; Effects Pedals; Loudspeakers; Organs

Further Reading

Faragher, Scott. 2011. *The Hammond Organ: An Introduction to the Instrument and the Players Who Made It Famous*. Milwaukee, WI: Hal Leonard Books.

Lenhoff, Alan, and David Robertson. 2019. *Classic Keys: Keyboard Sounds That Launched Rock Music*. Denton: University of North Texas Press.

Leslie, Donald J. 1945. "Rotatable Tremulant Sound Producer." U.S. Patent application, July 9. Accessed January 18, 2021. https://patents.google.com/patent/US2489653.

Limina, Dave. 2002. *Hammond Organ Complete*. Boston: Berklee Press Publications.

Vail, Mark. 2002. *The Hammond Organ: Beauty in the B*. San Francisco, CA: Backbeat Books.

Loops

The concept of loops involves the creation of repeating musical figures over which one might create and layer other musical material, rap, improvise, and so on. This is also a concept/category that is part of several of the other topics covered in entries in this volume, including video game music, generative music, DJ techniques, sampling, virtual studio software, and others. The basic concept goes back to the use of ostinato patterns that have been part of folk music and other genres around the world for years and years. An ostinato is a short pattern of pitches and/or rhythmic material that repeats over and over. Well-known examples of ostinato patterns include the keyboard bass lines in Doors songs such as "Light My Fire" and "When the Music's Over," the chord progression of Bob Dylan's "All Along the Watchtower," the background keyboard pattern in Mike Oldfield's *Tubular Bells* (the opening of which was used in the horror film *The Exorcist*), and numerous others. In 20th-century popular music, though, the use of ostinatos goes back further than these examples. For example, the left-hand part in piano pieces in the boogie woogie genre, which enjoyed popularity from the 1920s through the 1940s and influenced early rock and roll music, can be understood as an ostinato over which new melodic material appears (see, e.g., Meade Lux Lewis's 1927 recording of his composition "Honky Tonk Train Blues" at https://www.youtube.com/watch?v=iabfLNKjNQY, accessed January 15, 2021).

In terms of the loop as a part of the world of music technology, we might first consider the percussion pattern presets that were part of early drum machines that were integrated into electronic organs. In addition, the tape-based Mellotron, which became a staple of 1960s' psychedelic and progressive rock, was equipped with tape loops that allowed the user (albeit, not the psychedelic rock user) with the ability to create lounge-style accompaniments to singing or instrumental performance.

Some of the minimalist composers of the 1960s' and 1970s' avant-garde used tape loops as part of their compositions, such as Terry Riley on his 1969 piece *A Rainbow in Curved Air*. Other minimalists, including Steve Reich, created compositions based on the live performance of loop-like patterns that moved in and out of phase with each other. The use of the term, however, is probably more closely associated in popular culture with the use of loops beginning in the 1970s as part of the hip-hop age.

Early hip-hop DJs, such as DJ Kool Herc, used pairs of turntables to extract and loop snippets of music from popular R&B and rock hits to create pitch- and rhythm-based ostinatos to serve as a background for rap. As sequencing, sampling, and polyphonic sound synthesis became part of the world of synthesizers in the late 1970s and early 1980s with instruments such as the Roland MC-8 MicroComposer—which greatly expanded upon the sequencing power that previously had been available on synthesizers—and the Fairlight CMI—the first sampling keyboard—and their lower-priced successors, it meant that it was possible to create and store loops without using turntables and tape.

And, although loops usually are associated with the use of either turntables or digital samples, some were created by studio musicians. For example, the backing for the first major rap hit, the Sugarhill Gang's 1979 song "Rapper's Delight," is based on a snippet of "Good Times," which had been released by the group Chic earlier that same year and sounded like a sample, but the instrumental backing was newly recorded. This suggests the importance that looped samples played in the aesthetics of hip-hop music, even from its start. Although the use of loops to create a background base over which other new material and various samples are layered has greatly expanded and can be created entirely in a studio absent turntables, the work of the turntable-and-vinyl-record-based DJ has remained a vital part of the hip-hop aesthetic and part of live performances.

Ambient music of the late 20th century and early 21st century, including the generative electronic ambient music of Brian Eno, also makes use of loops. Although Eno's Generative Music is discussed in detail in an entry in this volume, suffice it to say that the *Bloom* app that Eno and Peter Chilvers introduced in 2008—and that continues to be expanded and updated at the time of this writing—is based on the use of ambient music-style loops that can be changed by the user or that the user can allow the application's algorithm to change.

The use of loops is not limited to hip-hop or ambient computer-generated music. As mentioned earlier, the drum machines that became popular in the 1980s could be programmed to loop a pattern that could be the basis of a recording or a live performance. In fact, the use of programmed (as opposed to sampled) loops was a regular part of the world of rock music in the 1980s and beyond. Today, even on low-priced digital synthesizers and in free and low-priced virtual studio software, users can easily create ostinato patterns and create loops. In fact, programs such as *GarageBand* seem to be designed to make this a process as effortless as possible. *GarageBand* includes percussion loops in a variety of rhythmic styles that can be plugged directly into the user's composition/recording. In *GarageBand* and other programs, users can also create percussion or pitched instrument loops by highlighting/selecting a piece of music and copying it over and over, or by looping it.

See Also: Generative Music; Tape Manipulation

Further Reading

Chang, Jeff. 2005. *Can't Stop Won't Stop: A History of the Hip-Hop Culture.* New York: Picador.

DJ Booma. 2017. *How to Be a DJ in 10 Easy Lessons: Learn to Spin, Scratch and Produce Your Own Mixes.* London: QED Publishing.

George, Nelson. 2012. "Hip-Hop's Founding Fathers Speak the Truth." In Murray Forman and Mark Anthony Neal, eds., *That's the Joint!: The Hip-Hop Studies Reader*, 2nd ed., 44–55. New York: Routledge.

Jones, Andy. 2018. "Brian Eno and Peter Chilvers Bloom: 10 Worlds Review." *MusicTech*, December 28. Accessed October 30, 2020. https://www.musictech.net/reviews/brian-eno-and-peter-chilvers-bloom-10-worlds-review/.

Schloss, Joseph. 2012. "Sampling Ethics." In Murray Forman and Mark A. Neal, eds., *That's the Joint!: The Hip-Hop Studies Reader*, 2nd ed., 610–630. New York: Routledge.

Loudspeakers

Loudspeakers, usually referred to simply as speakers, allow an electrical signal to be translated into and projected as sound waves. Although the materials that are used and other aspects of speaker technology have changed from the early 20th century, the basic electronic principals remain the same. An electrical signal is sent to the coil in the speaker. This creates an electromagnetic field that interacts with a permanent magnet that is located in the speaker. This interaction creates motion in the speaker's cone, and this motion creates the sound waves that are projected at the listener. Generally, this principle pertains to most speakers, be they massive speakers in large public address systems or small speakers in headphones.

When considering speakers, it is important to note that early listening, be it to cable broadcasts or early radio, was done using earpieces, essentially very small speakers that followed in the wake of Johann Philipp Reis's work of the 1860s and Alexander Graham Bell's 1870s development of the telephone. Werner Von Siemens and the electronics company that bears his name, and others, made refinements to these early telephone-based speaker designs; however, speakers remained quite small until the 1920s. In fact, one of the challenges early in the history of radio, even for the consumer market, was that listeners were essentially tethered to a fixed transmitter that was anything but portable. A major challenge that the inventors of the late 19th and start of the 20th century faced was the weakness of the electric currents that were involved with these early speaker designs.

At the end of the 19th century and beginning of the 20th century, there were several designs that aimed to produce speakers that could project sound beyond a distance that previously had been possible, with inventors such as Oliver Lodge, Peter L. Jensen, and Edwin Pridham, using the idea of a moving coil with a phonograph or Victrola-style horn to amplify sound. Despite the work of inventors such as these, the modern moving coil speaker was not developed until the mid-1920s, when Chester W. Rice and Edward W. Kellogg created a speaker that greatly improved on earlier designs by limiting the amount of noise generated. Rice and Kellogg filed for several patents through General Electric in the second half of the 1920s, and it is their principles that established the basic speaker design that is still with us today. Part of advantage that Rice and Kellogg had over some of their predecessors is that their speaker relied on amplification of an electronic signal by an amplifier that used vacuum tubes; the vacuum tube was not invented until the middle of the 1910s. With tube-based amplification of electronic signals, the need for large horns was eliminated.

After Rice and Kellogg's development of the modern nonhorn speaker in the 1920s, generally single speakers were sufficient for the demands of music reproduction in the home. Naturally, multiple speakers would be used for amplification in stadiums, auditoriums, and other large venues but because of the size of the venue and the need for greater amplification than a single speaker could handle. The point is that generally the electronic means with which music was transmitted to the public (e.g., AM radio, shortwave radio, 78-rpm records) had a limited frequency range through the 1940s and 1950s, and stereophonic records did not exist until the 1950s.

With the invention of vinyl records, development of stereo records, and the start of an audiophile subculture in the 1950s, however, that changed. Many home audio systems, either with stand-alone speakers or in console stereo systems in which everything was contained in one piece of furniture, were incorporating speaker systems that included at a minimum a woofer and a tweeter to cover the extremes of the frequency range. Some systems were based around the concept of a three-speaker setup, with the woofer and tweeter joined by a medium-sized speaker that accentuated mid-range frequencies. In systems with multiple sizes of speakers, manufacturers use electronic crossover systems so that the frequencies best suited for an individual speaker in the system reach that speaker, as opposed to all the frequencies.

One can also see the progression of speaker and speaker system technology by looking at the progression of car audio systems. Although car radios had been around since the early 1930s, the basic setup continued to be designed around monophonic AM radio into the 1970s. As a result, all that was needed was a single speaker that could reproduce the frequency range that AM radio signals carried.

As cassette and 8-track tape players became more commonplace, and as stereophonic FM radio became an increasingly popular source for music, left and right stereo speakers became more commonplace in automobiles. Even as some cars were still manufacturer with a single speaker, aftermarket speaker systems—either meant to be installed by the consumer or professionally installed—became available, either from specialty shops or from mass-market retailers. Many of these aftermarket stereo systems, as well as some of the systems installed by the auto manufacturers, had two speakers, often located in the panel behind the rear seats.

Because optimally speakers work best if they are more focused on just part of the frequency range, home audio and car audio systems started including woofers—large speakers for low frequencies—and tweeters—small speakers for higher frequencies. In automotive settings, ultimately this meant placing multiple speakers around the cabin. This surrounds the driver and passengers with sound, as well as allowing for the easier use of multiple speakers of different sizes. The subwoofer became a more-or-less standard part of some car audio systems in the 1990s. Because low frequencies are not perceived by the listener as being as directional as high frequencies, only one subwoofer was needed and often was situated in the car's trunk.

Subwoofers also found their way into home audio systems. Although they seem to be a natural fit with drum-and-bass and hip-hop styles, they perhaps are more closely associated with home theater systems. As they do in movie theaters, subwoofers allow the low-frequency sound effects and effects that are designed more to be felt than heard (e.g., the rumble of a train) in films to be experienced in a way that is impossible with conventional television speakers, or even with some soundbars.

Throughout the decades, numerous companies have made refinements and changes to speaker system design and speaker construction; however, for the most part, the basic principles developed in the 1920s remain. Some of the more intriguing and commercially well-known changes in speaker systems came from Amar G. Bose and the Bose Corporation. Although Bose's contributions to audio are detailed in a separate entry in this volume, it is worth mentioning some of Bose's developments here. Disappointed with the sound of an expensive system that Bose had purchased, in the mid-1960s, he developed a system that produced an experience more similar to the concert-hall listening experience in which the audience member hears sound that is directly coming at them but mostly reflected sound from the surfaces of the auditorium. The basic design paradigm at the time was for speaker systems to send sound directly at the listener. In addition to the array-based speaker system that Bose designed for the home, he also designed relatively small radio/cassette/CD Acoustic Wave shelf audio units that produced significantly more bass response than seemed possible, given the apparently small size of the unit and its speakers. As *Chicago Tribune* writer Rich Warren explained the technology, "They [Amar Bose and William Short] followed the laws of physics to where they bend. They discovered that placing a speaker at precisely the right place in a tube of precisely the right length and diameter results in stupendous bass reproduction, without distorting other frequencies" (Warren 1991). The secret to making this work in a small unit that could be placed on a shelf was that Bose and Short also discovered that this improved bass response was achieved even if the tube was bent and doubled back on itself.

There have been a variety of other approaches to speaker enclosures since the development of the loudspeaker in the 1920s. Some speaker systems have used closed boxes with baffles (e.g., openings) for the individual speakers. In these, in which the baffles may be completely open—exposing the speaker—or covered with a porous material with small openings, known as a grille, the sound generally is projected directly at the listener. Some speaker cabinets have incorporated openings, or ports, often in the rear, in order to allow the sound that is produced by the back of the individual speakers to be better audible for better bass response. In addition, today some manufacturers (e.g., Eden Acoustics) offer speaker systems that are designed around the concept of no enclosure. All these approaches produce sound with differing acoustic properties.

Over the years, companies have produced specialized speaker systems that were designed for specific purposes. One such invention was Donald Leslie's Leslie speakers, which have been produced since the early 1940s. Most closely associated with electronic organs—and specifically with Hammond organs—Leslie's speaker cabinets include an amplifier, a speaker, and a sound chamber that rotates in front of the speaker to produce a tremolo/vibrato effect. The Leslie speaker is covered in a separate entry in this volume.

See Also: Headphones; Leslie Speaker; Radio

Further Reading

The Editors of Bose.com. n.d. "Dream + Reach: The First 50 Years of Bose." Bose.com. Accessed April 29, 2020. https://www.bose.com/en_us/better_with_bose/dream_and_reach.html.

The Editors of the *San Diego Union-Tribune*. 2018. "From the Archives: September 20, 1919: 50,000 Hear President Wilson." *San Diego Union-Tribune*, September 20. Accessed January 18, 2021. https://www.sandiegouniontribune.com/news/150-years/sd-me-150-years-september-20-htmlstory.html.

Ferris, Robert. 2016. "How Amar Bose Used Research to Build Better Speakers." CNBC.com, March 24. Accessed January 18, 2021. https://www.cnbc.com/2016/03/24/how-amar-bose-used-research-to-build-better-speakers.html.

Warren, Rich. 1991. "Bose Improves Remarkable Acoustic Wave Music System." *Chicago Tribune*, December 20. Accessed January 18, 2021. https://www.chicagotribune.com/news/ct-xpm-1991-12-20-9104240238-story.html.

Mechanical Organs

A variety of mechanical musical instruments played roles in the popular culture of the late 19th century and early 20th century. Among these were player pianos, music boxes—and particularly the pay-to-play music boxes known as nickelodeons—and mechanical organs. Although this chapter contains entries for organs and theater organs, the unique roles played by mechanical organs make this classification of musical instruments deserving of its own entry. Although there are other types of mechanical organs, let us consider three types that historically have been the most visible: street organs, dance hall organs, and carousel organs.

The small portable (typically, they can be carried around the neck of the player or pulled on a small cart) street organs associated with organ grinders originated in Europe and made their way to North America beginning with immigration waves in the 19th century. These are pneumatic instruments that play encoded musical selections as the organ grinder turned a crank. At the time of this writing, viewers can find several amateur videos in which street organ aficionados explain the workings of street organs and demonstrate their instruments (see, e.g., https://www.youtube.com/watch?v=DSNAypbWe0g, accessed January 15, 2021).

Although the organ grinder could work alone as a street performer, sometimes the street organs would be part of an act put on by a troupe of street performers. However, perhaps most famously, some organ grinders worked with trained monkeys that would dance to the music and hold a tin cup in which tips from passersby would be deposited.

As implied by the name, the street organ's home was in an urban environment. These instruments and organ grinders—with or without monkeys—became fixtures in major European, U.S., and other cities to the extent that they became something of a stereotype of the urban landscape in films, cartoons, and literature. Street organs, which required no musical ability to play, provided employment for U.S. immigrants from various European countries, including Germany, the Netherlands, and Italy. Street organs and their players declined in number in large U.S. urban areas by the mid-20th century. Interestingly, however, the street organ and organ grinders remain a common sight in Mexico City. In 2016, Luis Román Dichi Lara, the leader of Mexico City's organ grinders' union, estimated that there were approximately 500 organ grinders working the streets of Mexico City (Ahmed 2016). This form of street entertainment arrived in Mexico from Europe in the 19th century with a wave of immigrants. Azam Ahmed's report on this musical phenomenon in Mexico that appeared in the *New York Times* also shows how the subculture of organ grinders and the source of their street organs have shifted over time from Germany to Central and South America. Today, although the number of cities with an active street organ culture are far fewer than in the first third of the 20th century, several workshops offer organ repairs, sales of used and new instruments, and sales of street organ kits. In addition, the street organ has come into the digital age, with new MIDI-compatible street organs being produced.

With regard to size, the mechanical organs on the opposite end of the spectrum from the street organs are the instruments that were part of dance halls of the early 20th century. However, size was not the only thing that differentiated dance hall organs, street organs, or the band organs that were part of the carousels of the early 20th century. Both the instruments and the arrangements that they played differed from those of other contemporary mechanical organs. For example, street organs were limited by the small number of pipes that they had. As a result, the encoding of music on paper strips or rolls was simple compared with that for instruments with numerous ranks of pipes, such as dance hall organs. The arrangements that were created for dance hall instruments tended to emphasize a solid rhythm with, for example, chords on the beats, to encourage dancing.

Dance hall organs were not only designed and built for their sound. Many of the instruments are elaborate with carved figures that are either stationary or animated. One can often find the figure of a conductor, for example, who is either stationary or who waves a baton in time with the music. A figure such as this connects the dance hall organ with the concept of the dance band or dance orchestra.

Although some dance hall organs continued to be produced, the heyday for these instruments generally ended in the mid-20th century. Some of the instruments that once played an important role in dance at standalone dance halls and dance halls at amusement parks and fairgrounds in the early 20th century are now housed in museums. For example, the large 1924 (modernized in the 1940s) Mortier Dance Organ that was built in Belgium is housed at the Grampian Transport Music in Scotland and can be heard by visitors to the museum (see, the Grampian Transport Museum's website for more information on the Mortier Dance Organ at http://gtm.org.uk/exhibitions/gtm-permanent-collection/mortier-dance-organ/, accessed January 15, 2021). Other museums in Europe and the United States house large organs of this type that also typically date from the early 20th century. Perhaps one of the best-known locations in the United States is the House on the Rock, a tourist attraction in Iowa Country, Wisconsin, that houses numerous mechanical musical instruments, some of which can be played briefly by visitors using a token.

Although they were not as large as some of the grandiose dance hall organs, mechanical organs that were integrated into outdoor fair and amusement park carousels (sometimes given as carrousel) are perhaps even more frequently seen today. These instruments, because they were located in busy amusement parks and outdoors, tended to be capable of playing more loudly than some other organs. Like theater organs and some dance hall organs, carousel organs also often included percussion—generally in the form of a snare drum, cymbals, and bass drum—which gave the carousel organ the tonal feel of a band; concert bands (e.g., that of bandleader and composer John Phillip Sousa) were popular attractions in the first third of the 20th century. Because of the aural connections these instruments made with the concert band, they are sometimes called band organs.

The great carousel band organs of the early 20th century are celebrated at museums such as the Herschell Carrousel Museum, which is located in North Tonawanda, New York. Herschell was one of the leading manufacturers of carousels, and its rides featured Wurlitzer band organs; Wurlitzer's headquarters and factory were located nearby. The Herschell Carrousel Museum includes an exhibit dedicated to Wurlitzer's band organs and the paper music rolls that they used and includes an example of an early 20th-century machine that cut the music-encoded rolls for the organs.

Some of the classic early 20th-century carousels and organs are still in use even into the third decade of the 21st century. For example, Tuscora Park, which is located in New Philadelphia, Ohio, boasts an operating 1925 Spillman Engineering Carousel, which includes a Wurlitzer band organ. Interestingly, Spillman, like Wurlitzer and Herschell, was also located in North Tonawanda, New York. The Wurlitzer instrument of the Tuscora Park carousel has been converted to MIDI compatibility, thus eliminating the need for paper rolls as well as for an operator to change rolls on a regular basis.

Although the band organ might seem like a museum-piece-like thing of the past, carousel band organs are still produced in the 21st century. One 21st-century manufacturer, Stinson, specializes in MIDI-compatible organs that the company makes in a variety of styles and sizes (http://stinsonbandorgans.com/catalog/index.html, accessed January 15, 2021). One of the rare late 20th-century carousels that was designed and built in the style of early 20th-century rides, the Richland Carousel Park ride in Mansfield, Ohio, includes a MIDI-operated Stinson band organ. The carousel made its debut in 1991.

The Hurdy-Gurdy

Because its usage has been so specialized, the hurdy-gurdy is an example of a music technology that is not necessarily significant or overarching enough to merit a full entry. It is, however, a particularly interesting technology, especially because the actual term has been used quite a bit over the decades; however, many readers might be unaware of exactly what the hurdy-gurdy is, what it does, or how it is played. The hurdy-gurdy dates back to the Middle Ages and is a mechanical instrument in which a wheel turns and serves the function of a bow on any of the orchestral stringed instrument: it engages the strings. However, instead of the player putting fingers down on a fingerboard or fretboard to select the pitches, the hurdy-gurdy uses a system of buttons, each corresponding to a different pitch. When the player pushes a button down, the desire pitch is selected, and the wheel engages with the correct string. As one might imagine, as long as the strings are tuned properly and the mechanism is in good working order, the hurdy-gurdy is absent the intonation challenges of, say, the violin. In addition to the strings that are played using the buttons, some instruments also have drone strings. Although hurdy-gurdy-like instruments were used in the Middle East, Europe, and elsewhere, it is the European type, especially as the instrument evolved through the Middle Ages and Renaissance, that seems to be the most frequently seen today in the West. At the time of this writing, there are numerous YouTube videos of hurdy-gurdy performances, both as a solo instrument and with various types of accompaniment, as well as videos that demonstrate how the instrument works (see, for example, https://www.youtube.com/watch?v=gYJg9cLklus, accessed May 13, 2020, and a TED Talk on the instrument by Caroline Phillips at https://www.ted.com/talks/caroline_phillips_hurdy_gurdy_for_beginners, accessed May 13, 2020).

Part of the reason that the nature of the instrument and the operation of the hurdy-gurdy is something of a mystery to many people is that the term has also been used somewhat confusingly for a portable mechanical street organ that has musical pieces programmed into it using a mechanism similar to that of a player piano. Although long considered an obscure centuries-old instrument primarily associated with various ethnic styles, the hurdy-gurdy came back into popular consciousness in 1968 with the release of Donovan's hit song "The Hurdy Gurdy Man," a recording that, ironically, featured prominent tambura and distorted electric guitar and not the hurdy-gurdy.

See Also: Music Boxes; Organs; Player Pianos; Theater Organs

Further Reading

Ahmed, Azam. 2016. "Mexico City's Organ Grinders, Once Beloved, Feel Shunned." *New York Times*, September 12. Accessed January 18, 2021. https://www.nytimes.com/2016/09/13/world/americas/mexico-city-organ-grinders.html.

Bush, Douglas E., and Richard Kassel, eds. 2006. *The Organ: An Encyclopedia*. New York: Routledge.

Ord-Hume, Arthur W. J. G. 1973. *Clockwork Music: An Illustrated History of Mechanical Musical Instruments from the Musical Box to the Pianola, from Automaton Lady Virginal Players to Orchestrion*. New York: Crown Publishers.

Seagrave, Shane. 2007. "Organ Figures." *Carousel Organ*, no. 33 (October): 15–22.

Mellotrons

Although it is not heard nearly as often in 21st-century popular music, the mellotron was a staple of mid- to late 1960s' and early 1970s' rock, particularly psychedelic and progressive rock. The instrument itself uses a keyboard to activate loops of magnetic tapes on which the sounds of acoustic instruments have been recorded.

The use of magnetic tape to record acoustic pitched and unpitched instruments for synthesizer-like playback certainly was not new when the Mellotron was developed in Birmingham, England, in 1963. American inventor Harry Chamberlain used magnetic tape to record a drummer playing 14 different rhythms that could be accessed with his 1949 Rhythmate. Similarly, Chamberlain's eponymous Chamberlain, produced in the 1950s, used magnetic tape loops of pitched acoustic instruments. The Mellotron came out of an attempt by Chamberlain's agents to market his instrument in Europe. Eventually, a legal agreement between both parties allowed for both Chamberlains and Mellotrons to be manufactured and sold, even though both instruments used the same basic core technology. The Mellotron was easier to mass-produce, and it, not the Chamberlain, became the iconic tape-loop-based keyboard instrument, remaining in production until the mid-1980s. At the time of this writing, Bell Tone Synth Works has a YouTube video posted in which the inner workings of the Mellotron M400 are shown and demonstrated (see, https://www.youtube.com/watch?v=ByD8gH7kYxs, accessed January 15, 2021).

In another video on YouTube, Paul McCartney explains and demonstrates how the mellotron works, the instrument's early associations with lounge music—based on some of the Mellotron's original 1960s' presets—and how his band, the Beatles, used it on the song "Strawberry Fields Forever" (see, e.g., https://www.youtube.com/watch?v=tTKPW92ndUA, accessed January 15, 2021). The lounge-style preset that McCartney demonstrates in the video suggests a strong connection back to Harry Chamberlain's Rhythmate and to the rhythm presets associated with various electronic organs of the 1950s and 1960s.

The Beatles' famous use of the Mellotron in "Strawberry Fields Forever" is entirely different from the lounge-like rhythm accompaniment presets that were associated with the original incarnation of the Chamberlain and the Mellotron and apparently what their manufacturers envisioned for the instrument. In this well-known 1967 song, the Beatles used the flute preset to suggest a live acoustic flute; however, the Mellotron's pitch slide feature makes it clear to the listener that the opening of the song is a product of an electronic instrument. As a British-built instrument, it is perhaps natural that British rock bands were the 1960s' and 1970s' musicians, most closely associated with the Mellotron.

As evidence of the importance of the mellotron in music of the "Summer of Love," 1967, even a band associated with psychedelic music for a very short time, the Rolling Stones, used the Mellotron in notable ways. In particular, Brian Jones played the instrument on "We Love You" and "Dandelion," the A and B sides, respectively of a 1967 single, as well as on "She's a Rainbow," perhaps the band's best-remembered recording from the Rolling Stones' brief psychedelic period. Perhaps the most interesting of these Rolling Stones' use of the Mellotron is the emulate of a brass section on "We Love You." In his book *Old Gods Almost Dead: The 40-Year Odyssey of the Rolling Stones*, Stephen Davis quotes recording engineer George Chkiantz as saying, "*You* try playing a Mellotron . . . Just *try*. There's a horrible time lag, depending on how many notes you push down, and most people, even great musicians, screwed it up. A terrible instrument—dreadful, very hard to play, impossible to maintain tempo—unless you were Brian Jones. *Nobody* else could have gotten anything like that" (Davis 2001, 210, italics in the original).

More so than the Rolling Stones, who pursued psychedelic rock for only a brief period, or the Beatles, the Moody Blues was the band that perhaps is most appropriately associated with the Mellotron. Beginning with the band's 1967 album *Days of Future Passed*, through the 1972 album *Seventh Sojourn*, the Moody Blues' keyboardist Mike Pinder made liberal use of the Mellotron, particularly in providing string lines to the group's songs. At the time of this writing, there is a YouTube video of Pinder demonstrating some of the Mellotron figures that he played with the Moody Blues (see, https://www.youtube.com/watch?v=llhy6GJcyt4, accessed January 15, 2021). Pinder also made use of the similarly designed Chamberlain on some of the group's recordings. In addition to the Moody Blues, other progressive rock bands, including King Crimson, used the Mellotron on their recordings and in their live performances during this period. The Mellotron also provided string lines in David Bowie's well-known 1969 song "Space Oddity." Although the Mellotron fell out of production in the mid-1980s, it was later revived, and today in the third decade of the 21st century, one can purchase digital versions of the Mellotron as well as Mellotron patches that are available for a variety of digital synthesizers.

See Also: Loops; Reel-to-Reel Tape

Further Reading

Davis, Stephen. 2001. *Old Gods Almost Dead: The 40-Year Odyssey of the Rolling Stones*. New York: Broadway Books.

Margotin, Philippe, and Jean-Michel Guesdon. 2016. *The Rolling Stones: All the Songs: The Story behind Every Track*. New York: Black Dog & Leventhal Publishers.

Metronomes

One of the challenges that musicians, particular in the milieu of Western art music, have faced for hundreds of years is how to find the appropriate tempo at which a piece of music should be performed. In the 17th and 18th centuries, it was common for Western art music—what the general public calls "classical" music—scores to contain general terms that indicated the feel and tempo of the piece. Originally, these terms were used by Italian composers, and as a result, Italian terms such as Allegro (fast), Adagio (slow), and Andante (at a walking tempo) became the norm. The challenge is that one person's Allegro might not be the same as another musician's Allegro, and they might both be different than what the composer originally had in mind. From the mid-19th century through the present—particularly as nationalism grew in the 19th century—other languages, especially German, French, and English, were increasingly used to provide performers with an idea of the approximate tempo.

The metronome, which can be set to make a steady ding, ping, click, or beep at a user-specified number of times per minute, originally came of the same mechanical tradition as clocks, such as grandfather's clocks. Working in Amsterdam, Dietrich Nikolaus Winkel invented the metronome in 1812. This might be surprising to some readers, because often the name Johann Nepomuk Maelzel is more frequently associated with the device. Maelzel took Winkel's invention and, in 1816, made improvements to it and began manufacturing metronomes under his own name. Part of the reason for the long-standing association of Maelzel's name with Winkel's invention is that traditionally, from the early 19th century, metronomic indications in printed music carried the designation "M.M.," for Maezel's Metronome.

Composer Ludwig van Beethoven knew Maelzel and was the first major composer to include numerical metronome markings in some of his scores. Some music historians and conductors speculate that these early metronomes were not entirely accurate, as some of Beethoven's metronome markings are believed to be higher/faster than what was probably the tempo in which the pieces were meant to be performed. Some composers after Beethoven, who died in 1827, included metronome markings in their scores; however, many of the compositions of the last

3/4 of the 19th century—known as the Romantic Period—were meant to be performed with subtle, expressive changes of tempo, known as rubato, so the use of the metronome to give performers a definitive representation of a composer's tempo intensions for an entire piece, an entire movement, or entire sections of a movement was not as common as they would become in the 20th century.

Some 20th-century composers were creative in their use of metronome markings. Paul Hindemith, for example, included metronome markings in his scores that would give musicians important information about what rhythmic unit (e.g., quarter note and eighth note) he intended to be used for the feeling of the composition's pulse. Sometimes the pulse units that Hindemith included in his metronome markings were not those that a performer might intuitively assume, so Hindemith's markings played an important role in providing an idea not just of the tempo but also about the feeling of pulse (e.g., two slow pulses per measure, as opposed to four quicker pulses per measure).

One of the challenges posed by metronomes in general is that unlike general and open-to-interpretation designations such as "Allegro," "Andante," "Moderately Slow," and so on, they are sometimes presented as absolutes. The danger in this is that different acoustical environments (or such things as the technical expertise—or lack thereof—of the performers) are not taken into account. To put it another way, one performer's "Moderately Fast" in a given acoustical environment might not be another performer's "Moderately Fast" in a different environment; M.M.=92, on the other hand, is not as open to interpretation.

The traditional pendulum-based metronomes are susceptible to other problems as well. Because their function is based on the use of a weighted pendulum, they must be placed on a flat-level surface in order for them to tick evenly. And like mechanical spring-wound clocks or spring-wound gramophones or phonographs, over-winding the spring can cause failure of the device. Interestingly, weighted pendulum-based metronomes are still available today.

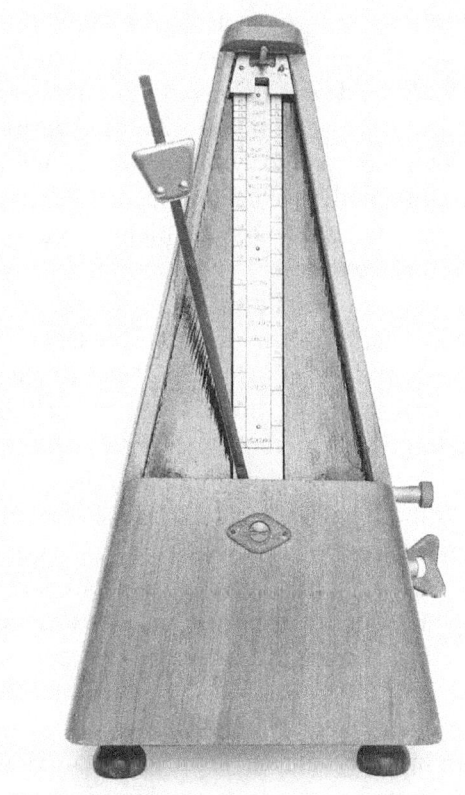

Although metronomes are still available in this classic design, the devices have entered the electronic age. Several companies offer electronic metronomes and metronome apps for tablets and smartphones. (Slavapolo/Dreamstime.com)

Even a mass market company generally associated with more high-tech products such as Amazon markets an "Amazon Basics" mechanical metronome with a shape that is clearly based on that of Johann Maezel over two centuries earlier. Although it is roughly the same shape and uses essentially the same wound-spring mechanical technology of pre-electric metronomes, the Amazon Basics metronome includes one feature not common on early metronomes: it can produce a differently toned sound every 2, 3, 4, or 6 pulses, which is helpful when one practices music that is in duple, triple, quadruple, or sextuple meter along with the metronome.

In the 20th century, some companies began manufacturing electric metronomes. One of the leading early manufacturers of these devices, Frederick Franz, held patents related to his "electrically driven metronome" that date back to the 1930s (see, e.g., https://patents.google.com/patent/US2150967A/en, accessed January 15, 2021). Although some of Franz's patents are later than U.S. patent 2,150,967 (patent applied for in 1937), the Franz Manufacturing Company continued to make metronomes that were based on the basic principles of the 1930s and 1940s into the 1990s. One of the features of electric metronomes that was not possible on mechanical metronomes was the inclusion of a light that flashed in time. A musician could activate the light and deactivate the audio click and find a tempo or practice along with the device using only visual cues.

In the last several decades of the 20th century, electronic metronomes became available that would provide a different downbeat tone for various meters. Some electronic metronomes, too, could be programmed for changing meters. Unlike their mechanical predecessors, electronic metronomes tend to be highly accurate and do not have to be placed on a level flat surface in order to click or ping evenly. As is the case with mechanical metronomes, electronic metronomes are still widely available in the third decade of the 21st century. This is despite the fact that musicians with smartphones on various platforms, including Android and Apple's iOS, have free and low-cost metronome apps available. Some of these, such as *MusiciansKit*, can easily be programmed to give a different downbeat for meters that were not commonly possible with earlier electronic metronomes. *MusiciansKit*, for example, allows the user to hear a different downbeat tone for meters with up to 16 pulses per measure. This app can also provide up to 8 subdivisions per pulse. Another feature available on some electronic metronomes and metronome apps is a tap mode. This feature allows the user to tap a pulse on the metronome/smartphone screen, and the metronome or app displays the numerical metronome marking for that pulse.

The continuing miniaturization of metronomes and integration of a metronomes into multifunction devices can also be seen in smartwatches for musicians, such as the Soundbrenner Core. The vibration feature on the Soundbrenner Core allows the user to feel the pulse. This is particularly helpful for musicians who find an audible metronome tone to be obtrusive during their practice sessions. This watch also includes a contact tuner that picks up the frequency of an instrument from its vibrations, and a decibel meter function, and also functions as a timepiece.

See Also: Smartphones

Further Reading

Davis, Stephen. 2001. *Old Gods Almost Dead: The 40-Year Odyssey of the Rolling Stones.* New York: Broadway Books.

Margotin, Philippe, and Jean-Michel Guesdon. 2016. *The Rolling Stones: All the Songs: The Story behind Every Track.* New York: Black Dog & Leventhal Publishers.

Microphones

Microphones date back to the late 19th century, long before they were used for sound amplification in public address (PA) systems or for audio recording. Arguably, the first microphone was that invented by Alexander Graham Bell for use in his telephone in the 1870s. Other inventors improved on microphone design, including Thomas Edison a decade after Bell's invention.

Although microphones initially were associated with telephone communication, by the end of the second decade of the 20th century, microphones were an essential part of military and civilian radio, as well as of early PA systems. For example, when U.S. president Woodrow Wilson addressed a crowd that was estimated to be of at least 50,000 people at a San Diego, California, sports facility in 1919, it was a microphone into which he made his remarks. Microphone technology has continued to be refined into the 21st century; however, some of the designs that were part of the recording industry and live performance amplification as early as the 1940s are still highly valued today. Eventually, many different types of microphones appeared, based on how they functioned electronically and the pattern of the sound they picked up. Let us first examine some of the sound input patterns.

Perhaps the best-known sound response patterns in microphone technology are omnidirectional, unidirectional, and cardioid. Omnidirectional microphones pick up sounds generally uniformly in a 360-degree pattern around the device. As one might imagine, this makes omnidirectional microphones particularly useful for applications such as video conferencing, in which participants might be sitting around a conference table. In musical applications, they generally have less wind noise and are less susceptible to popping when enunciating consonants. However, in live performances or in the recording studio, these microphones will allow for several instruments/voices to be heard in a given channel. On the negative side, in live performance applications, they can also be susceptible to feedback, because they can pick up not only the performers but also potentially input from nearby speakers, regardless of whether the speakers are in front of or behind the microphone.

Unidirectional microphones are designed to pick up sound directly in front of the microphone. They tend to be more susceptible to popping as well as to a buildup of bass response and boominess because of a phenomenon called proximity effect. To put it another way, unidirectional microphones tend to emphasize lower frequencies to their detriment in some applications. However, these microphones possess some distinct advantages over omnidirectional microphones; in particular, they are less susceptible to feedback and their narrow field of pickup makes it easier to isolate voices and instruments.

The other principal direction/field of input type of microphone is known as cardioid. The *Merriam-Webster Dictionary* provides a concise definition of the cardioid microphone, as follows: "a microphone having approximately uniform response over 180 degrees in front and minimum response in back, a polar curve representing its directional response being a cardioid" (https://www.merriam-webster.com/dictionary/cardioid%20microphone, accessed January 15, 2021). The cardioid to which the dictionary refers might less technically be pictured as a heart-shaped pattern. Cardioid microphones work particularly well in live performance situations. The fact that they pick up a minimal amount of sound from the back means that audience sound is minimized. It also means that in live applications, they can be less susceptible to feedback than, say, omnidirectional microphones, because the cardioid microphones' response pattern means that they will pick up less of the amplified sound of the musicians, provided that the speakers are placed closer to the front of the stage, effectively in back of the microphone. Because the cardioid pattern is in between those of unidirectional and omnidirectional microphones, some of the challenges associated with unidirectional microphones (e.g., a tendency in some toward boominess) and omnidirectional microphones (e.g., bleed-through and feedback) are minimized.

In addition to the different response/field patterns, microphones pick up and send their electrical signals in several ways. The earliest and least complex are dynamic microphones. The German microphone company Neumann provides the following definition for this type of microphone: "In this context [the microphone], it [the term 'dynamic'] refers to the kind of electromagnetism that happens for instance inside your bicycle's dynamo: Then an electrical conductor moves in a magnetic field, and electric current is induced. Dynamic microphones, thus, are microphones that convert sound into an electrical signal by means of electromagnetism" (https://www.neumann.com/homestudio/en/what-is-a-dynamic-microphone, accessed January 15, 2021). Basically, what occurs in a dynamic microphone is that a coil inside the device moves as activated by sound waves. The moving coil, being in an electromagnetic field, creates the electrical current that is sent to the amplifier or recording device.

Ribbon microphones work in a similar way and are often grouped along with coil-based dynamic microphones in the same category. The main difference between the coil type and ribbon microphones is that in ribbon-based microphones, a thin metallic ribbon is activated by sound, with the resulting vibrations in the electromagnetic field creating the electrical current that is sent to the amplifier or recording device. Because ribbon microphones generally are more fragile and do not provide as full a range of frequency response as coil-based microphones, coil-based microphones are the most common type of dynamic microphones in use.

Condenser microphones work in a fundamentally different way. These microphones require external power, either through the mixing board or through another device into which they are plugged or in some cases batteries that are inside the microphone. As audio and studio recording equipment company PreSonus defines condenser microphones in a technical article on the corporate website: "Condenser microphones are essentially a highly specialized capacitor. For those of you without an EE degree, a capacitor is a passive electrical component that is designed to temporarily story energy in an electric field.. . . The diaphragm of a condenser microphone is what makes these microphones so well-known for superior sound quality, especially when picking up minute details" (https://www.presonus.com/learn/technical-articles/what-is-a-condenser-microphone, accessed January 15, 2021). The reference to "minute details" comes from the fact that a coil in a dynamic microphone requires more sound input in order for the coil to move than the diaphragm in a condenser microphone. It should be noted that condenser microphones are built with different-sized diaphragms, categorized as large and small. Large diaphragm condenser microphones, the first type of condenser microphones to be used in the 1940s, tend to minimize noise, while small diaphragm condenser microphones generally have a more focused and well-defined pattern to the sound they pick up. Condenser microphones can be used for live sound amplification, particularly of small acoustic ensembles and solo performers, but perhaps the most common use of condenser microphones is in studio recording applications, particularly in which a wide dynamic (volume) range is used. Although condenser microphones possess certain advantages over dynamic microphones, it is possible to damage the diaphragm with too much volume, and they can be more susceptible to distortion than dynamic microphones.

Another type of microphone that gained popularity in the 1970s and 1980s was the contact microphone. These devices, also widely known as piezo microphones, pick up vibrations directly from the instrument and are often attached to the body of the instrument. They are small in size and in some applications are less susceptible to feedback than microphones that are activated by sound waves from the air. Some musicians argue that contact microphones do not, however, reflect the subtleties of the instrument's tone as well as traditional microphones. The technology behind piezo microphones has been widely used in the 21st century for tuners that can clip onto instruments, perhaps most widely used by guitarists.

Aside from the general types of microphones with their differing electronics and input patterns, microphones have also become more specialized and—in some cases—miniaturized. Wireless microphones, which transmit an electronic signal to a receiver, had been available and used sporadically since the 1950s (see, https://www.shure.com/en-US/about-us/history for a brief description of the first wireless microphone, the Shure Vagabond 88, which made its debut in 1953, accessed January 15, 2021). However, beginning in the late 1970s with the one-off wireless headset microphone that singer Kate Bush used on a brief tour, the wireless headset microphone increasingly became popular for singers. In 1990, Madonna made the wireless headset microphone a popular culture icon, and numerous artists have used similar devices since.

In addition to wireless technology and the miniaturization event in many of the headset microphones available since the 1990s, microphones have become increasingly specialized. Instrument microphones designed are for guitars, resonator guitars, woodwind instruments, and brass instruments. Although piezo, or contact, microphones are still used for these acoustic instruments, microphones that pick up sound waves from the air but that clip onto instruments arguably allow for a better representation of the natural sound of the instrument. Higher-end specialized instrument microphones are designed and built to optimize the acoustical traits of the instruments. For example, those that clip onto the bell of brass instruments, such as the trumpet, are made to withstand the high volume level of which brass instruments are capable.

See Also: Electrical Recording

Further Reading

The Editors of the *San Diego Union-Tribune*. 2018. "From the Archives: September 20, 1919: 50,000 Hear President Wilson." *San Diego Union-Tribune*, September 20. Accessed January 18, 2021. https://www.sandiegouniontribune.com/news/150-years/sd-me-150-years-september-20-htmlstory.html.

MIDI

MIDI—Musical Instrument Digital Interface—is arguably one of the most significant music technologies of the late 20th century and something that today is interwoven into numerous facets of music making, musical composition, musical notation, and so on.

The goal behind what came to be known as MIDI was to establish a standard method of communicating musical information between devices. The technology was anticipated by a paper that Dave Smith and Chet Wood of Sequential Circuits presented at the 1981 meeting of the Audio Engineering Society. However, at the same time, industry leaders in sound synthesis, such as Roland's Ikutaro Kakehashi, were also working on a standardized way of connecting the new digital synthesizers with computers. Smith and Wood's "Universal Synthesizer Interface" was rolled out within two years of their paper as MIDI. Part of the significance of MIDI is that its adoption made it possible for synthesizers and computers using different systems to connect and communicate despite their differences. To put it another way, because of the immediate adoption of MIDI by synthesizer and computer makers and programmers, conflicts between various noncompatible proprietary technologies that had plagued the early recording industry (e.g., various disc speeds and cylinders versus discs) and even the home video industry (e.g., Betamax versus VHS video cassettes) were to a great extent avoided.

The first incarnation of the MIDI standard was immediately adopted by synthesizer manufacturers, with the Sequential Prophet 600 and the Roland Jupiter-6 coming to market in late 1982 and early 1983. The first computers that supported the MIDI standard were introduced at around the same time. The year 1983 also saw the introduction of the first MIDI-compatible drum machine: the Roland TR-909. MIDI keyboards dropped in price, and the interfacing of computers to digital musical instruments became commonplace by the late 20th century. From the late 20th century to the present, other electronic instruments have also been used as MIDI controllers, including electronic wind instruments, electronic percussion instruments, and so on.

Although it is likely that many readers will associate MIDI solely with the types of instruments and machines described earlier—electronic synthesizers of various kinds and desktop, laptop, and tablet computers—MIDI technology also plays roles in the playing and programming of otherwise acoustical instruments. For example, since at least the late 1990s, some pipe organ manufacturers have offered MIDI interfaces. This allows, to cite just one example, a church organist who has to miss a service to prerecord organ accompaniments for hymns. The MIDI recordings can then be played back on the instrument in the organist's absence by anyone who can learn what buttons to push, even if they have no musical background or experience. Likewise, MIDI interfaces on pipe organs make transposition relatively simple, as well as making it possible for the organist to play a duet, either at the console or perhaps for MIDI-controlled organ and live piano.

Another less visible MIDI application related to pipe organs can be found in some of the preserved mechanical carousel organs from the early 20th century that have been retrofitted with MIDI interfaces. This enables these instruments—which originally played music encoded on rolls of paper, similar to those used with player pianos—to play digitally encoded music, thus eliminating the danger of the paper rolls breaking, eliminating the need to change rolls, and enabling numerous musical selections to be saved on a flash drive or some other small mass storage device.

MIDI technology has played a significant role in composition, arranging, music notation, and publishing since the 1980s. The adoption of the MIDI standard was closely followed by the release of notation software that allowed users to produce scores and parts that eventually—with the arrival of relatively inexpensive and high-quality inkjet and laser printers—rivaled the quality of professionally published music. With this capability, MIDI-based notation software made an impact in self-publishing work of composers and arrangers. Publishers have also turned to MIDI technology to offer MIDI files to consumers, both so that pieces can be heard and, in some cases, so that the consumer can import the MIDI file into a notation program and transpose a score and/or parts when necessary.

Another aspect of computer-based music notation software that is important to note is the development of Optical Music Recognition (OMR) software. Although considerably more complicated than its alpha-numeric-based counterpart, OMR is somewhat analogous to Optical Character Recognition (OCR). There has been software created at academic institutions and several open-source programs, as well as commercial applications such as *PhotoScore*, *SmartScore*, and *PlayScore*. The best-known and most frequently found are probably *PhotoScore* and *SmartScore* because of packaging of versions of these programs that has been done with versions *Sibelius* and *Finale*, respectively, over the years.

MIDI also led to several fundamental changes in how musicians used keyboards and computers. As the capabilities of computer-based notation and sequencing programs grew, it became less and less necessary for musicians who primarily worked in the studio or in-home recording to rely solely on expensive large keyboards. Smaller MIDI-controller keyboards did not need to have all the synthesis bells and whistles, because computer-based virtual synthesizers offered presets and ways to design and manipulate sounds, and MIDI provided the means for controllers to access and communicate with the computer software.

Although MIDI was operational and became a recognized standard in 1983, it was not static. Most recently, the MIDI Manufacturers Association announced MIDI 2.0 in January 2020. As the MIDI Association, a group of MIDI designers, users, and manufacturers, defines the new standard, "MIDI 2.0 takes the specification even further, while retaining backward compatibility with the MIDI 1.0 gear and software already in use." The organization continues by describing the differences between MIDI 1.0 and MIDI 2.0 as a change "from a monologue to a dialog" in which MIDI 2.0 devices will be able more fully to communicate with each other (The MIDI Association n.d.a.).

See Also: Digital Synthesizers; Drum Machines; Music Notation Software

Further Reading

Abernathy, David. 2015. *The Prophet from Silicon Valley: The Complete Story of Sequential Circuits*. Auckland, New Zealand: AM Publishing.

Anderton, Craig. n.d. "Craig Anderton's Brief History of MIDI." MIDI Association. Accessed January 18, 2021. https://www.midi.org/midi-articles/a-brief-history-of-midi.

McKee, Ruth, and Jamie Grierson. 2017. "Roland Founder and Music Pioneer Ikutaro Kakehashi Dies Aged 87." *Guardian*, April 2. Accessed January 18, 2021. https://www.theguardian.com/music/2017/apr/02/roland-founder-and-music-pioneer-ikutaro-kakehashi-dies-aged-87.

The MIDI Association. n.d.a. "Details about MIDI 2.0TM, MIDID-CI, Profiles and Property Exchange." MIDI.org. Accessed November 20, 2020. https://www.midi.org/midi-articles/details-about-midi-2-0-midi-ci-profiles-and-property-exchange.

The MIDI Association. n.d.b. "The History of MIDI." MIDI.org. Accessed January 18, 2021. https://www.midi.org/midi-articles/the-history-of-midi.

Mixing Boards

The basic concept behind the mixing board is that it allows multiple inputs—from instruments, microphones, and so on—to be controlled for volume, equalization, and so on and then for the fuller mixed result to be output to a public address (PA) system or recording device. Although we might think of mixing boards as being such firm staples of live concert venues and recording studios, they must have been around since the early days of PA systems or audio recording. In actuality, the modern mixing board came about because of the requirements and opportunities offered with the advent of multi-track recording.

After the establishment of electrical recording in the 1920s, multiple microphones were sometimes used in the recording studio. However, recordings were still live, as there was no multi-track recording. As a result, the inputs from the various microphones would be mixed together, but the mixing was rudimentary compared with what would later be the case after the development of multi-track recording.

As multi-track tape recorders grew from four to eight tracks and then well beyond even starting as early as the late 1950s when stereophonic albums first appeared, the need arose for a console apart from the controls on the tape machine itself to control input volumes, equalizations, reverberation, and other effects. Mixing boards are also used to control such things as placement of the instruments and/or voices in the output across the stereophonic soundscape. In fact, it is this aspect of multi-track recording and what can be done on the stereo master tape that became a central part of the production/composition process during the psychedelic period of rock music between 1966 and 1968. Perhaps the best-known example of the role of the mixing board in the multi-track recording and mastering process is the 1967 Jimi Hendrix's song "Purple Haze." In fact, the recording, which was produced by Chas Chandler and engineered by Eddie Kramer, includes the use of several technologies detailed in this volume, including Hendrix's use of various effects pedals on his electric guitar parts. In considering the impact of the mixing console, however, perhaps the primary aspects of the stereo release of "Purple Haze" to consider are the placements of Hendrix's voice and the various instruments across the stereo space, as well as the panning across the stereo soundscape.

At least in part because of the role of the mixing console in this recording, the mix itself and the motion around the stereo space become part of the composition, as suggested by Hendrix's contemporary, rock guitarist and songwriter/composer Frank Zappa, who wrote, "On a record, the overall timbre of the piece (determined by equalization of individual parts and their proportions in the mix) tells you, in a subtle way, *WHAT* the song is about" (Zappa with Occhiogrosso 1989, 188, italics and capitalization in the original). Later writers, such as Gracyk (1996) and Zak (2001) are even more emphatic than Zappa about the importance of the mix and recording effects in defining a recording as a musical composition. This is important to keep in mind regarding the mix and production in "Purple Haze" as this recording appeared on various millennial lists as being one of the greatest pieces of the 20th century. For example, National Public Radio's "The 100 Most Important American Musical Works of the 20th Century," as selected by the network's listeners, included "Purple Haze," and the network included a segment on the song that details the work of producer Chas Chandler and recording engineer Eddie Kramer on the iconic recording of the song (see, Wegman 2000).

Although multi-track recording in professional studios continued to expand during the 1980s and 1990s, there was a simultaneous interest in home multi-track recording. Some of the 4-track cassette-based recorders of the period incorporated sliders for the individual tracks as well as effects such as reverb that were offered on large professional consoles. As digital technology entered the home studio market, cassette-based recorders with mixers were replaced by mixers that offered more tracks, and over the past several decades, these mixers have dropped in price. At the time of this writing, various online musical instrument and equipment retailers offer 16–48-track digital mixing boards for between approximately $800 and $4,000.

However, one of the outcomes of the mass computerization of music that began with the introduction of MIDI in the early 1980s was the introduction over the years of a number of virtual recording studio programs and tablet and smartphone apps. Although some of these programs offer only rudimentary editing and mixing capabilities, several, including such well-established free downloads as *Audacity* and *GarageBand*, include mixing capabilities. In particular, *GarageBand* is set up so that the mixing process is reminiscent of using an actual mixing console. Other programs, such as *Davinci Resolve*, include video and audio editing. Although some of these programs might not offer all the capabilities of a professional mixing console or even the home studio hardware that is available, the fact that many of the things that once could only be done by using a professional mixing board in a studio can now be done by musicians using a laptop, tablet, or even a smartphone has greatly democratized the music production process.

Although we might reasonably think that the most common usage of a mixing board or mixing console might be in a recording studio, whether it be a huge professional studio or in multi-track home recording, we must consider that the mixing board also plays a significant role in live concerts. In these settings, the board often is used to control levels on inputs from microphones and instruments that are directly plugged in. The board then is used to control outputs through PA system speakers.

Despite what might at first appear to be a relatively static use of a mixing board, we can see the importance of the mixing process and how it can go far beyond just setting the levels of all the inputs and outputs and the equalization and so on during the soundcheck and then leaving the mixing board pretty much alone during a concert in the live work of the band Roxy Music. When Brian Eno played keyboards with the band in the early 1970s, he performed on the group's recordings;

Mixing boards allow sound technicians and engineers to precisely balance the volume, tone color, and placement of vocal and instrumental sounds both in the recording studio and in live performance venues. (Andris Andrejuks/Dreamstime.com)

however, at concerts he manned the mixing board. This gave Eno not only major role in shaping the mix of the vocal and instrumental sounds but also a role in adding effects to the voices and instruments. Essentially, then, Eno played the role of a studio record producer and recording engineer, but in a live performance setting.

Live performance mixing boards have moved into the digital age in some interesting ways. Perhaps most significantly, some boards that are available today can be controlled by a tablet computer or smartphone. This allows the technician to check the sound in various parts of the auditorium or theater and make appropriate adjustments. This can enhance the audience members' experiences even in a small theater that might include several loudspeakers. However, it is also particularly helpful for live music productions in large venues. For example, at an outdoor venue that includes a covered amphitheater with additional lawn seating, typically the mixing console might be located in middle or back of the audience area in the covered area. A sound technician who essentially is tethered to the mixing board might not necessarily be aware of or be able to make appropriate adjustments to the output to the audience members seated on the lawn. Bluetooth or WI-FI-based control over the mixing board from a portable device, such as a tablet or smartphone, allows the technician to walk out into the lawn seating area to monitor and adjust the output there, which may sound entirely different than what the audience members in the covered pavilion hear.

See Also: Amplification; Electrical Recording; Multi-Track Recording

Further Reading

Coryat, Karl. 2005. *Guerrilla Home Recording: How to Get Great Sound from Any Studio (No Matter How Weird or Cheap Your Gear Is)*. San Francisco, CA: Backbeat Books.

Gracyk, Theodore. 1996. *Rhythm and Noise: An Aesthetics of Rock*. Durham, NC: Duke University Press.

Huber, David M. 2018. *Modern Recording Techniques*. 9th ed. New York: Routledge.

Owsinski, Bobby. 2017. *The Mixing Engineer's Handbook*. 4th ed. Burbank, CA: Bobby Owsinski Media Group.

Wegman, Jesse. 2000. "The Story Behind 'Purple Haze.'" National Public Radio (NPR), September 18. Accessed January 18, 2021. https://www.npr.org/2000/09/18/1088122/jimi-hendrix-purple-haze.

Zak, Albin J. III. 2001. *The Poetics of Rock: Cutting Tracks, Making Records*. Berkeley: University of California Press.

Zappa, Frank, with Peter Occhiogrosso. 1989. *The Real Frank Zappa Book*. New York: Poseidon Press.

Moog Synthesizers

Robert Moog's early interest in electronics and particularly in oscillator-based electronic musical instruments is borne out in the Theremins he constructed. In fact, one of Moog's early publications is a 1961 article on building a transistorized Theremin that appeared in *Electronics World* (see, Moog 1961). Moog rolled out his first synthesizer, a monophonic analog instrument, in 1967. The Moog synthesizer quickly became the most frequently heard synthesizer in commercial popular music. In fact, country singer-songwriter Buck Owens is widely acknowledged as being the second person to purchase one of Moog's commercially produced instruments. The first electronic music composer Paul Beaver collaborated with several popular musicians and played a significant role in bringing the Moog synthesizer into the world of popular music.

Paul Beaver and Bernie Krause demonstrated the early Moog synthesizer at the Monterey Pop Festival in the summer of 1967, the "Summer of Love." Beaver then collaborated with the Doors in programming the Moog for their use on the title track from their 1967 album *Strange Days*. Beaver also collaborated with other prominent rock musicians as a programmer.

Writer Mark Brend credits the Byrds with being the first pop musicians to release a single that used the Moog synthesizer: "Goin' Back." Brend states, however, that Paul Beaver's performance on the Moog for the band was recorded "during a session on 9 October [1967]" (Brend 2012, 164). Although the Byrds' release was on October 20, 1967, earlier than the Monkees' release of "Daily Nightly" as a single, according to Andrew Sandoval's extensive liner notes for the Monkee's four-CD collection *The Monkees Box Set*, the sessions for "Daily Nightly" took place in June 19, 1967 (Sandoval 2001, 89), before the Byrds' sessions.

In the case of the Monkees, the Moog was put to use most notably on the aforementioned "Daily Nightly," a track on the 1967 album *Pisces, Aquarius, Capricorn & Jones Ltd*. Although penned by Monkees guitarist Mike Nesmith, drummer Micky Dolenz sang the lead vocals and performed the Moog synthesizer part; Paul Beaver, who seemed to be everywhere there was a Moog synthesizer being recorded in 1967, is credited with programming the instrument for the Monkees. "Daily Nightly" was included as a music video in two episodes of the television series *The Monkees*, "Fairy Tale" (broadcast January 8, 1968) and "Monkees Blow Their Minds" (broadcast March 11, 1968) (Lefcowitz 1985, 94, 95). In addition to "Daily Nightly," the album *Pisces, Aquarius, Capricorn & Jones Ltd*. also included a prominent Moog synthesizer part on the final track, "Star Collector," with Paul Beaver performing the synthesizer part.

The year 1968 saw the release of *The Notorious Byrd Brothers* by the Byrds. The album integrated the Moog synthesizer into the arrangements of several songs. However, easily the most famous, complex, and influential use of the Moog synthesizer in 1968 came in the form of Wendy Carlos's album *Switched-On Bach*. For this recording, Carlos interpreted complex polyphonic works of the famous Baroque period composer Johann Sebastian Bach using the monophonic Moog synthesizer—and numerous overdubs—in an impressive use of the multitrack recording studio.

AllMusic's Bruce Eder wrote that, despite the fact that *Switched-On Bach* offended some in the classical music world, it "became the first classical music LP ever to be certified for a Platinum Record Award, by selling to hundreds of thousands of mostly younger listeners wo didn't normally buy classical recordings." Eder describes Carlos's use of the Moog synthesizer as being "highly musical in ways that ordinary listeners could appreciate, itself a first in the use of the instrument" (Eder n.d.).

The technical side of the composition and production of *Switched-On Bach* is covered in part in a 1989 BBC documentary on the album and Wendy Carlos's work on it in which Carlos demonstrates the Moog synthesizer and how the various tone colors were created on it (see, https://www.youtube.com/watch?v=Z3cab5IcCy8, accessed January 15, 2021). Carlos's other major contribution on Moog synthesizer at the time was her soundtrack work for the 1971 Stanley Kubrick film *A Clockwork Orange*.

In 1969, the Beatles made a more subtle and more conventional musical use of a Moog IIIp synthesizer that had been purchased by George Harrison on the band's album *Abbey Road*, the last Beatles album to be recorded (*Let It Be* had been recorded earlier in 1969 but was not released until 1970). Compared with, say, the Monkees' use of the Paul Beaver-programmed Moog synthesizer for "Star Collector" and "Daily Nightly," the Beatles' use of the Moog on songs such as "Maxwell's Silver Hammer," "Because," and "Here Comes the Sun" was subtler, more melodically inspired. As such, it anticipated the ways in which synthesizers would be interwoven into the texture of pop songs in the 1970s. To put it another way, the sound of the Moog synthesizer on *Abbey Road* was less of the sound of a scientific novelty than it was another tone-color possibility that could be shaped in whatever way made the most musical sense to the artists.

Designed by Robert Moog, the analog synthesizer that bears his name was one of the first widely heard synthesizers in popular music in the late 1960s and early 1970s. Not limited to that time period, this Minimoog Voyager synthesizer was used by hip-hop artist J Dilla. (Collection of the Smithsonian National Museum of African American History and Culture, Gift of Maureen Yancy)

The complexity of programming and playing the early Moog synthesizers detailed previously in the discussion of *Switched-On* Bach is confirmed by the Abbey Road Studios maintenance engineer, Ken Townsend, who said of the Beatles' first use of the Moog synthesizer in the *Abbey Road* album sessions, "It was 'Because,' and the Moog was a bit of a marvel instrument. To get that French horn sound it took a whole set of flightcases full of jack plugs and filters" (Babiuk 2002, 246). The album's engineer, Alan Parsons, said of Paul McCartney's playing of the Moog using the instrument's ribbon controller, over which the performer could slide their finger to play the notes on "Maxwell's Silver Hammer," "Paul did 'Maxwell' using the ribbon, playing it like a violin and having to find every note—which is a credit to Paul's musical ability" (Babiuk 2002, 246).

Because of the length of the Moog synthesizer solo and the synthesizer's prominence late in the song, perhaps one of the best-remembered uses of the instrument in popular music came in the form of the 1970 single and album cut "Lucky Man," by Emerson, Lake & Palmer. Written by the trio's guitarist and bassist Greg Lake, the group's song "Lucky Man" was an almost accidental release. As *AllMusic* reviewer François Couture writes, the group was short one song for its debut album, so they turned to "Lucky Man," a song that Lake had written when he was 12 years old, and "stuck at the 11th hour, the group started to record, improvising the arrangements they laid down track after track. Emerson's landmark Moog solo at the end was his first take (and he wasn't even aware the tape was running). A set of circumstances and the song of a child became ELP's long-lasting hit" (Couture n.d.).

Interestingly, some writers credit "Lucky Man" with being the first significant Moog synthesizer solo on a popular recording, despite the fact that Paul McCartney's Moog solo on "Maxwell's Silver Hammer" had been recorded and released well before Emerson, Lake & Palmer recorded "Lucky Man." Still, "Lucky Man" was seen by some of Robert Moog's associates as a key ingredient in the popularity of Moog synthesizers. For example, keyboardist, music educator, and early supporter and marketer of Moog synthesizers David Van Koevering is quoted as saying, "Here's the big breakthrough. . .'Lucky Man' shows up, and then there's a Keith Emerson tour, followed by a Rick Wakeman [of the progressive rock band Yes] tour. I mean, we knew where these artists were by the cities that were calling [to order Moog synthesizers]" (Pinch and Trocco 2002, 248).

Several years later, an often-heard well-known use of a Moog analog synthesizer was Mike Post's use of a Minimoog Model D for the instrumental theme music for *The Rockford Files*, a television program that ran from 1974 to 1980. After the show became a hit during its first season, MGM Records issued Post's recording of the theme music, a piece that Post had cowritten with Pete Carpenter. The single was a hit and stayed on *Billboard* magazine's pop singles chart for 16 weeks. In *The Rockford Files* theme, the Minimoog plays the melody in the first section; harmonica is featured in the piece's bridge section. Although later versions of the theme music as the series continued its run place more of a focus on electric guitar, the Moog synthesizer sound continued to define the opening of the piece.

The Minimoog continued to play a prominent role in popular music, particularly during the 1970s and the early 1980s. For example, the innovative and influential German electronic band Kraftwerk used the instrument, as did the American experimental new wave rock band Devo. This happened even though Moog completed with more and more analog synthesizer brands and models as the 1970s went on. Even with the advent of digital synthesizers with the Casio VL-1 in 1979, the sound programming flexibility of Moog's analog synthesizers was still popular with the public. In fact, in addition to manufacturing synthesizers under its own name, Moog made the Realistic Concertmate MG-1 for the Radio Shack chain of electronics stores beginning in the early 1980s.

At the time of this writing, patches are available for digital synthesizers that emulate some of the classic tones that were associated with the analog Moog synthesizers back in their heyday.

See Also: Multi-Track Recording; Patches; Theremins

Further Reading

Babiuk, Andy. 2002. *Beatles Gear: All the Fab Four's Instruments, from Stage to Studio*. Revised ed. San Francisco, CA: Backbeat Books.

Brend, Mark. 2012. *The Sound of Tomorrow: How Electronic Music Was Smuggles into the Mainstream*. New York and London: Bloomsbury.

Couture, François. n.d. "Emerson, Lake & Palmer: 'Lucky Man.'" *AllMusic*. Accessed January 18, 2021. https://www.allmusic.com/song/lucky-man-mt0053497938.

Douglas, Adam. 2019. "The Synths behind 8 Classic TV Theme Songs: *Miami Vice*, *Doctor Who*, and More." *Reverb*, June 11. Accessed January 18, 2021. https://reverb.com/news/the-synths-behind-8-classic-tv-theme-songs.

Eder, Bruce. n.d. "Wendy Carlos: *Switched-On Bach*." *AllMusic*. Accessed January 18, 2021. https://www.allmusic.com/album/switched-on-bach-mw0000976916.

Lefcowitz, Eric. 1985. *The Monkees Tale*. San Francisco, CA: The Last Gasp.

Moog Music, Inc. 2011a. *A Brief History of the Minimoog*, Part 1. Accessed January 18, 2021. https://www.youtube.com/watch?v=sLx_x5Fuzp4.

Moog Music, Inc. 2011b. *A Brief History of the Minimoog*, Part 2. Accessed January 18, 2021. https://www.youtube.com/watch?v=xh4Ok0ex2vU.

Moog, Robert A. 1961. "A Transistorized Theremin." *Electronics World* 65 (January): 29–32, 125.

Pinch, Trevor, and Frank Trocco. 2002. *Analog Days: The Invention and Impact of the Moog Synthesizer*. Cambridge, MA: Harvard University Press.

Sandoval, Andrew. 2001. "The True Story of the Monkees" (liner booklet for the *The Monkees Music Box*). Four CD set. Rhino Records R2 76706.

Motion Picture Soundtracks

Although the technology for synchronizing sound and visual images in movies had already been in existence for years, it was not until the late 1920s that things coalesced and musical soundtracks that included both diegetic and nondiegetic music became the norm.

The 1927 film *The Jazz Singer* is routinely credited with being the first "talkie," a movie that included synchronized sound and visual images. There were earlier films that synchronized sound and visuals, but *The Jazz Singer* garnered considerably more attention when it was released. For one thing, the film seems designed to draw attention to the technology. This is particularly evident when the audience experiences a transition from the silent-film parts of the movie, in which characters are not heard speaking and a condensed transcription of what dialogue appears to be transpiring appears on screen, to the scenes in which actor Al Jolson's singing clearly exhibits audio and visual synchronization. In fact, in the sequence in which Jolson sings "Toot, Toot, Tootsie," Jolson incorporates so many quick-paced ad libs, handclaps, and so on that it seems as though the arrangement and performance of the song were designed as a showpiece for the audio-visual synchronization. Likewise, when Jolson's character performs the Irving Berlin standard "Blue Skies" for his mother while seated at the piano, the quick cuts between audible dialogue and singing make it clear to the audience that they are in entirely new territory than they were during the silent-film era.

Within two years of *The Jazz Singer*, films such as *The Hollywood Revue of 1929* featured completely synchronized soundtracks. Ironically, Hollywood motion picture studios brought out so many movie musicals in the late 1920s and start of the 1930s that, as film historian Thomas S. Hischak puts it, "by 1931 there were too many musicals on the market, the novelty had worn off, and audiences started to avoid movie musicals" (Hischak 2013, 617). However, even at that time, dialogue and orchestral background music were already being included in the soundtracks of nonmusicals.

One of the technical challenges faced in some of the movie musicals of the past was in synchronizing the audio with the audience's view of the character singing because in that particular film, a particular character had their vocals overdubbed by a more competent singer. The soprano Marni Nixon, in particular, was one of the more frequently heard overdubbed voices in films such as *The King and I*, *My Fair Lady*, *West Side Story*, and others. Some of the ghost-singer efforts of movie directors of the mid-20th century worked better than others and were subject to the technical limitations of the day. In her autobiography, Marni Nixon discussed the particular challenges that she faced in ghost singing for Natalie Wood in *West Side Story*, which caused Wood's apparent lip-synching to be slightly out of synchronization with Nixon (Nixon and Cole 2006, 133–134).

As mentioned earlier, the use of a movie soundtrack was not limited to movie musicals. In fact, one of the best-known pre–World War II movie soundtracks, despite the film's ongoing controversy, is Max Steiner's elaborate orchestral score for the 1939 film *Gone with the Wind*. In contrast to movie musicals, in which diegetic music is the norm (in fact, the characters not only hear the music but they also play a role in making the music), Steiner's score is largely nondiegetic. Not only does the score set the mood for the film's action but it also includes an element of the 19th-century aesthetic that German composer Richard Wagner pursued in his operas, in which the various characters have their own musical themes. In fact, one of the best-known melodies from *Gone with the Wind* is "Tara's Theme," which musically symbolizes the plantation mansion of the story.

More recent composers have expanded on the work of Steiner and other mid-20th-century film composers. For example, in the original *Star Wars* trilogy, composer John Williams provided all the central characters with thematic material. Williams wove the various melodies throughout the score in such a way as to highlight the moods and relationships of the characters, sometimes providing hints of insights into the audience that the action and dialogue alone did not fully capture.

Even in films in which the music and characters, buildings, and locations are not as closely integrated as in the original *Star Wars* trilogy or in *Gone with the Wind*, over the years, film scores have continued to play essential roles in establishing mood, building audience expectations (which sometimes deliberately are not met for dramatic effect), and manipulating the audience's sense of time and pacing.

Putting purely musical connections between film and sound aside, one of the technical innovations that helped to make both music and sound effects more spectacular use of surround sound. From its use in the 1940 film *Fantasia*, in which a bumblebee sounds as though it is flying around the theater, to the extra-large visual and audio spectacles of IMAX films (the 1974 film *Man Belongs to the Earth*, with its rhythmically complex score by Don Ellis is a particularly impressive early example of an IMAX documentary), various technologies for surrounding the audience with sound have played a major role in putting the audience member in the drama in a way that was not possible with the sound coming at the audience from only the front of the theater.

The use of Dolby noise reduction, which is covered in a separate entry in this volume, as well as Dolby Lab's later surround-sound system, has provided more immediate and clearer sound for dialogue, sound effects, and music. Similarly, the THX sound certifications that began in the early 1980s and are still used in the 21st century have allowed the moviegoer's audio experience to be as close to that intended by the film's production staff as possible. Some moviegoers might perceive THX movies that make full use of Dolby noise reduction encoding as being louder than other movies; however, that is probably because THX movies typically have a wider frequency range and more even response across that frequency range than non-THX movies.

As a sort of sidebar to consideration of movie soundtracks, it is interesting to note that some record producers recognized the advantages that encoding music on film had over recording on magnetic tape. Violinist, bandleader, and producer Enoch Light founded Command Records in 1959, and for the next half decade, the label produced a number of innovative albums that highlighted some of the new techniques of the recording industry, for example, heavily exploiting stereophonic sound. Light also produced a series of albums on Command in which the multi-track recordings were made directly onto 35mm film instead of onto tape. The fact that film was fed by means of sprockets meant that it was absent the wow and flutter that tended to create audio challenges with the use of magnetic tape.

See Also: Dolby Noise Reduction; Quadraphonic and Surround Sound

Further Reading

Hischak, Thomas S. 2013. "Musical Film." In Charles H. Garrett, ed., *The Grove Dictionary of American Music*, 2nd ed., vol. 5, 617–620. New York: Oxford University Press.

Light, Enoch, producer. 1967. Liner notes for *Kites Are Fun* by the Free Design. LP, Project 3 PR5019SD.

Nixon, Marni, and Stephen Cole. 2006. *I Could Have Sung All Night: My Story*. New York: Billboard Books.

MP3s

For all intents and purposes, and particularly for the general public, music entered the digital age with the release of commercial compact discs in the early 1980s. At that time, the management of large digital audio files was being dealt with on several fronts. The mp3 file format itself grew out of the desire of the dissertation advisor to Karlheinz Brandenburg—widely known as the "father of the mp3"—to use ISDN phone lines for the transmission of music. According to Brandenburg, in 1982, his advisor "composed and submitted a patent application that ISDN digital phone lines should be used to transmit music and the patent examiner told him, 'This is impossible, we can't patent impossible things'" (Rose and Ganz 2011). Brandenburg began working on the problem, and by 1986, the technology had advanced to the point at which several aspects of the what would come to be mp3 technology were patentable.

In 1987, the German institution Fraunhofer-Gesellschaft established the Fraunhofer Institute, which began researching high-quality low-bit-radio audio encoding; Karlheinz Brandenburg's funding came from the institution. The following year, the Motion Picture Experts Group, of which Brandenburg was part, was established. One of the principal findings of the group was that digital music files could be compressed a significant amount if the audio content was limited to the range of human hearing. It was this group's name (Motion Picture Experts Group) and the compression format's designation as Audio Layer III that led to the name mp3. By 1992, the International Standards Organization recognized the mp3 compressed file format as international standard. As a result, the mp3 was widely adopted and was free from the kinds of competitive technology wars of the past, such as the battle between Thomas Edison's wax cylinders and Emile Berliner's flat discs for domination of the early commercial record industry, or between VHS and Beta for home video.

As mentioned earlier, a significant reason for the success of the mp3 format was that digital audio files could be greatly reduced in size by limiting the frequency range of the recordings to the limits of human hearing. On the surface, that might sound like a win-win situation, and to a great extent it was. However, some listeners to mp3 compressions of music might find that because some of the higher overtones are missing from the tone of certain instruments, they lack the brilliance that can be heard in noncompressed versions of the same recording. This might be particularly noticeable with instruments such as crash cymbals or particularly high-pitched instruments. Generally, though, mp3 compression resulted in a musical product that exceeded the frequency spectrum and clarity of, say, FM radio. And compression of audio information into mp3 files meant that music could be saved on a computer's hard drive, an mp3 player, or, later, a USB drive or SD card or any other mass storage device.

By the late 1990s, consumers could choose from a variety of mp3 players and software for their home computers. Although many of the lower-cost mp3 players had limited capacity, particularly compared with the iPods and smartphones that would be introduced early in the new millennium, they helped to make personal music more portable than earlier technology, such as Sony's Walkman and Discman. Using these players and the software (e.g., the Windows program *MusicMatch Jukebox*) that could "rip" compact discs (e.g., convert the CD's files into mp3 files) might have been seen as little more than a digital extension of what individuals had been doing with recording their vinyl albums and singles onto cassette tapes for portability, except for two significant developments. First, although *MusicMatch*'s program was completely legal and involved royalty payments to the record industry, some coders developed mp3 programs that circumvented the music industry by not paying use rights. At almost the same time, coder Shawn Fanning and entrepreneur Sean Parker rolled out their peer-to-peer file-sharing network Napster. Napster allowed users to share digital files (e.g., mp3 files) without going through a central server. Although Napster in its original form was only active for a short time, it created a significant amount of controversy, and some studies (e.g., Michel 2006) suggest that Napster and successor file-sharing programs made an impact on the sales of compact discs. Anecdotally, other writers state unequivocally that Napster and other peer-to-peer file sharing networks "were actually bringing the entire music business to its knees" (McIntyre 2018).

Even after these controversies at the very end of the 1990s and start of the 2000s and as mp3 files became more mainstream and legitimate, the files remained one of the main ways in which consumers received downloaded digital music. For example, companies such as Amazon and Apple's *iTunes* made mp3 versions of albums and individual tracks available for purchase in the format. In fact, one of the more interesting marketing devices that Amazon used for a period was to make mp3 versions of albums available free to consumers who purchased CDs of certain albums.

By the end of the second decade of the 21st century, Advanced Audio Coding (AAC), also known as MPEG-4, or mp4, had become an established standard for compressed music. This successor to the mp3 standard allows files with higher-quality audio fidelity to be compressed to approximately the same size as mp3 files. A 2017 article in *Tech Times* (Velasco 2017) suggested that because of the improvements that mp4 offered, the mp3 is "dead"; however, the mp3 is still used. It is easy to edit and can easily be incorporated into videos using applications such as Apple's *iMovie*.

See Also: iPods and Other Portable Digital Music Players

Further Reading

Coleman, Mark. 2004. *Playback: From the Victrola to MP3, 100 Years of Music, Machines, and Money*. Cambridge, MA: Da Capo Press.

McIntyre, Hugh. 2018. "The Piracy Sites That Nearly Destroyed the Music Industry: What Happened to Napster?" *Forbes*, March 21. Accessed January 18, 2021. https://www.forbes.com/sites/hughmcintyre/2018/03/21/what-happened-to-the-piracy-sites-that-nearly-destroyed-the-music-industry-part-1-napster.

Michel, Norbert J. 2006. "The Impact of Digital File Sharing on the Music Industry: An Empirical Analysis." *Topics in Economic Analysis & Policy* 6, no. 1. This article is available for download from the Recording Industry Association of America (RIAA) at https://www.riaa.com/wp-content/uploads/2004/01/art-the-impact-of-digital-file-sharing-on-the-music-industry-michel-2006.pdf.

Moreno, Ashley. 2013. "It's Still All about Peer-to-Peer Connectivity: Alex Winter Raps about Napster with Shawn Fanning and Sean Parker." *Austin Chronicle*, March 12. Accessed January 18, 2021. https://www.austinchronicle.com/daily/sxsw/2013-03-12/its-still-all-about-peer-to-peer-connectivity/.

Rose, Joel, and Jacob Ganz. 2011. "The MP3: A History of Innovation and Betrayal." *NPR*, March 23. Accessed January 18, 2021. https://www.npr.org/sections/therecord/2011/03/23/134622940/the-mp3-a-history-of-innovation-and-betrayal.

Velasco, Carl. 2017. "The MP3 Is Dead: Here's a Brief History of MP3." *Tech Times*, May 13. Accessed January 18, 2021. https://www.techtimes.com/articles/207213/20170513/the-mp3-is-dead-heres-a-brief-history-of-mp3.htm.

Winter, Alex, director. 2013. *Downloaded*. Documentary film. New York: VH1.

MTV and Music Videos

Although one might think of the start of the music video age as beginning with the first broadcast of the Buggles' video for "Video Killed the Radio Star" on MTV on August 1, 1981, the history of the music video goes well back before that often-cited event. The mature music video would become a unified work in which the visual imagery, acting, visual effects, and the music would work in conjunction with one another to produce an integrative experience. In other words, music videos needed to go beyond merely showing musicians lip-synching to their work.

Some of the sequences in Elvis Presley's films, most notably the "Jailhouse Rock" sequence in the film of the same name, and some of the songs in the Beatles' 1964 film *A Hard Day's Night*, most notably the "Can't Buy Me Love" sequence, anticipated some of the kinds of effects and production that would mark mature music videos. Some of the song sequences in *The Monkees* television program, which aired between 1966 and 1968, too, presented something approaching and integration of storytelling and music; however, some of the promotional films that other artists produced so that they would not need to appear live on music television programs of the 1960s also qualify and direct antecedents to the mature music videos of the MTV age. Perhaps as well as the work of any other 1960s' rock artists, the films of the Beatles in 1967 and 1968 show the range of promotional films of that time period. The Beatles' 1968 promotional film for "Hey Bulldog" (see, e.g., https://www.youtube.com/watch?v=M4vbJQ-MrKo, accessed January 15, 2021) goes beyond a mere studio performance; however, it does not integrate a story. The Beatles' 1968 promotional film for "Revolution" (see, e.g., https://www.youtube.com/watch?v=BGLGzRXY5Bw, accessed January 15, 2021) is even closer to what we might term the old-school approach to music film and video making, showing the band lip-synching to the music. Interestingly, the band's 1967 promotional films for "Penny Lane" and "Strawberry Fields Forever" had been significantly closer to what would later define mature music video (see, "Penny Lane," for example, at https://www.youtube.com/watch?v=S-rB0pHI9fU, accessed January 15, 2021, and "Strawberry Fields Forever" at https://www.youtube.com/watch?v=HtUH9z_Oey8, accessed January 15, 2021).

A one-time member of the Monkees, Michael Nesmith, was one of the individuals who had the idea of a music video–based network and television evolved from a broadcast base of three major networks and public television into the cable TV age, providing the idea that Warner Cable developed as MTV. During the 1970s, Nesmith had been developing the concept of what would become the mature music video, and his company, Pacific Arts, was a noted producer of music videos even before MTV became part of popular culture. Even just a few months into the life of MTV, Nesmith recognized the potential impact of the music video while sensing that the old guard of the entertainment industry did not understand the medium. Nesmith is quoted as saying, "It's frustrating to talk to people in the entertainment here who aren't aware of how potent music video is. Moving some people into a position where the understand is like pulling teeth. I have no axe to grind. It's not hurting me. Let 'em sit by the side of the road as the bandwagon goes by" (Massingill 1997, 177).

The 1980s were the heyday of the MTV music video, with videos for Madonna's "Like a Virgin" and "Material Girl"; Prince's "Cream," "U Got the Look," "Purple Rain," and "Let's Go Crazy"; the Pretenders' "Brass in Pocket"; Eurythmics' "Sweet Dreams (Are Made of This)"; Devo's "Whip It!"; the Specials' "A Message to You, Rudy"; Culture Club's "Karma Chameleon"; and Michael Jackson's "Beat It," "Billie Jean," and "Thriller" being just some of the more memorable highlights. Some of these videos, particularly Jackson's "Beat It," involved elaborate productions with dozens of actor/dancers and elaborate Hollywood-like sets, locations, and sound stages.

Early in its life, MTV included more esoteric videos in its rotation than what would later be the case, perhaps because there was so much time to fill and relatively few music videos. One of the more unusual examples, "O Superman," by singer, composer, and performance artist Laurie Anderson, became a surprise pop hit.

The impact of MTV and the music video medium has been suggested by a number of anecdotal events. Perhaps one of the more interesting was the effect on the life and career of British scientist Magnus Pyke, who appeared in the video for Thomas Dolby's "She Blinded Me with Science." The sampled Pyke exclaims the word "Science!" numerous times in the video. The scientist reported to Dolby that throughout a lecture tour to the United States (in which Pyke was not as well known as he was in Britain) that took place around the time of the wave of popularity of the Dolby song and music video, "Every time I walked down the street, someone would come up behind me and shout 'SCIENCE!' It frightened me out of my skin. Your MTV video is better known than my body of academic work" (Majewski and Bernstein 2014, 149–150).

In the 1990s and early 2000s, MTV and music videos in general increasingly generated controversy over sexual, violent, and drug images and lyrics. Other music videos of the period included what were perceived as satanic imagery, which added to the controversies surrounding the video medium. At the same time, MTV itself increasingly focused on programming so-called reality shows and programs such as *Beavis and Butt-Head*. Although music was still part of MTV, to some viewers who remembered the musical focus of the network in the 1980s, however, it seemed that the *M* part of MTV was disappearing. By the end of the 2010s, it appeared that the role of music networks on television was increasingly giving way to other media. Although it is not necessarily indicative of overall interest in music videos, it must be noted that 2017, 2018, and 2019 all saw drops in the audience share for MTV's *Video Music Awards* program. In fact, the 2019 show had the lowest viewership in the history of the awards program (Kelley 2019).

One of the aesthetic issues raised by the prominence of the music video, whether they be on television, social media, streaming, or downloaded, largely as a result of the success of MTV and its principal competitor, VH1, is that of the listener's opportunity to interpret a piece of music, particularly a song and its lyrical intent. This is because the music was more closely tied to the visual imagery in mature music videos than in the promotional music films of the 1950s and 1960s. In the music videos of the 1980s through the present, often the musical artist and the writer, producer, and director of the video present the audience member with visual imagery that is so memorable and, in the most successful music videos, so impactful that it cements one interpretation of the song in the audience member's mind. This is exacerbated by the fact that in the world of popular music, it seems as though the effect of MTV and the music videos of the 1980s is that virtually every new song has to have a video attached to it. As a result, it can be argued that

MTV ushered in the age of the music video with its debut on August 1, 1981. (Raffaele1/Dreamstime.com)

MTV and its success in the 1980s not only changed how consumers view their music but also how they listen to and interpret it.

In recent years, another aspect of the music video has emerged: consumer/fan-produced videos. Even a cursory search of sites such as YouTube shows that for popular music—particularly for that of the pre-music video age—for which promotional films or videos were not officially released, enterprising fans have prepared and published their own videos. In some cases, these consist of little more than a film of a record playing, while some others include a photo montage of the artist or artists. The more enterprising video producers include a variety of material, such as cuts to live performances by the artist or artists, photos, perhaps the lyrics of a song, and so on. Some of the more intriguing fan-produced videos synchronize the visual images from a live performance with the music from a studio recording.

Beginning with its rollout in China in 2016 and its worldwide rollout in 2018, TikTok has provided another form of music video, based on brief posts. Although arguably these are not really comparable to the best and most elaborately produced fan-made videos of YouTube—not to mention official professionally produced Vevo videos and the like—at the time of this writing, these short-form videos are quite popular among young people. Despite its popularity, at the time of this writing it remains to be seem whether or not TikTok can survive its controversies (e.g., foreign ownership, attempts to ban it in the United States, its information-gathering capabilities).

See Also: iPods and Other Portable Digital Music Players

Further Reading

Kelley, Caitlin. 2019. "VMAs Ratings Plunge to All-Time Low This Year." *Forbes*, August 30. Accessed August 7, 2020. https://www.forbes.com/sites/caitlinkelley/2019/08/30/vmas-ratings-plunge-to-all-time-low-this-year.

Majewski, Lori, and Jonathan Bernstein. 2014. *Mad World: An Oral History of New Wave Artists and Songs That Defined the 1980s.* New York: Abrams Image.

Massingill, Randi L. 1997. *Total Control: The Michael Nesmith Story.* Mesa, AZ: FLEXquarters.

Williams, Cameron. 2017. "How MTV Changed the World with Its Industry of Cool." Special Broadcast Services (SBS) Australia, February 13. Accessed January 18, 2021. https://www.sbs.com.au/guide/article/2017/02/13/how-mtv-changed-world-its-industry-cool.

Multi-Track Recording

When commercial audio recording began in the late 19th century, singers and instrumentals were recorded live. The musicians played and sang into a large horn that was part of a mechanism that transferred their sound directly onto a disc, or in the case of Edison's cylinder-based recordings, directly onto a cylinder. There were no overdubs, and there was no way to isolate individual instruments or voices, short of moving them closer to the horn. The live performance, or direct-to-disc, paradigm continued to be the norm into the electrical recording era. With the advent of tape recording and the capability of magnetic recording tape to be separated into several tracks, the recording industry underwent a fundamental—albeit incremental—change.

During the 1950s, the principal force in making the full use of the new tape-based possibilities in the recording studio was the duo of guitarist Les Paul and singer Mary Ford. Especially germane to our consideration of multi-track recording, Paul's instrumental recordings and the duo's recordings made use of overdubbing, which was possible because of multi-track tape machines. In the late 1950s, producer Tom Dowd of Atlantic Records used the multiple tracks available on tape to isolate the instruments of the rhythm section (e.g., bass and drums) on separate tracks. This allowed Dowd to balance these instruments in a way that had not been possible before the advent of multi-track recording.

American composer-guitarist Frank Zappa wrote:

> On a record, the overall timbre of the piece (determined by equalization of individual parts and their proportions in the mix) tells you, in a subtle way, *WHAT* the song is about. The orchestration provides *important information* about what the composition *IS* and, in some instances, assumes a greater importance than *the composition itself* [italics and capitalizations in the original]. (Zappa with Occhiogrosso 1989, 188)

Zappa's context is the music of his time, roughly the 1960s through the 1980s, and his statement suggests that listeners consider a musical work to be synonymous with the best-known recording of the work. Later writers such as Gracyk (1996) and Zak (2001) concur with Zappa and, in fact, are even more emphatic about the point that the recording essentially equals the piece for listeners of rock music, music recorded during the period in which advanced studio techniques such as timbral effects, tape manipulation, and creative use of the stereophonic space came into their own. What these writers suggest is that although making full use of the possibilities of multi-track recording was something of a novelty in the 1950s, by the time musicians such as Zappa were recording in the mid-1960s and beyond, extensive overdubs, balancing individual voices and instruments, stereophonic panning from channel to channel, various other studio effects, and so on were already becoming part of the defining and essential aspects of popular music compositions (as much as melody, harmony, lyrics, arrangement, and so on). One could argue, then, that the way in which musicians and producers used multi-track recording became part of the compositional process.

Eight-track machines were used in some studios in the second half of the 1950s. Eventually, 16-track, 32-track, and even 48-track machines have been used. In some studios for many years, 4-track machines were the standard even though more tracks had been available. Perhaps most famously, the Beatles' highly acclaimed 1967 album *Sgt. Pepper's Lonely Hearts Club Band*, which was produced by George Martin, was recorded using 4-track machines. Even when 4-track and 8-track recording were being done, though, multi-track recording was not limited to four or eight separate instruments or voices to be recorded. One of the techniques that gave even 4-track machines the ability to handle complex recording was bouncing tracks. One could, say, record two separate tracks (e.g., electric bass and drum set), move those two tracks to a third track, and then rerecord on the first two tracks that were used. This process could continue, and multiple 4-track machines could be connected to further increase the possibilities for including multiple tracks in a recording.

For the home, multi-track recording for many years was limited to owners of reel-to-reel tape recorders. That changed with the introduction of multi-track cassette recorders. The TASCAM Portastudio, which was introduced as the Teac 144 in 1979, was the first well-known example. The Teac 144 was a 4-track studio that used standard cassette tapes. With various effects and the ability to bounce tracks and so on, in many ways this machine and others that subsequently came out (e.g., multi-track machines from Fostex, Yamaha, Akai, Marantz, and others) brought many of the capabilities of professional studios (such as 1967-era Abbey Road Studios at which the Beatles' *Sgt. Pepper's Lonely Hearts Club Band* was recorded). It should be noted that a machine such as the Teac 144 did not necessarily record with the same signal-to-noise ratio as professional studio machines, but it and others like it democratized multi-track recording. And it was not just aspiring musicians who used machines such as the Teac 144; Bruce Springsteen recorded his album *Nebraska* on a Teac 144. No longer cassette-based, home studio hardware continued to expand in capabilities, including in the number of tracks that could be recorded. The machines also became more economical. For example, at the time of this writing, TASCAM's DP-24SD 24-track Digital Portastudio has a list price of $599 but is available from at least one online-retailer for $499. Eight-track digital home studios are even less expensive.

More recently, computer applications such as *GarageBand*, *Audacity*, and other software packages allow for multi-track digital recording. And, unlike cassette-based multi-track recorders, digital applications such as *Audacity* also allow the user to do such things as change the speed of a recording without changing pitch (which was not possible in analog recording), easily record backward, and add numerous effects beyond what was possible on a consumer-level multi-track recorder during the early 1980s' heyday of 4-track cassette-based machines made by TASCAM, Akai, Yamaha, Vestex, and other companies. However, it should be noted that a program such as *Audacity* that does not necessarily default to synchronized recording of tracks is perhaps better suited to recording live performances and to doing final editing than multi-track tape recorders or synchronization-default computer programs such as *GarageBand* and other virtual recording studio software.

Koninklijke Philips N.V. (Royal Philips)

Based in the Netherlands, and better known in the United States simply as Philips, this company has played significant roles in music technology for decades. Although formed in 1891 by Frederik Philips to make incandescent light bulbs, Philips soon moved into a variety of other electronics areas, including developing x-ray technology. In the area of music, the company was responsible for the experimental multimedia Philips Pavilion at the 1958

Brussel's World's Fair, in which an 8-minute electronic music composition by Edgard Varèse, *Poème électronique*, was heard by an estimated 2 million visitors during the exposition's run. During the 1970s, Philips was one of the premiere record labels for classical music. Philips collaborated with Sony Corporation in the development of the compact disc (CD), and the first commercial CD release, of pianist Claudio Arrau's recording of Chopin Waltzes, was on the Philips label. Because the boombox evolved dramatically in the 1970s and 1980s, what might be lesser known is that Philips is credited with manufacturing the first boombox, the Norelco 22RL962, which made its debut in 1966. Today, Philips's focus is on healthcare technology, an outgrowth of its pioneering early 20th-century work on x-ray technology.

See Also: Cassette Tapes; Reel-to-Reel Tape; Tape Manipulation; Virtual Studio Software

Further Reading

Coryat, Karl. 2005. *Guerrilla Home Recording: How to Get Great Sound from Any Studio (No Matter How Weird or Cheap Your Gear Is)*. San Francisco, CA: Backbeat Books.

Gracyk, Theodore. 1996. *Rhythm and Noise: An Aesthetics of Rock*. Durham, NC: Duke University Press.

Morgan, Robert P. 1972. Liner notes for *Edgard Varèse: Offrandes, Intégrales, Octandre, Ecuatorial*. Reissued on CD as Elektra/Nonesuch 9 71269–2.

Murnane, Kevin. 2017. "'Sgt. Pepper's' Was a Perfect Storm of Musical and Recording Creativity." *Forbes*, June 3. Accessed January 18, 2021. https://www.forbes.com/sites/kevinmurnane/2017/06/03/sgt-peppers-was-a-perfect-storm-of-musical-and-recording-creativity.

National Museums Liverpool. n.d. "Emergence of Multi-Track Recording." Accessed August 4, 2020. https://www.liverpoolmuseums.org.uk/emergence-of-multitrack-recording.

Zak, Albin J. III. 2001. *The Poetics of Rock: Cutting Tracks, Making Records*. Berkeley: University of California Press.

Zappa, Frank, with Peter Occhiogrosso. 1989. *The Real Frank Zappa Book*. New York: Poseidon Press.

Music Boxes

It seems likely that the experience of most citizens of the 21st century with music boxes might be with cheap adornments to jewelry boxes and the like that play perhaps one piece of music. In the days before digital music, before radio, and before sound recordings, however, music boxes were one of the more elaborate and technologically advanced mechanical music devices around.

Still, given the advances that would be made in mechanical musical instruments—or automatic musical instruments—particularly from the late 19th century through the present—music boxes might seem relatively low-tech. Music boxes represent an example of what musicologist David Fuller describes as "the simplest automatic instruments." According to Fuller, in such instruments, "pins on the cylinder or projections from the disc engage the teeth of a metal comb, sometime through the intermediary of a star wheel, causing them to vibrate and make a sound" (Fuller 1986, 62).

Early music boxes were made in Europe probably beginning sometime around the 1770s; however, the principle of encoding music using cylinders or discs can be found in some carillons, water organs, and other instruments going back to the Renaissance era. In any case, the basic technology of a rotating cylinder or disc closely relates to the workings of mechanical clocks. Despite the existence of mechanical carillons and water organs before the late 18th century, it is the principle of "pins on the cylinder or projections from the disc" engaging with differently tuned metal teeth described previously by Fuller (1986) that became the technology that we most likely associate with pre-player piano mechanical musical instruments, as well as with the nonelectric mechanical musical instruments that we are still most likely to encounter.

The early music boxes, those of the late 18th century, originally were miniature, approximately the size of a pocket watch of the day. They soon became devices that were designed more for use in the home, as opposed to portable as the early music boxes had been. The music box, once it had literally become a wooden box in the early 19th century, was generally approximately 13 inches long with 96 teeth and an interchangeable brass cylinder that allowed different pieces to be played (*Encyclopædia Britannica* 2020).

As anyone who has experienced music boxes and mechanical clocks knows, there is in many cases a significant difference between the two mechanisms: although the hands of a clock will maintain a more-or-less constant pace as the spring winds down, inexpensive music boxes, such as those integrated into late 20th-century and early 21st-century jewelry boxes and the like, tend to slow down as the spring winds down. Such particularly is the case with inexpensive music boxes because many do not include a governor to regulate speed as part of their mechanisms. Other devices, including clocks and nonelectric phonographs, Victrolas, and so on, have a governor, which keeps their clock hands, turntables, and so forth moving at the more-or-less constant speed, at least until the spring is close to being fully unwound.

In many respects, the second half of the 19th century represented the culmination of the era of the music box. For one thing, some of the machines of this time period were quite ornate so that they not only provided one of the rare forms of home musical entertainment that did not require family members, staff, or any other people to perform but they were also aesthetically pleasing visually.

Perhaps the ultimate music boxes that were built in the late 19th century and early 20th century were manufactured by the Leipzig, Germany-based Polyphon Musikwerke. The company's machines used a disc, as opposed to the more familiar cylinders. One of the advantages of the discs was that they could easily be changed; therefore, unlike the music boxes with which many of us are familiar today, a single machine could play multiple different pieces, like the record players such as phonographs and Victrolas that were their contemporaries. In fact, the home models of the disc-based music boxes that were produced in the first decade of the 20th century might appear more like the record players of the mid- to late 20th century than the phonographs and Victrolas of the period; because the sound is produced by the metal plectrums on the disc activating the tuned metal teeth, there is no need for the large horn that record players used before electrical amplification became the norm. Although some earlier music boxes allowed the user to play different musical selections, the flat discs made storage more convenient, and the joint marketing of music boxes and discs of Polyphon Musikwerke actively encouraged the purchase of multiple discs. Incidentally, this connection of player and medium would continue into the early age of sound recording when several companies manufactured and marketed players and proprietary records that could only be played on the company's players.

Although the Polyphon was invented around 1870, it was not until near the end of the 1880s that Polyphon Musikwerke was set up to sell the machines and discs commercially. The most impressive models were not the home machines but the multi-disc coin-operated changers that in the 1890s were found in public locations such as train stations. The user could select one of six discs to play for their coin. Although this is a more limited range of musical selections, it is easy to understand the coin-operated multi-disc changer music box as a predecessor of the jukebox. The Dutch YouTube program *Music Machine Mondays* included a 2017 episode that examined the Polyphon in the context of other mechanical musical instruments of the past and present. In addition to providing a historical overview of the device, the program provides the viewer to see how the Polyphon works (see, https://www.youtube.com/watch?v=LyXY4Bz3PLs, accessed January 15, 2021).

Ultimately, the heyday of the music box ended with the increasing prominence of the record industry and as radios became part of the landscape of homes in the 1920s. With these technologies, listeners could hear vocal music, jazz bands, blues singers, and orchestral music, things that went beyond the capabilities of even the most advanced mechanical music instruments, including music boxes. The stock market crash of October 1929 and the ensuing Great Depression effectively put the final nail in the coffin of the music box as a major player in commercial music.

Today, historical music boxes can be found at several museums, including The Music House Museum in Traverse City, Michigan. Perhaps one of the most impressive collections of mechanical musical instruments, which includes music boxes, automatic pianos, automatic organs, and a variety of others, is part of the House on the Rock in Spring Green, Wisconsin. Several of the instruments at the House on the Rock can be played by visitors using tokens that can be purchased on site. Likewise, the mechanical music gallery at the Phoenix, Arizona, Musical Instrument Museum includes music boxes as well as mechanical musical instruments that are extensions of music box technology. The music box is not, however, an example of music technology with roots in the European clockmaking tradition that today is related to inexpensive instruments readily available in stores and online, or the most elaborate examples from decades ago that are housed in museums. As evidenced by some price guides to collectibles, there is also a collector's market for music boxes, particularly the elaborate 19th-century devices (see, e.g., Miller 2018, 267).

See Also: Cassette Tapes; Reel-to-Reel Tape; Tape Manipulation; Virtual Studio Software

Further Reading

Bowers, Q. David. *Encyclopedia of Automatic Musical Instruments.* Lanham, MD: Vestal Press.

The Editors of *Encyclopædia Britannica.* 2020. "Music Box." *Encyclopædia Britannica.* Accessed May 8, 2020. https://www.britannica.com/art/music-box.

Fuller, David. 1986. "Automatic Instrument." In Don M. Randel, ed., *The New Harvard Dictionary of Music,* 61–63. Cambridge, MA: The Belknap Press of Harvard University Press.

McLerran, Marina. 2017. "A Brief History of the Music Box." *McLerran Journal,* June 4. Accessed January 18, 2021. https://www.mclerranjournal.com/technology-1/2017/6/4/a-short-history-of-the-music-box.

Miller, Judith. 2018. *Collectibles Handbook & Price Guide 2019–2020.* London: Miller's.

The Music House Museum. 1994. *The Music House.* 2nd ed. Traverse City, MI: The Music House.

Ord-Hume, Arthur W. J. G. 1973. *Clockwork Music: An Illustrated History of Mechanical Musical Instruments from the Musical Box to the Pianola, from Automaton Lady Virginal Players to Orchestrion.* New York: Crown Publishers.

Ord-Hume, Arthur W. J. G. 1995. *The Musical Box: A Guide for Collectors.* Atglen, PA: Schiffer Publishing.

Werner, Karen, ed. 2012. *MIM: Highlights from the Musical Instrument Museum.* Phoenix, AZ: Musical Instrument Museum.

Music Notation Software

Although the printing of music using movable type dates back to the late 15th century, mechanical printing of music generally was limited to music publishers well into the 20th century. Composers, arrangers, and other musicians who were not affiliated with publishing houses generally had to rely on their own skills at music calligraphy or on music copyists, who would use pen and ink to write out scores and parts. And some publishing companies also used music copyists to produce master copies of scores and parts that could then be duplicated. Some highly skilled copyists could produce results that rivaled the precision of typeset music, and some companies, such as Keuffel & Esser, manufactured music drafting systems that were similar to the lettering systems that they made for professional cartoon letterers, draftsmen, and so on. The development of music notation software was a natural fit for MIDI when MIDI (Musical Instrument Digital Interface) was rolled out in 1982, and the development and subsequent refinement of MIDI-based notation software eventually brought near-professional-quality or professional-quality music printing to musicians in a way that was never before possible.

As performing musicians who lived through the introduction of computer-produced music notation remember, some of the early music that was printed by composers, arrangers, and even publishers was less than fully satisfactory for a number of reasons, not the least of which was because some of the sheet music was produced using dot-matrix printers. As ink-jet and laser printers became the norm, generally the sheet music became easier for the singer or player to read.

Although some of this printing technology was available at the same time as dot-matrix technology, the high cost of laser printers tended to exclude them from the mass consumer market until into the 21st century. Once these machines became available to the consumer, professional-quality music printing had made it into the hands of working musicians.

However, just as printing music using movable type or employing a professional copyist to write out scores and parts by hand that are then reproduced is not an infallible process, such is also the case with software-based musical notation. One of the challenges posed by some of the programs, especially problematic when producing large scores with many parts, is associating dynamic marks and other editorial markings with the correct instrument or voice. However, unlike ink-on-paper music copying or published typeset music, in which errors might continue to be published for decades, music notation software made it possible to correct scores and parts quickly and easily.

Although many musicians associate music notation software primarily with the last decade or so of the 20th century through the present, notation software actually has a longer history. Stanford University professor Leland C. Smith developed the first computer-based notation program, *SCORE*, which was first released in 1967. The first iterations of *SCORE* and other notation programs released before the mid-1980s generally ran on mainframe computers and worked with the complex plotter printers of the day. As a result, they generally were used at well-funded educational institutions and tended to be associated with new music and computer music programs.

Computer graphics for the home market made a significant leap forward with the introduction of Apple's Macintosh computers in 1984. The graphics capabilities of Macintosh computers made these machines a natural fit for the first MIDI-based notation programs that began appearing later in the decade. In fact, during the first phase of computer notation software in the late 1980s and through the 1990s, the norm was for software developers to focus on the Macintosh operating systems, as opposed to DOS or the early incarnations of Microsoft's Windows.

Arguably, the best-known MIDI-based music notation program of the 1990s through the present has been *Finale*, which MakeMusic first released in 1988. Although there were other notation programs available in the late 20th and early 21st centuries, *Finale* and perhaps its nearest competitor, *Sibelius*, have been the most visible notation applications. Professional versions of the programs might be priced out to the range of some musicians—and particularly music students—however, "light" versions of these professional programs made a more limited music notation software significantly more readily and economically available. The less expensive versions tend to limit the size of scores; however, the quality of the notation for single instruments, instrument or voice with piano, small chamber groups, and so on is quite good, and standard notational items such as slurs, ties, articulation and dynamic markings, glissandi, trills, and so on are included with the less expensive versions of programs such as these.

The importance of this software can be found in several areas. Perhaps most obviously, parts for new unpublished arrangements could now be given to performers with a clarity of notation that was easier to decipher than handwritten notation. And, in cases in which a copyist—or in this case someone entering the pitches, rhythms, accents, dynamic changes, and so on using a computer or MIDI keyboard—made a mistake, it was easier to make the correction and print out a new part or a new page as needed.

Arrangers and composers also benefited from the software in several ways. For one thing, they could hear their scores played back. Some software packages could identify notes that are out of the range of the instrument indicated in the score, and some programs will allow the composer or arranger to further limit ranges so that they do not exceed the standard easily playable range of the instruments, a feature that is helpful for arrangers writing for, say, school bands and orchestras.

As with any technology, there can be dangers with musicians relying too heavily on the technology as a be-all and end-all. One of the complaints that practicing musicians have about composers and arrangers who rely solely on their ears to tell them would sound good and on notation software telling them what literally is impossible to play because of range limitations is that scores are produced and sometimes published that contain awkward and sometimes physically impossible passages. Another challenge that the MIDI/notation software combination poses is in the volume balance between parts in MIDI playback. An arranger who relies on MIDI playback can produce a score that poses unintended and unnecessary dynamic challenges for the performers, much like listening to, say, an orchestral score from a film or television show can give one an incorrect impression of dynamic level balance that is practical or possible outside a recording studio and post-recording mixing setting.

Perhaps one of the more valuable aspects of notation software is that transposition, both of individual parts from a score and of the entire score, if necessary, is far easier, less time consuming, and generally less prone to error than when done by hand. The transposition of an entire score using a notation program, in particular, can be especially helpful in ensembles of a wide variety of sizes that include singers and instrumentalists. If a vocal soloist needs to sing an opera aria, for example, a half-step down from how it is written in the original score, if the score is a MIDI file or a file format that is proprietary to one of the notation software packages, the score and all the parts can be instantly transposed, with the new parts printed out using a standard computer printer.

Despite the challenges that notation software and MIDI pose, clearly MIDI-based notation programs have opened up the world of music publishing to small companies and to the composers and arrangers themselves. In fact, it can be argued that MIDI helped to democratize notated music in a way perhaps never previously seen in history. Among the options that MIDI notation programs have helped to open up to publishers and performing musicians is the ability to print on demand or to tailor published music to specific ensembles and specific situations. For example, some publishers send purchasers of a score a MIDI-compatible file version of the music. This can then, for example, be easily transposed if, say, a vocal soloist needs to sing a piece a whole step lower than the key in which it was originally published.

Another aspect of computer-based music notation software that is important to note, although to date it has not been developed to the extent that other aspects of computer-based notation have been, is the development of Optical Music Recognition (OMR) software. Although considerably more complicated than its alphanumeric-based counterpart, OMR is somewhat analogous to Optical Character Recognition (OCR). There has been software created at academic institutions and several open-source programs, as well as commercial applications such as *PhotoScore*, *SmartScore*, and *PlayScore*. The best-known and most frequently found are probably *PhotoScore* and *SmartScore* because of packaging of versions of these programs that has been done with versions *Sibelius* and *Finale*, respectively, over the years.

As is the case with other types of software, music notation software is now available for tablets and smartphones. Because notation programs are not necessarily as effective on the small screen of a smartphone, let us consider tablets, including the iPad. Several notation programs, including *Notion*, can be purchased for a fraction of the cost of laptop and desktop computer-based programs such as *Finale*, *Sibelius*, and others. *Notion* for iPad, for example, is priced at $14.99 at Apple's App Store at the time of this writing. The low-cost tablet programs might not all necessarily be able to handle large scores and might not have all the features of the expensive programs; however, for musicians who need to have a highly legible performance part available quickly, or who need to quickly produce a small group arrangement for a gig, these inexpensive notation programs are more than adequate.

See Also: iPads and Other Tablets; MIDI

Further Reading
Williams, David B., and Peter R. Webster. 2008. *Experiencing Music Technology*. Updated 3rd ed. Boston: Schirmer.

Music Printing with Movable Type

One of key points in human achievement that often appears in history books and generally is well known to the public is Johannes Gutenberg's introduction of the printing press in Europe in 1439. Gutenberg's innovations in printing with movable type made it possible to print more copies, faster and with less effort. Gutenberg's printing of the Bible, the first printed with movable type, remains famous all these centuries after his work in the 1400s. And although Gutenberg's most famous publication was of a typeset Bible, his invention helps to bring the notation, publication, and preservation of literature out of the virtual sole domain of the church and bring publishing into the secular world.

Interestingly, but perhaps understandably, Gutenberg's work did not immediately result in the mechanical printing of music using movable type. In fact, it was not until 1501 that Ottaviano Petrucci printed and issued the first set of polyphonic music printed with movable type, the publication *Harmonice Musices Odhecaton*. The work that Petrucci printed was a set of polyphonic chansons. There were some predecessors; however, *Harmonice Musices Odhecaton* was the first example of a completely movable typeset music publication in which the music was scored for more than one part. According to scholar Mary Kay Conyers Duggan, "The earliest book with music notes and staves printed from movable type that is actually dated is the *Missale Romanum* of Ulrich Han, issued at Rome in 1476 with sixteen leaves of Roman plainchant. That priority cannot seriously be challenged by an earlier dated book that appeared in 1473 with only one printed line of music notes without a staff" (Duggan 1992, 13). The earlier work of Han is not nearly as well remembered today; it is Ottaviano Petrucci who is credited with fully ushering in the age of music printing with movable type.

As one might imagine, printing music with movable type was more complicated than printing text alone. For one thing, music printing involved the staff—or staves in the case of multipart music—and, in the case of vocal music (which was the type most frequently printed when printing with movable type was developed), aligning text with the musical notation. Although the printing of music using movable type has a long history, it must be noted that it did not necessarily become a universal in the publishing world, and not entirely because of the early challenges just described. In particular, handwritten music was the norm, until the development and refinement of music notation software, in musical theater and opera circles. In these genres, it was typical for overtures, songs, arias, and so on to be written, arranged, reorchestrated, and transposed right up to the opening of the production. As a result, the norm here was for professional copyists to handwrite the parts for the orchestra from the composer or arranger's score. In the 21st century, more and more parts for musical theater works are produced using notation software, and the rights holders of some of the old standbys of the world of Broadway musicals (from whom scores and parts typically are rented) have been converting the old handwritten and reproduced parts into software-produced versions. Another innovation that has occurred recently in the publishing world is an increasing move to distribution of pdf files, thus requiring the musician, music director, or other person on the consumer end to print out the sheet music. Typically, these files, too, come from computer-generated notation, thus continuing the move beyond printing with movable type.

Be that as it may, the use of movable type and the move away from sole reliance on the music scribes before the 16th century helped to fundamentally change music and music publishing in important ways. By a wide margin, the music that was written out and copied by hand was the product of monks. As one might reasonably expect, the vast majority of notated music of the Middle Ages and into the early Renaissance was religious in nature. And this music was collected and codified carefully. The secular music of the time was considerably more ephemeral; there may have been hundreds or thousands of popular dance tunes and secular songs that simply were never written down. As a result, we might find references to them today but have no idea what they sounded like.

So, it is important to return to that first completely typeset musical publication: Ottaviano Petrucci's *Harmonice Musices Odhecaton*, a set of polyphonic chansons, or multipart songs. Although the scribes of the Middle Ages and

Renaissance sometimes notated secular music, the printing and publication of typeset music freed music from the necessity of having ties to the clergy. Therefore, this new process played a significant role in elevating secular music. It lost some of its earlier ephemeral, transitory nature and increasingly was able to become part of the musical canon of its time. By the late 16th and early 17th centuries, Renaissance madrigals in collections were being published using movable type. Instrumental music, too, probably the least well-preserved music of the era that completely relied on scribes, came into publication in typeset editions in the late Renaissance and early Baroque periods, with significant works such as Giovanni Gabrieli's *Sonata pian'e e forte* appearing in print.

The use of mechanical musical printing arguably reached a high point in the late 19th century and first several decades of the 20th century, as a new popular culture developed. Although this can be seen in other parts of the Western world, it was perhaps most famously connected with the United States and the development of the Tin Pan Alley publishing industry. Popular songs and popular instrumental pieces that could be played at home on the piano (e.g., ragtime hits such as Scott Joplin's *Maple Leaf Rag* and post-ragtime hits, such as Zez Confrey's *Kitten on the Keys*) and others became widely available because of this publishing industry. The way in which popular music and the growing sense of an American popular culture permeated the publishing industry is suggested by the fact that well-known pieces, such as some of John Phillip Sousa's marches, were published for bands, orchestras, and even arrangements for piano, guitar solo, and various combinations of stringed instruments that one might have at home, including instruments such as the banjo and the zither. The mass printing of thousands of copies of popular hits was only possible because of machine-driven enhancements of what pioneers such as Ottaviano Petrucci had started hundreds of years before.

The prominence and importance of the Tin Pan Alley publishing industry in the United States can also be seen in the elaborate and colorful sheet music covers that, along with songs, were being mechanically printed by the thousands and thousands. However, by the 1940s, popular singers increasingly dominated the record charts and public consciousness. At the same time, record players became more numerous, and increasingly, radio continued to be available in homes, automobiles, and so on. Likewise, specifics of recorded arrangements increasingly defined what members of the public thought of as to the definition of songs, something that increasingly became the case during the rock era beginning in the 1950s. As a result of all these developments, the singer increasingly took the spotlight, or as the title of the 1953 Audrey Erskine Lindop's novel, the title of a 1961 film, or the 1965 Rolling Stones song, the music industry was now more about "the singer, not the song." The appetite for published sheet music did not die, but the heyday for typeset published sheet music, as well as for singing in the home and public group singing, was gone.

Although computers were used to notate music earlier, the rollout of MIDI in 1983 meant that the need for typesetting for music publishing went away. Some publishers continued to issue typeset music, but increasingly, computer-based music notation has played a significant role in the music industry. In fact, some publishers today supply scores and parts as pdf files, which the purchaser must print themselves.

See Also: Music Notation Software; Musical Notation; Sheet Music

Further Reading

Duggan, Mary K. 1992. *Italian Music Incunabula*. Berkeley and Los Angeles: University of California Press.

Samuel, Harold E. 1986. "Printing of Music." In Don Randel, ed., *The New Harvard Dictionary of Music*, 655–656. Cambridge, MA: The Belknap Press of Harvard University Press.

Musical Notation

Although musical notation at first glance might seem like a decidedly low-tech or even a dubious connection of music and technology, neumatic notation, as it developed over hundreds of years, formed the basis for modern music notation. Although there is historical evidence of musical notation systems from the ancient world—whether or not we can actually decipher exactly what they mean—modern Western music notation began as a sort of short-hand reminder to people who had already learned to sing the melodies by rote. It then developed from a descriptive system to a prescriptive notation as it became systematized and standardized to the extent that one could use it to accurately sing music that the singer had never heard before.

When studying the history of Western musical notation, one must keep in mind that there was a wide divide between sacred and secular music at least through the Middle Ages. Secular music, particularly secular instrumental music, tended to be much more ephemeral. Sacred music, as it was associated with the Roman Catholic Church and its monasteries, convents, cathedrals, and so on, was significantly more deeply rooted in tradition and study. To put it another way, one did not simply improvise chant melodies. Certain chant melodies were intrinsically tied to certain prayers.

The challenge came with the creation of new melodies and just the sheer amount of musical material that had to be memorized. The earliest aids to memory included markings above the texts that indicated general melodic shape in some manuscripts. This became more formalized through the 9th century, and relative reference points indicated by lines—the predecessors of today's staff lines—began to appear. The lines provided singers with more information about not only melodic direction but also relative distance between melodic intervals.

The notational symbols themselves, known as neumes, eventually were square in shape and indicated not only the relative direction of pitches but also melodic interpretation when more than one pitch was to be sung per syllable. Pitch was relative, in other words there was no C, E-flat, or so on that was defined such as what we can find from a piano or pitch pipe. Over time, however, additional levels of detail were added, including additional reference lines, until the notation included the five lines and four spaces that we recognize today as the modern staff. The music theorist Guido d'Arezzo is credited with the five-line staff, as well as with the system of syllables that came to be called the solfege system, which is still used in ear-training and sight-singing study today. The reader might be familiar with the solfege system through the well-known song "Do-Re-Mi" from the Rodgers and Hammerstein musical *The Sound of Music*. Clefs, too, provided additional reference information, similar to our treble, bass, and other staffs used today.

One important feature that modern music notation provides was absent from neumatic notation: rhythm. So long as the notation was still of melodies associated with monophonically sung prayers, it is generally believed that performance might have been rather smooth melodically, perhaps with reference to speech rhythms. With the development of polyphonic music, particularly as it involved more than two parts, it was necessary to include more information about rhythm and, eventually, meter into musical notation. Eventually, too, the square notation that had been inherited from neumatic notation was largely dropped for rounded note head shapes; however, even today in some collections of Roman Catholic chant melodies, the square notation is still used.

It was not until the late 16th century that the Italian composer Giovanni Gabrieli helped to make musical notation even more prescriptive in his *Sonata pian'e e forte*. Not only did Gabrieli's score provide dynamic markings (piano, or soft, and forte, or loud) but it also specified what brass instruments were to be used. Previous publications of music for instrumental ensembles tended to leave the choice of instruments to play the parts up to the performers. To put this piece in perspective, the rhythmic and pitch notation, along with the scoring for specific instruments and dynamic markings of *Sonata pian'e e forte*, provides performers virtually all the information that is provided by today's Western music notation.

During the 17th and 18th centuries, a system of tempo (speed) markings, based on the Italian language, continued to be standardized; however, one performer's interpretation of Allegro—or, fast—might differ from another performer's interpretation. More specificity in this area came with the development of the metronome. Although credited to Johann

A manuscript showing neumes on a five-line staff, the predecessor of modern musical notation. (R Martin Seddon/Dreamstime.com)

Nepomuk Maelzel, because of his patent, but apparently actually invented but not patented by Dietrich Nikolaus Winkel, this mechanical device utilized principles of the clocks of the early 19th century and allowed composers to specify tempo in terms of the number of beats, or pulses, per minute, as opposed to the open-to-interpretation Italian relative terms that were the norm. Ludwig van Beethoven was the first well-known composer to use numerical metronome markings in his manuscript scores.

In addition to developing the system of translating letters of the alphabet into raised tactile dots, the young Louis Braille also developed a system of tactile music notation for unsighted musicians. Braille's system essentially broke down the various components used in traditional print music notation so that musicians using his system would be able to read pitches, rhythms, relationships of the various parts, dynamics, and so on. Interestingly, the system that Braille developed in the first half of the 19th century resembled the kinds of markup-language notation used to program music into computers until the development of MIDI and Optical Character Recognition (OCR). The computer markup-language notation that was developed throughout the second half of the 20th century addressed certain issues that had limited the traditional visual, for-the-live-performer notation that preceded it: by using numbers, it was possible to specify pitches in between the 12 discrete pitches per octave used in Western music, and it was possible to code rhythmic events that would be impossible or nearly impossible to write using the fraction-based rhythmic notation that has been used for hundreds of years.

With the development of MIDI (Musical Instrument Digital Interface) as a standard, several computer notation programs were published. Today, arguably, *Finale* and *Sibelius* are the best known, although there are other powerful notation programs, and unlike the numbers-based programs described earlier, the vast majority of notation programs use traditional Western musical notation. In the 21st century, there have been numerous notation applications published for tablet computers, including the iPad. In some cases, these apps cost a fraction of what some of the best-known desktop and laptop notation programs cost. Although there are some limitations to these relatively inexpensive apps, some are powerful enough to use for chamber music scores, as well as piano, guitar, voice-and-piano, and a variety of other scores. Tablet applications and notation programs for larger machines offer the composer, arranger, and performing musician the ability to make quick changes to scores and parts. In addition, the use of notation software for scores and parts that traditionally were handwritten and then photocopied (e.g., pit orchestra parts for musical theater works) allows for parts that are more legible.

Another significant change in how music notation is made available to and used by the performer—albeit not necessarily a change in the notation itself—is through computer applications that allow sheet music to be saved as pdf files that can be quickly accessed on iPads and other tablets by performing musicians. One of the programs that became popular during the 2010s is *forScore*, which allows musicians to make notations in their parts, set up a series of pages as a whole (which can be reordered), and which can incorporate an optional foot pedal that is used for turning pages forward and backward.

See Also: Desktop and Laptop Computers; iPads and Other Tablets; Music Notation Software

Further Reading

Floros, Constantin. 2011. *The Origins of Western Notation*. Revised and translated by Neil Moran. Frankfurt am Main, Germany: Peter Lang AG.

Muzak

By definition, background music of the type provided by Muzak and other piped-in music services is designed to be minimally noticed, if noticed at all. Although the original musical style employed by Muzak has been replaced by smooth jazz, recordings of name artists performing their hit songs, and satellite radio services, among others, the story of Muzak really is one of the more intriguing stories of the connection of music and technology of the 20th century.

Something of the proverbial Renaissance man, George Owen Squier was a U.S. Army major general, an aviation pioneer, and inventor. One of Squier's principal contributions related to music was his work on wired and wireless communications. Several of Squier's patents played roles in the early 20th-century development of the radio industry. As a matter of fact, it was Squier's work on wired transmission of information that led to the delivery of music to homes before wireless transmission became widespread. Squier's developments in the transmission of information signals over electrical wires also led to Muzak. With advances in wireless radio, Squier took the wired delivery of music in a new direction, with his late-in-life venture: Muzak, a concept that was aimed at commercial customers and provided background music.

The idea of background—or, environmental—music was not new in the mid-1930s. French composer Erick Satie wrote his *musique d'ameublement* (usually referred to as "Furniture Music" in English) in 1917, and American composer/inventor Thaddeus Cahill invented his telharmonium at the start of the 20th century. In fact, Cahill's instrument, which he envisioned would provide background music to department stores, is a conceptual predecessor of Muzak.

Muzak differed from other earlier attempts to create and transmit background music in several important ways. For one thing, until Muzak began using original-artist recordings of songs and popular instrumental pieces in the 1980s, the service (which has gone through multiple ownership changes) provided instrumental music of popular material, recorded by studio musicians who were regular members of significant symphony orchestras in the United States. The focus of the arrangements and orchestrations was on clarity, but, because this was meant to be background music, typically Muzak arrangements did not have wide changes of dynamic levels. Because of the background nature of Muzak, the quality of the musicianship might be easy to overlook. Anecdotally, when I was an undergraduate student—before Muzak moved into employing more vocal music and original-artist recordings of popular songs—Capital University clarinet professor David Hite suggested to his students that they pay attention to Muzak in order to hear good examples of classical instrumental phrasing within the context of arrangements of popular songs.

The other, perhaps most notable feature of Muzak was how the piped-in music was programmed. The selections were organized in 15-minute segments, with each segment generally increasing in tempo from piece to piece. Although the effectiveness of this approach has been widely debated over the decades, the idea was that the pace change through each segment would increase customers' purchases at retail stores and workers' productivity in offices and other such settings.

Muzak was not the only company that offered piped-in background music in commercial venues since the 1930s, but it was the best known. In fact, the word "muzak" came to be used generically for the background music one might hear in an elevator; easy listening music recorded by artists such as Mantovani and His Orchestra, the Ray Conniff Singers, and so on; and music that was actually produced, programmed, and distributed by Muzak. As Paul Allen Anderson wrote in *Critical Inquiry*, Muzak "sits alongside Kleenex, Q-Tip, and Prozac as both a protected trademark and an informal name for a whole category of products" (Anderson 2015, 811).

Muzak's marketing of itself for years and years as a potential boost to retail sales and worker productivity raises the question, was it successful? Milliman (1982) presents information that suggests that earlier surveys and studies that sought to answer that questions were not necessarily conclusive, for a wide variety of reasons. His study, however, focused on the use of piped-in background music in a supermarket setting and found that slow music tended to result in shoppers spending more time in the store shopping. One might reasonably assume that by spending more time in the store, these shoppers might be more likely to make more purchases than individuals who shop at a faster pace. In fact, the study found that "the average gross sales increased from $12,112.35 for the fast tempo music to $16,740.23 for the slow tempo music . . . a 38.2% increase in sales volume" (Milliman 1982, 90).

The ownership and programming style of Muzak evolved over the years. With the 2011 acquisition of the company by Mood Media, Muzak took on the name Mood. Mood took a different approach than the old Muzak by offering "curated playlists for corporate clients" (Lazarus 2017). The service's approach now allows for clients such as the Target store chain to better define their brand by defining it musically for shoppers. The new version of Muzak also takes cues from Milliman's and subsequent studies that link both mood and tempo to sales. So, Muzak, in a sense, lives on but not with the proprietary, nearly exclusive instrumental focus it had for decades.

The entire concept of piped-in background music has come under attack in the late 20th and early 21st centuries. In the early 1990s, a movement known as Pipedown emerged in the U.K. Eventually, the movement did result in some retail stores and other establishments removing piped-in music. However, this has not entirely been a U.K. phenomenon; in 2017, there was a movement afoot in Ann Arbor, Michigan, to eliminate piped-in background music at bars, restaurants, and various locations on the University of Michigan campus. Interestingly, most of the statements from leaders of both movements focused on the physical and psychological impacts of excessive noise—in other words, the loudness itself rather than the manipulative aspects of piped-in music (see, Remy 2017; Watts 1999). This could be taken as a suggestion that the so-called Muzak of the 1990s through the present is not perceived to be as psychologically manipulative as the piped-in music of the years when such music was proprietary and programmed in such a way as to encourage worker productivity and consumer spending. Interestingly, there are suggestions in the marketplace that the new guises under which Muzak operates, and its rivals, which now include strategic use of Pandora, Spotify, iHeart Radio, and satellite radio to set mood and generate either more lingering (e.g., in retail stores) or quicker throughput (e.g., in bars and restaurants), might be more effective in terms of dollars and cents, and therefore even more potentially manipulative than the original 1930s' incarnation of Muzak. In his *Critical Inquiry* article on the evolution of mood-based music programming, however, Anderson (2015) establishes a clear connection between these new services and Muzak, suggesting that the legacy of Muzak is alive and well in the 21st century.

See Also: Pandora and the Music Genome Project; Radio; Satellite Radio; Streaming Audio Services

Further Reading

Anderson, Paul A. 2015. "Neo-Muzak and the Business of Mood." *Critical Inquiry* 41, no. 4 (Summer): 811–840.

Clark, Paul W., and Laurence A. Lyons. 2014. *George Owen Squier: U.S. Army Major General, Inventor, Aviation Pioneer, Founder of Muzak*. Jefferson, NC: McFarland.

Farkas, Remy. 2017. "Quiet Ann Arbor Looks to Limit Piped Music." *Michigan Daily*, December 5. Accessed May 26, 2020. https://www.michigandaily.com/section/ann-arbor/quiet-ann-arbor-looks-limit-pipped-music.

Kennelly, Arthur E. 1938. *Biographical Memoir of George Owen Squier, 1865–1934*. Washington, DC: National Academy of Sciences of the United States of America.

Lazarus, David. 2017. "Whatever Happened to Muzak? It's Now Mood, and It's Not Elevator Music." *Los Angeles Times*, July 7. Accessed January 18, 2021. https://www.latimes.com/business/lazarus/la-fi-lazarus-store-music-20170707-story.html.

Milliman, Ronald E. 1982. "Using Background Music to Affect the Behavior of Supermarket Shoppers." *Journal of Marketing* 46, no. 3 (Summer): 86–91.

Parkinson, Justin. 2016. "What Is Shop Music Doing to Your Brain?" *BBC News Magazine*, June 1. Accessed January 18, 2021. https://www.bbc.com/news/magazine-36424854.

Watts, Janet. 1999. "No Thank You for the Muzak." *Guardian*, December 1. Accessed January 18, 2021. https://www.theguardian.com/society/1999/dec/01/guardiansocietysupplement2.

Ondes Martenots

Invented in 1928 by French musician and inventor Maurice Martenot, the ondes Martenot (roughly translated as Martenot's Waves) used similar principles with regard to electronic oscillators as did Leon Theremin's slightly earlier theremin but was considerably more practical than Theremin's instrument. Whereas the theremin required the player to literally pull the pitches out of the air (the player does not actually touch the theremin but moves one's hand closer and further away from an antenna to find a pitch along a continuum), the player of the ondes Martenot controls pitch by means of a keyboard and ribbon device for slides between pitches. Like the theremin, the ondes Martenot is a monophonic instrument, capable of producing single pitches. Because of its use of electronic oscillators and a keyboard, the ondes Martenot is one of the early 20th-century predecessors of the monophonic analog synthesizers of the 1960s and 1970s.

After its invention, the instrument was soon taken up by experimental composers such as Edgard Varèse and Olivier Messiaen. For example, Varèse included two ondes Martenots in the score of his 1934 work *Ecuatorial*. Ironically, the text of this piece, which is scored for bass voice, four trumpets, four trombones, piano, organ, the aforementioned two ondes Martenots, and five percussionists, is based on the *Popol Vuh*, the book of ancient Mayan culture, history, and beliefs. Varèse's use of a new electronic instrument and in particular the fact that the ondes Martenots are the final "voices" in the piece that revolves around the ancient Mayans not only give the piece an eerie feeling of irony but also suggest the composer's view of an electronic future. Messiaen included the ondes Martenot in the scores of both his 1944 *Trois petites liturgies de la presence divine* and his 1948 *Turangalîla-symphonie*.

Much more recently, the band Radiohead has used the ondes Martenot in the studio and in concert. Although there are numerous examples of Radiohead's Jonny Greenwood playing the instrument, videos of the song "How to Disappear Completely" show Greenwood's technique and allow the viewer to hear the ondes Martenot in a rock music context (for a live performance of the piece that shows some of the techniques used for playing the instrument, see, for example, https://www.youtube.com/watch?v=chE1_g3GAWw, accessed January 15, 2021). Greenwood played and discussed his use of the ondes Martenot in a section of the 2012 Caroline Martel film *Wavemakers*. At the time of this writing, the segment is available on YouTube (see, e.g., https://www.youtube.com/watch?v=B92ZgRSM2tI, accessed January 15, 2021).

Although the ondes Martenot has been used by relatively few composers and performers, especially compared with the theremin (which can easily be built in kit form), perhaps the instrument's most important legacy comes in Martenot's fitting electronic oscillators to a piano-like keyboard. This set the stage for the electronic organs, electric pianos, and analog synthesizers that would follow the ondes Martenot over the course of the four decades following its invention.

See Also: Moog Synthesizers; Organs; Theremins

Further Reading
Beaumont, Rachel. 2017. "The Ondes Martenot: The Eeriest Instrument Ever Invented?" Royal Opera House (U.K.). Accessed April 8, 2020. https://www.roh.org.uk/news/the-ondes-martenot-the-eeriest-instrument-ever-invented.

Orchestral String Instruments

Today's orchestral string instruments, the violin, viola, violoncello (often abbreviated cello or cello), and double bass, have been in use since the Baroque period, having largely superseded the members of the bowed, but fretted, viol family, which primarily were popular in Europe in the Renaissance but that continued to be used even as the bowed, nonfretted strings became more popular. The fact that the modern orchestral string instruments were well established centuries ago is borne out by the fact that players today still use prized instruments made by Antonio Stradivari (1644–1722), members of the Amati family (active from the mid 16th century through the mid-18th century), and members of the Guarneri family (active from the mid-17th century through the mid-18th century. Although instruments from centuries ago have been reinforced and use more robust strings than those for which they were intended, the basic principles behind the instruments remain largely unchanged. In fact, some instruments made today are based on the designs of the Italian masters of the 16th through 18th centuries.

One of the few changes to the orchestral string instruments that is easily visible to audiences is the addition of a fifth string—commonly called a C extension—on the double bass. This addition extended the range of the instrument down the interval of a major third to C1, the C that sounds three octaves below middle C on the piano keyboard. For practicality, the C extension commonly has keys that the player presses down to play the notes between C1 and E1, which is the lowest note that is playable on the normal fourth string. Some bassists use instruments with a low fifth string that is tuned down to B0. Few works in the standard orchestral repertoire require either the C extension or the low B string, with most being compositions from the mid-19th century through the present. These modifications to the standard 4-string double bass are also helpful for earlier works, in which the double bass is frequently called upon to sound an octave below the cello, whose lowest string is normally tuned to C2.

With the use of the orchestral string instruments in jazz, rock, commercial country, and avant-garde art music of the 20th and 21st centuries, electronics have played an increasing role. One of the challenges faced when trying to use a small microphone attached to any of these acoustic string instruments is controlling feedback. Contact microphones were particularly popular for use with string instruments because of their compact size and because some were particularly good at controlling feedback because they picked up the vibrations of the instrument itself rather than sound vibrations in the air in the instrument's resonating chamber or outside the instrument near the strings. The problem with some of the lower-end microphones of this type with any acoustic instrument is that the instrument-body vibrations alone do not define the unique tone of the instrument because of the harmonics/overtones that one can hear but that are not necessarily easily picked up through the structural parts of the instrument itself. More recent developments in electric string instruments—discussed later—have gone a long way in solving these issues, as well as the issue of unwanted feedback.

One of the recent developments in string instrument construction technology that has been heralded by some performers and is visible to audience members is the use of carbon fiber as a construction material. The technology was pioneered by cellist Luis Leguia and Steve Clark, who founded Luis and Clark in 2000. Leguia and Clark received their patent for the design and technology in 2001. Luis and Clark have expanded its design and production into the rest of the orchestral string family. This includes the 15.25 viola, a smaller-than-standard instrument that is reputed to produce the tone of a full-sized viola but is considerably lighter than even a 3/4-size traditional wooden viola. Since Luis and Clark's introduction of carbon fiber string instruments, several other companies have produced instruments made out of similar, non-wood-based materials. Among the best known is Glasser Bows, which makes carbon composite orchestral string instruments, as well as carbon fiber bows.

The principal advantage of materials such as carbon composite and carbon fiber is that the instruments are considerably less susceptible to cracking caused by temperature changes. Similarly, the instruments are less susceptible to breakage upon impact from dropping or careless handling when traveling. Glasser's carbon composite instruments are available in acoustic-electric models, with built-in pickups. Other companies have gone further in focusing on electrifying what for

A Parisian musician playing an electric cello, one of the recent developments that have brought orchestral strings into the electronic age. (Minacarson/Dreamstime.com)

centuries were entirely acoustic instruments. Yamaha Corporation, for example, manufactures a line called the "SILENT Series," which includes electric violin, electric viola, electric cello, and electric double bass in the series' orchestral string category. Yamaha's "SILENT" string instruments are not completely silent; however, the instruments rely on electronic circuitry and amplification to produce enough sound to be heard by an audience. In this regard, they somewhat resemble the electric guitar. And because they are plugged into an amplifier and can be run through the amplifier's or any external effects units, they are capable of producing tones that range from a close approximation of the sound of a regular acoustic string instrument to the limits of a player's wildest experimental imagination.

Several notable performers have made a special point of using electric string instruments for the flexibility of sound and the freedom of movement that wireless technology offers. For example, Alison Lynn of the Moxie Strings plays an electric cello. Not only can the instrument be played like a standard orchestral cello but it can also be worn with a strap so that the performer can play the instrument sitting, standing, or even walking around. In her performances with the Moxie Strings, Lynn also uses various electronic effects, including digital looping, with her NS Design CR Electric Cello. Perhaps the most highly visible performer associated with electric string instruments, however, is Mark Wood, formerly of the Trans-Siberian Orchestra. Wood regularly travels around the United States performing with high school and college orchestras, boosting interest in school string programs with music and arrangements that make strong use of his electric violin in his "Electrify Your Stings" program. Wood is also active as a manufacturer of electric violins that feature a patented design that allows the instruments to be more easily held.

In addition to more expensive and well-made electric stringed instruments such as Yamaha's offerings or the instruments of smaller companies such as NS Design and Wood Violins, internet sales sites such as Amazon.com and others have seen a proliferation of electric string instruments that at the time of this writing sell for a fraction of the price of Yamaha's, Wood's, and NS Design's instruments. Various YouTube demonstrations and reviews of the low-priced electric stringed instruments available through online retailers show instruments purported to have actions (the distance between the fingerboard and the strings) that make it necessary for the performer to move the string significantly further to get it onto the fingerboard (see, e.g., https://www.youtube.com/watch?v=Ns9xiFjKtrI, accessed January 15, 2021), or inexpensive electric instruments that purportedly had to have their electronic components replaced because the original components did not function properly (see, e.g., https://www.youtube.com/watch?v=p6jQw-tT4g8, accessed January 15, 2021).

See Also: Moog Synthesizers; Organs; Theremins

Further Reading

Benade, Arthur H. 1960. *Horns, Strings, and Harmony.* Garden City, NY: Anchor Books.

Sandys, William, and Simon A. Forster. 2006. *History of the Violin.* Mineola, NY: Dover Publications.

Organs

Widely known as "the king of the instruments," the organ is one of the most complex musical instruments. It is also an instrument with an exceptionally long history. Here, we are concerned with virtually all types of organs, except mechanical (e.g., self-playing) organs and theater organs, as they are detailed in other entries in this volume. The basic principle behind traditional air-operated organs is that some sort of mechanism is used to force air through a pipe or a set of pipes that produce sound. The means of setting the air into motion have changed over the years, and in the 20th century, electronic organs became commonplace in a wide variety of settings, even replacing pipe organs in some of their traditional places of highest usage: places of worship.

The earliest instruments that are generally considered to be the predecessors of the mature organ were the water organs that go back to ancient Greek and Roman times. These instruments used pipes—much like present-day pipe organs—however, water organs used water pressure, either from a nature source or from manually operated pumps, to provide the pressure to move air. At the time of this writing, a brief performance on a reproduction of an early hydraulic organ can be found on YouTube (see, https://www.youtube.com/watch?v=bP2u8NBI5m8, accessed January 15, 2021).

Before the advent of electric motors that could be used to provide the moving air needed to create sound through the organ pipes, post-hydraulic organs featured bellows that were operated manually by an assistant. Small pump organs—sometimes called parlor organs—used a bellows mechanism that the player operated with their feet. Famed American organist Diane Bish appeared in a television program in which she explores and plays the oldest remaining playable organ in the world, a Swiss instrument that dates back to the 15th century. The portion of the program that is available at the time of this writing on YouTube provides rare access to how pre-Baroque instruments functioned and sounded (see, https://www.youtube.com/watch?v=jxwZTfILJDA, accessed January 15, 2021).

Arguably, the Baroque period (approximately 1600–1750) was the high point for organ compositions. Composers such as Girolamo Frescobaldi, Johann Jakob Froberger, Dietrich Buxtehude, and Johann Sebastian Bach successively raised standards for the instrument and for the complexity of compositions for it. During the Baroque period, many of the standards for subsequent pipe organs were established.

The largest cathedral organs and even small parlor organs often have several different sounds/voices of which they are capable of producing. These are often called stops, based on the fact that pushing their associated knobs in the organ console stops the air from reaching that set of pipes. Pulling out the stop (knob) allows the pressurized air to go to those pipes as the keys are depressed on the organ's keyboard. Hence, the phrase "pulling out all the stops," sometimes used in conjunction with such nonmusical activities as an athlete's performance at a sporting event, literally refers to an activity that would allow pipes associated with all the voices of the organ to sound simultaneously, thus producing the fullest possible sound.

The traditional way in which organ keys and the pedals of the pedalboard activate the valves that allow air to flow to the pipes is with a mechanical linkage. Even today, some pipe organs, called tracker organs, are built in this way. Some organs from the 20th century through the present, however, use an electronic connection. In these instruments, depression of a key on the keyboard makes an electrical contact, with the signal being sent through wiring to the valve. As one might reasonably imagine, the tracker mechanism tends to require more forceful depression of the keys and pedals, and the touch can vary based on how many stops are pulled. Keyboard response on organs that use electrical contacts does not change with the number of stops pulled.

Electronic organs of various types gained popularity starting in the mid-20th century, with the first introduced by Baldwin in the mid-1940s. This class of organs covers a wide range, including electronic instruments that can offer as many stops/voices as large cathedral pipe organs. Allen became one of the best-known companies in the field of electronic church organs. However, electronic organs were also made for the home market and, because they are more portable than instruments that use pipes that have been installed in a building (e.g., church organs and theater organs), could be transported to performances in various venues. The Hammond organs of the mid-20th century became popular in the jazz world and later became associated with gospel, soul, and rock music, including hits such as Booker T. and the M.G.'s' "Green Onions." Organs by companies such as Lowrey and Thomas made for the consumer market included such features as color-coded keys, which would illuminate to show how to play particular chords. These organs also were equipped with percussion presets that allowed the organist to include a percussion accompaniment to their playing. The presets used in organs of the 1950s and 1960s reflected some of the popular dance rhythms of the mid-20th century, many drawn from various Caribbean and Central and South American dance styles (e.g., rhumba and bossa nova). Although other rhythm presets were offered, the focus on Latin dance styles that were primarily popular in the United States during the 1950s and early 1960s dates instruments from that era. It should be noted, however, that the concept of percussion presets set the stage for the later drum machines of the 1970s, 1980s, and beyond. Unlike the earlier electronic organs, though, the later drum machines would be programmable and not limited to presets. One of the curiosities of some of the electronic organs that were popular in the second half of the 20th centuries is that the pedalboards do not line up with respect to the keyboard in the same way as do those of pipe organs. Arguably, this makes them ill-suited for practice purposes for an organist who performs on a traditional pipe organ or on an electronic organ that uses the pipe organ's keyboard to pedalboard relationship.

Although rock and roll of the 1950s was largely guitar based, with a focus on the piano in the work of stars such as Little Richard and Jerry Lee Lewis, and with the more-than-occasional tenor saxophone solo, the electronic organ entered the rock genre by the early 1960s. This was anticipated by Dave "Baby" Cortez's instrumental hit of 1959, "Happy Organ," which used the Hammond B-3. In the early to mid-1960s, other electronic organ-based hits appeared, including Chris Montez's recording of "Let's Dance"; however, the real heyday of the rock electronic organ came with the adoption of the transistorized Vox Continental organ, an instrument that made its debut in 1962. This British-designed instrument was used by British and American rock bands alike, particularly between 1964 and 1968. Keyboardists who made extensive use of the Vox Continental include Paul Revere of Paul Revere and the Raiders, Alan Price of the Animals, Mike Smith of the Dave Clark Five, Ray Manzarek of the Doors, and others. During the new wave rock era of the late 1970s and early 1980s, some of the instrumental sounds of early 1960s surf music and British Invasion rock reentered the pop music mainstream. With regard to the electronic organ, the B-52's' Kate Pierson used a Farfisa Compact Deluxe organ, which can easily be seen in live videos of the band, to re-create the retro sound of early 1960s' rock and roll.

The electronic organ came into the digital age that began in the 1980s. So, new technologies such as MIDI compatibility also entered the world of the electronic organ. However, it should also be noted that at the same time, digital keyboards, which often included at least one pipe and/or electronic organ preset in addition to all the other presets they offered, decreased the need for electronic organs for home and even for some concert accompanying. Although these keyboards did not and do not have pedal boards like organs, arguably, they are better suited to pianists who need to play an instrument that sounds like an organ than organs themselves.

It was not solely through electronic organs that the King of Instruments came into the electronic and digital ages. After the rollout of MIDI technology, some pipe organ manufacturers added MIDI capability to these acoustic instruments. As a result, organists can play and store pieces in electronic memory to be played via MIDI on the instrument. This is helpful in instances such as when the organist might miss a church service due to vacation and so on. The electronic technology on some pipe organs also allows the player to automatically transpose what they are reading and playing into a different key. This is helpful, for example, in cases in which a song or aria might be too low or too high for a vocal soloist. By equipping pipe organs with this kind of technology, pipe organs built or retrofitted from the late 20th century through the present essentially can combine the best features of traditional pipe organs and electronic organs.

Even with the technological advances that have been made with the pipe organ and the development of electronic organs, the traditional pipe organ continues to be used primarily in churches around the world. One particular church organ continues to capture the attention of visitors and should, theoretically, continue to do so for hundreds of years as it plays John Cage's *ORGAN2/ASLSP—As SLow aS Possible*, a work being performed in Halberstadt, Germany. According to the composer's score, the entire performance should last for 639 years, not ending until the year 2640. In order to sustain the sound, a special organ was built for the performance, with the keys maintained in depressed position by means of weights. So that there are not significant breaks in the sound, chord changes are accomplished by keeping the keys depressed and replacing the pipes that need to be changed for the transitions from one chord to the next. Readers can find several YouTube videos of the most recent, September 2020, chord change; however, a 2019 video available at the time of this writing provides a sampling of a small fraction of the composition as well as an explanation of the piece and how it will continue to be performed (see, https://www.youtube.com/watch?v=6yUzaZDq3Hg, accessed January 15, 2021).

See Also: Drum Machines; MIDI; Theater Organs

Further Reading

Bush, Douglas E., and Richard Kassel, eds. 2006. *The Organ: An Encyclopedia*. New York: Routledge.

Faragher, Scott. 2011. *The Hammond Organ: An Introduction to the Instrument and the Players Who Made It Famous*. Milwaukee, WI: Hal Leonard Books.

Lenhoff, Alan, and David Robertson. 2019. *Classic Keys: Keyboard Sounds That Launched Rock Music*. Denton: University of North Texas Press.

Limina, Dave. 2002. *Hammond Organ Complete*. Boston: Berklee Press Publications.

Shannon, John R. 2009. *Understanding the Pipe Organ: A Guide for Students, Teachers and Lovers of the Instrument*. Jefferson, NC: McFarland & Company.

Vail, Mark. 2002. *The Hammond Organ: Beauty in the B*. San Francisco, CA: Backbeat Books.

P

PA Systems

The need for amplification came with the popularity of large theaters and large public events, sporting events, and so on. However, until the development of the vacuum tube and amplifiers that used tubes, speakers did not produce a great deal of sound, and early speakers were unwieldy, especially with the horns that they used to provide amplification that electronics alone could not provide. All that changed in the 1910s.

Loudspeakers and amplifiers such as those designed by Peter Jensen and Edwin Pridham of Magnavox in the 1910s made the projection of sound even in large arenas possible. Most famously, Jensen and Pridham's PA system made it possible for an audience of at least 50,000 people to hear an address by U.S. president Woodrow Wilson in San Diego, California, at what would later be known as Balboa Stadium.

The PA systems that began appearing in music halls and theaters negated the necessity of relatively quiet singers in vaudeville shows using megaphones—a standard feature for some, including, most famously, Rudy Vallée—however, some theaters did not include such systems well into the 1930s.

Theatrical and music hall PA systems took a significant step forward in 1947 with the introduction of Altec Lansing's Voice of the Theatre speaker systems. Although originally used in indoor venues, the Voice of the Theatre systems later found usage at outdoor rock concerts, particularly from the late 1960s into the 1990s. Perhaps most famously, Voice of Theatre speakers can be seen in film footage from the Woodstock Festival of 1969.

Over time, PA systems have become more sophisticated, with systems being tuned to the rooms in which they are used and for the types of musical and spoken-word applications. The tuning of PA systems entered the computer age with computerized mixing boards and, more recently, with the use of devices such as tablet computers. In fact, some of today's systems are controlled by sound technicians with iPads. Because of the portability of these devices, they can be carried to various locations in the hall, and the sound balanced throughout the entire audience area. Even in outdoor or combined indoor-outdoor venues, such as the Cleveland Orchestra's summer home at the Blossom Music Center, one can see an audio technician checking the sound from various parts of the venue. Volume, balance from across the stereophonic spectrum, equalization, balance from microphone to microphone, and so on can be controlled remotely using these devices. This technology has proven to be an improvement over the standard way in which PA system controls have been designed: with the mixing board in one fixed location, either somewhere in the audience space in the venue or in a sound booth that, by the nature of its location, does not allow the audio technician to hear what the audience member hears in any particular part of the auditorium.

Some of the technology that is used for audio induction—or, more commonly called, hearing loop, systems that are used as an adjunct to PA systems in churches, auditoriums, and other venues—dates from the 1930s. These systems, which can be thought of as PA systems for the hearing impaired, became more commonplace beginning in approximately the 1990s. Some earlier in-auditorium audio assistance systems (still in use in some venues) used FM transmissions or infrared technology and required that the user wear a headset; however, increasingly, one can find hearing loop systems that can send an electronic signal directly to a user's hearing aid, provided that they are sitting in a seat in the venue that is within the loop and provided that the audience member's hearing aids have the necessary wiring to pick up the signal; some hearing aids are so small that the necessary electronic components will not fit. For those users with the compatible hearing aids, however, research suggests that hearing aid–based loop systems are perceived by users to produce significantly better, clearer sound (Kaufmann et al. 2015, 300).

See Also: Amplification; iPads and Other Tablets; Microphones; Mixing Boards; Motion Picture Soundtracks

Further Reading

The Editors of the *San Diego Union-Tribune*. 2018. "From the Archives: September 20, 1919: 50,000 Hear President Wilson." *San Diego Union-Tribune*, September 20. Accessed January 18, 2021. https://www.sandiegouniontribune.com/news/150-years/sd-me-150-years-september-20-htmlstory.html.

Kaufmann, Thomas, Juliette Sterkens, and John M. Woodgate. 2015. "Hearing Loops: The Preferred Assistive Listening Technology." *Journal of the Audio Engineering Society* 63, no. 4 (April): 298–302.

Pandora and the Music Genome Project

One question that users of the 21st-century service Pandora might reasonably ask is, how does the company decide what music to include on, say, the Kanye West channel, if it does not solely consist of a playlist built from the recordings of Kanye West? The answer lies in the thing that distinguished Pandora from other early 21st-century streaming audio services (and the principal reason that Pandora has a separate entry in this volume): a proprietary system known as the Music Genome Project, a registered trademark of the company.

Pandora was first rolled out as Pandora Internet Radio in 2000. Founder Tim Westergren, himself a musician, and his associates developed a set of criteria by which songs and instrumental recordings would be analyzed, categorized, and grouped. Interestingly, the Pandora Music Genome Project procedures continue to combine human analysis of music and computer analysis of listeners' habits, likes, and dislikes. According to Pandora's website, each recording under consideration "is analyzed using up to 400 distinct musical characteristics by a trained music analyst." These analysts typically have "a four-year degree in music theory, composition or performance, has passed through a selective screen process and has completed extensive training in the Music Genome's rigorous and precise methodology" (https://www.pandora.com/corporate/mgp.shtml, accessed January 15, 2021).

Although Pandora uses only human listening and analysis to evaluate the hundreds of criteria, computer-based algorithms are used as listeners develop their own "channels" or like or dislike individual tracks on established Pandora channels. For example, if a listener selects the Jean Sibelius channel and dislikes Sibelius's *Finlandia*—as unlikely as that might be, considering that it is the composer's best-known and most famous work—that dislike is logged. Presumably, then, *Finlandia* might be removed from the Sibelius channel the next time the listener goes to that channel. The reason for the mixture of human-based analysis and computer-based continuing development of channels is that, despite the fact that the Music Genome Project's proprietary analytical criteria are numerous and rigorous, listener and even widespread public likes and dislikes can involve intangible factors.

Despite the apparent rigor and time-intensive nature of the Music Genome process, by October 2009, approximately a decade after Pandora's initial rollout, over 700,000 pieces of music by 80,000 artists had been analyzed (Walker 2009). And now, more than a decade later, the numbers continue to grow, and the repertoire and genre bases continue to expand.

Despite how interesting the concept is, during the first decade of its existence, Pandora encountered numerous business challenges. As *Inc.* magazine's Stephanie Clifford wrote in 2007, "It's undeniably cool and completely addictive, but Pandora has never quite found its footing as a business. Indeed, the company has been through an almost unbelievable number of setbacks, a series of blows that would make the most determined entrepreneur throw in the towel. Westergren has run out of money, which forced to him to lay off his entire staff (except for those willing to work for free). He's been rejected some 350 times by venture capitalists. He has faced bankruptcy, haggled with anxious creditors, and been sued by employees. Deal after deal has fallen through at the last minute" (Clifford 2007). Despite additional fits and starts, stock-price drops, and other ups and downs, in 2019, *Inc.* magazine reported that along the way Westergren and Pandora had found and built a viable business model, as evidenced by SiriusXM's 2019 purchase of Pandora for $3.5 billion (Ledbetter 2019).

In developing a detailed process that analyzed music, categorized it, and then offered it in what are essentially like curated packages (channels) for listeners, Pandora anticipated the plethora of curated experiences that arose in the late 2010s, including chef-curated meal kits, designer- and fashion consultant-curated clothing lines, and so on. At the time of this writing, Pandora does not actually use the term "curated" in its description of the Music Genome Project (see, The Editors of Pandora.com n.d.); however, if one were to translate the project's impact on the listener using the terminology that became trendy in the late 2010s, then one could label the service as providing a curated listening experience.

See Also: Internet Radio; Satellite Radio; Streaming Audio Services

Further Reading

Clifford, Stephanie. 2007. "Pandora's Long Strange Trip: Online Radio That's Cool, Addictive, Free, and—Just Maybe—a Lasting Business." *Inc.*, October 1. Accessed May 11, 2020. https://www.inc.com/magazine/20071001/pandoras-long-strange-trip.html.

The Editors of Pandora.com. n.d. "About the Music Genome Project." Pandora.com. Accessed May 11, 2020. https://www.pandora.com/about/mgp.

Eschner, Kat. 2017. "John Philip Sousa Feared 'The Menace of Mechanical Music': Wonder What He'd Say about Spotify." *Smithsonian Magazine*, November 6. Accessed January 18, 2021. https://www.smithsonianmag.com/smart-news/john-philip-sousa-feared-menace-mechanical-music-180967063/.

Flanagan, Andrew. 2017. "Pandora Co-Founder and CEO Tim Westergren to Step Down." *NPR*, June 27. Accessed January 18, 2021. https://www.npr.org/sections/therecord/2017/06/27/534542303/pandora-co-founder-and-ceo-tim-westergren-to-step-down.

Ledbetter, James. 2019. "How Tim Westergren Steered Pandora from 'the Brink of Shutting Down' to a $3.5 Billion Exit." *Inc.*, June 6. Accessed January 18, 2021. https://www.inc.com/james-ledbetter/pandora-founder-tim-westergren-fast-growth-tour-san-francisco.html.

Walker, Rob. 2009. "The Song Decoders." *New York Times Magazine*, October 14. Accessed November 30, 2020. https://www.nytimes.com/2009/10/18/magazine/18Pandora-t.html.

Patches

Even a cursory perusal of online or print catalogs of synthesizers and plug-ins for synthesizers will reveal references to patches. Although today this reference appears in advertisements and specifications related to digital synthesizers, its roots go back to the earliest analog synthesizers.

Early monophonic analog synthesizers, such as the first Moog synthesizers of the second half of the 1960s, as well as the larger oscillator-based synthesizers that were primarily found in academic music departments with computer music programs, required the use of numerous patch cords in order to program various tone colors. The programmer used cords that were somewhat like short electric guitar cables, to make connections to various filters that might, for example, filter out part of the harmonic spectrum—or overtone series—from a root sound wave form. Programmers also used patch cords to control the attack and decay of tones, as well as other aspects of the sounds. One can think of this as an extension of what human telephone operators once did when they had to use cords to connect one phone to another from a central console, except that the number of tonal aspects that could be controlled might make the telephone operator's console look like it was controlling a complex multi-user conversation.

As one might imagine, the process for making these patch cord connections on the early analog synthesizers was slow and complicated. For example, according to audio engineer Ken Townsend, who worked on the Beatles' 1969 *Abbey Road* album, "to get that French horn sound [on the song 'Because'] it took a whole set of flightcases full of jack plugs and filters" (Babiuk 2002, 246). Some of the early experts on analog synthesizers, such as Paul Beaver, were called in by numerous record producers and artists to program these unwieldy instruments.

Some analog synthesizers later employed internal circuitry, which allowed for easier and quicker connections; however, analog synthesizers continued to require programming. So, although patch cords began to disappear to some extent, their function remained part of the world of synthesizers and the design of tone colors. Later, when some users of digital synthesizers wanted to re-create digitally the sounds that earlier analog synthesists had designed (e.g., the aforementioned French horn tone on "Because"), the pre-programmed synthesizer settings became known as patches, in keeping with the original way in which synthesizers had been programmed manually using patch cords.

In more recent digital synthesizers, plug-ins have been used that re-create the classic patches of the past. In addition, some patches are available for download for virtual synthesizers. And patches are not limited to re-creating classic analog synthesizer tone colors; patches are available that allow digital synthesizers to emulate the classic tone colors offered on mellotrons and other now-difficult-to-find instruments.

See Also: Digital Synthesizers; Mellotrons; Moog Synthesizers; Virtual Synthesizers

Further Reading

Babiuk, Andy. 2002. *Beatles Gear: All the Fab Four's Instruments, from Stage to Studio*. Revised ed. San Francisco, CA: Backbeat Books.

Pinch, Trevor, and Frank Trocco. 2002. *Analog Days: The Invention and Impact of the Moog Synthesizer*. Cambridge, MA: Harvard University Press.

Sweetwater Sound. 2004. "Virtual Analog Synths vs. Analog Synths." Sweetwater.com, April 22. Accessed January 18, 2021. https://www.sweetwater.com/insync/virtual-analog-synths-vs-analog-synths/.

Pedal Steel Guitars

Beginning in the 1940s, the pedal steel guitar became a standard part of the sound of country music. Born out of the earlier popularity of the Hawai'ian guitar, lap steel guitar, and the square-neck resonator guitar, the pedal steel guitar solved many of the tuning and adaptability problems associated with those instruments. It is acknowledged, however, as one of the more complex and difficult-to-play-well instruments still in regular use in vernacular music.

The history of the modern pedal steel guitar begins with the introduction of the standard Spanish-style acoustic guitar to the Hawai'ian Islands in the 19th century. Instead of the standard guitar tuning, Hawai'ians began using what was known as slack-key tuning, because some of the strings were slackened so that the entire six-stringed instrument was tuned to a major chord. Guitarist Joseph Kekuku is credited with turning the slack-key guitar on its side so that it would rest—strings on top—on the player's lap. Kekuku raised the instrument's strings and used a steel bar to get the pitches by sliding the bar along the strings. The term "steel guitar" comes from this use of a steel bar.

From Kekuku's development of the steel guitar in the 1870s and into the early 20th century, Kekuku's students and others continued to define a distinctly

Hawai'ian approach to the guitar. In the first several decades of the 20th century, Hawai'ian music became one of the biggest musical fads on the mainland and featured the sound of ukulele and slack-key guitars and lyrics about sunny beaches. The 78-rpm records of Hawai'ian music and music in the Hawai'ian style sold by thousands and thousands. Hawai'ian slack-key guitarists such as Sol Hoopi'i became well known, and eventually Hoopi'i began performing in a Western swing band, generally thought to be the entrance point of the steel guitar into country music.

In the 1920s and early 1930s, the Dopyera Brothers and others were pioneering both the resonator guitar and early incarnations of the electric guitar. The next step in the developments that would lead to the pedal steel occurred when a pickup was installed on the Hawai'ian-style guitar. Electrification of the guitar meant that a large body was no longer required for sound resonance and projection, so the lap steel guitar was born. The fact that the electric lap steel guitar could be smaller than a conventional guitar and the fact that none of the strings were actually fretted (a finger was not used to push the string down to the fretboard behind a fret, because the steel bar was used to play the notes) meant that lap steel guitars could be made with more than six strings.

One of the challenges faced by all players of guitars tuned to an open major chord was how to play minor chords. Another was that certain tonalities—or, keys—were easier to play in than others, based on the chord to which the guitar was tuned. Although tuning a Hawai'ian acoustic guitar or lap steel to a four-note chord that had both major and minor qualities (e.g., the popular lap steel C6

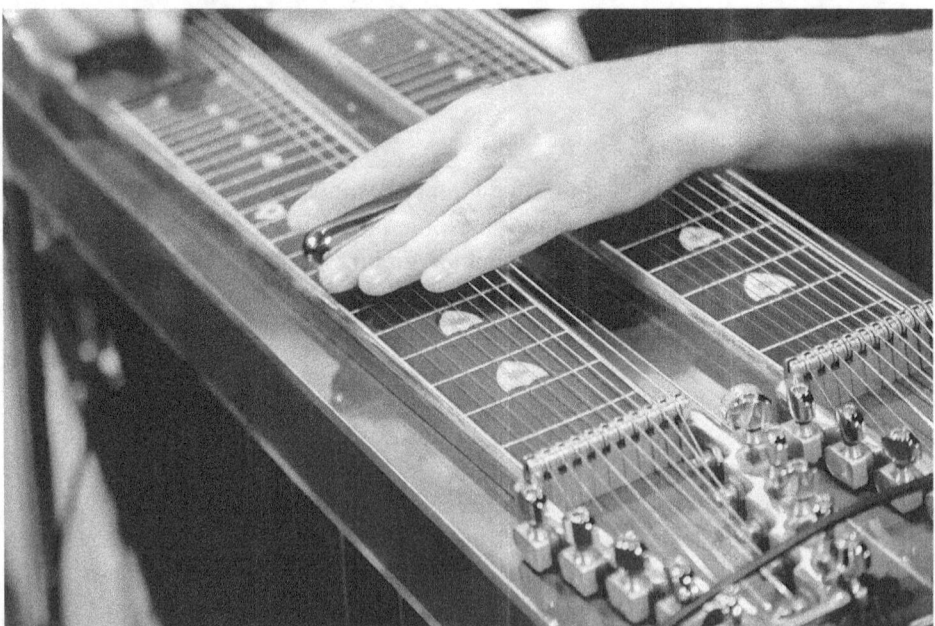

Most closely associated with country music, the pedal steel guitar has its roots in Hawai'ian-style steel guitar and in the electric lap steel guitar. (Tomo Nogi/Dreamstime.com)

tuning, which includes the three notes of the C major chord—C, E, G—and the three notes of the A minor chord—A, C, E), it does not address all the challenges. Some lap steel guitars were fitted with two or three necks and multiple sets of strings, each in a different tuning.

The pedal steel guitar took the basic principal of the electrified lap steel guitar and added a series of pedals that functioned much like the pedals on the orchestral harp. The pedals change the pitches of sets of strings, such that chords can be changed without use of the bar and without having to use the tuning pegs. Although patents continued to be issued for various aspects of different pedal steel designs for decades, the guitarist Alvino Rey is generally credited with being the impetus behind the development of the pedal steel guitar in the late 1930s. Gibson, one of the major corporations in the guitar design and manufacturing field, introduced its first pedal steel guitar in 1940.

The pedals on the first instruments raised the open-tuned strings by a half step by stretching them. Zane Beck, a pedal steel player who had performed with several country music stars during his career, added a mechanism operated with the knees in 1952. Beck's mechanism lowered the strings by a half step, providing the player even more tuning flexibility. Other pedal steel players made additional modifications and refinements throughout the 1950s, including Bud Isaacs and Buddy Emmons. Emmons and machinist Harold Jackson formed the Sho-Bud company in 1957. This company differed from Fender, Gibson, and some of the other major manufacturers of pedal steel guitars by focusing exclusively on the instrument. It also put many of the developments that had been made since the start of the 1940s and put them in one package. The company expanded into offering acoustic guitars in the 1970s and eventually became part of Gretsch. However, today instruments are no longer produced under the Sho-Bud.

Pedals that must be put down or lifted with precise timing, knee-operated levers, multiple fretboards, and a method of finding pitches (e.g., using the metal bar) that is inherently imprecise make the pedal steel guitar one of the most complex and difficult to play instruments; however, all the complexities also make for an instrument that sounds like no other and is capable of playing musical figures that simply cannot be played on any other member of the guitar family.

See Also: Electric Guitars; Resonator Guitars

Further Reading

Jarnow, Jesse. 2020. "The Endless Potential of the Pedal Steel Guitar, an Odd Duck by Any Measure." NPR, January 7. Accessed January 18, 2021. https://www.npr.org/2020/01/07/793989801/the-endless-potential-of-the-pedal-steel-guitar-an-odd-duck-by-any-measure.

Ross, Michael. 2015. "Pedal to the Metal: A Short History of the Pedal Steel Guitar." *Premier Guitar*, February 17. Accessed January 18, 2021. https://www.premierguitar.com/articles/22152-pedal-to-the-metal-a-short-history-of-the-pedal-steel-guitar.

Shah, Haleema. 2019. "How the Hawaiian Steel Guitar Changed American Music." *Smithsonian Magazine*, April 25. Accessed January 18, 2021. https://www.smithsonianmag.com/smithsonian-institution/how-hawaiian-steel-guitar-changed-american-music-180972028/.

Peer-to-Peer File Sharing

The sharing of music and the bootlegging of music were nothing new at the end of the 20th century. The copying or dubbing of a friend's cassette tapes was well established, and the illegal copying and sale of music on bootleg vinyl discs, cassettes, and especially on 8-track tapes had been going on for years and years. The digital age brought the opportunity for sharing and bootlegging music to an entirely different level, because now copies could be made that were exact digital clones of the information (music) of the original. With tape-based copying, particularly as copies were made of copies and so on, some audio fidelity was lost from generation to generation.

One of the challenges of digital file sharing in the 1980s and 1990s was that audio files tended to be very large, so in the days before cloud computing and before storage devices such as USB drives, the means of sharing were limited. Certainly, sharing over the internet was difficult because of the size of files, even compressed music files in mp3 format. Peer-to-peer file sharing was developed at the end of the 20th century. The process was popularized with Napster, a program developed by Shawn Fanning (the coder) and Sean Parker (the entrepreneur). Napster users accessed a central computer server that contained information about what users owned particular pieces of music as digital files. The service then essentially allowed the users to connect in order to share the desired music. The name of the process, peer-to-peer, comes from the fact that the two users' computers were connected to each other for the direct online transfer of the digital files.

Although other peer-to-peer networks emerged, Napster was the most famous, or perhaps more correctly, the most infamous. It is still difficult to assess the exact financial impact of Napster and peer-to-peer file sharing on the music industry, but it certainly is safe to say that for the couple of years Napster operated in its original format—as a community for music sharing—sales of commercially recorded music decreased, and the music industry itself blamed Napster and the similar peer-to-peer networks that went online in its wake.

In a 2013 conversation with Ashley Moreno of *The Austin Chronicle*, Napster cofounder Sean Parker stated, "Just about every interesting, meaningful, large-scale shift introduced [on the Internet] has something to do with connecting people to one another and to this idea of global community. You see it in Napster. You see it in Facebook. It's all about how we connect people in ways that either approximate real-life connections or [are] completely novel, new ways we've never seen before. Napster is really the first company that harnessed the underlying true potential of the Internet as a peer-to-peer communication medium" (Moreno 2013).

Although the quote from Parker suggests that something positive came out of Napster—a demonstration of the connective power of the internet—he does not address the damage to the recording industry of the widespread sharing of music. *Forbes* magazine's Hollywood and Entertainment contributor Hugh McIntyre described himself as being in 1999 "one of millions of people who downloaded music without paying for it via a number of programs, all of which seemed legitimate to somebody who didn't know better," without the realization that these peer-to-peer file sharing services "were actually bringing the entire music business to its knees" (McIntyre 2018).

In the article "The Impact of Digital File Sharing on the Music Industry: An Empirical Analysis," published in the journal *Topics in Economic Analysis & Policy*, Norbert J. Michel cites statistics that appear to suggest a link between computer usage and lack of expenditure on commercially released compact discs. For example, in this study, which was initially part of Michel's doctoral dissertation, the author found that "the year Napster went online (1999). . . computer owners' mean CD expenditures increased $0.68 (1 percent), a small, statistically insignificant change. On the other hand, non-computer owners' mean CD expenditure increased $3.30 (20 percent) in 1999" (Michel 2006, 4). Michel includes other statistical analyses of CD expenditures, computer usage, broadband access, and so on and concludes that "our micro-level data test results suggest that file sharing may have reduced album sales (between 1999 and 2003) by as much as 13 percent for some consumers" (Michel 2006, 11).

Ultimately, it was the widespread copyright infringement that Napster enabled and the fact that users made their initial connections with one another through Napster's central server that led to the service being shut down in 2001. As Yochai Benkler wrote in the book *The Wealth of Networks: How Social Production Transforms Markets and Freedom*, "peer-to-peer file sharing networks are an excellent example of a highly efficient system for storing and accessing data in a computer network" (Benkler 2006, 83–84). Benkler continues, "For fairly obvious reasons, we usually think of peer-to-peer networks, beginning with Napster, as a 'problem.' This is because they were initially overwhelming used to perform an act that, by the analysis of almost any legal scholar, was copyright infringement" (2006, 84).

Copyright infringement was at the core of the lawsuits that brought down Napster. Perhaps the most famous case was the 2000 suit brought by heavy metal band Metallica against Napster, Indiana University, the University of Southern California, and Yale University that alleged that the band's copyrights were being violated because of Napster's central server and because students at the universities named in the lawsuit were using their student accounts through the institutions' IT departments to illegally obtain copyrighted music. The next generation of peer-to-peer networks further complicated the legal questions and further broadened the possibilities for file sharing by connecting various end users directly to one another; there was no need to go through a Napster-like central server. Limewire, Kazaa, and Gnutella were among this next generation of peer-to-peer networks. Some of the post-Napster networks allowed for sharing other types of files, including movies, video games, and so on. However, Napster remains the most infamous of these late 20th-century and early 21st-century networks. Writing in the wake of the first version of Napster being shut down in 2001, *Wired* magazine's Brad King described it as "the software application that ignited the music file-trading frenzy" (2002). At the conclusion of the lawsuit that ultimately led to Napster being shut down in 2000, U.S. district judge Marilyn Hall Patel stated that the service encouraged "wholesale infringement" against copyrights (ABC News 2000).

Napster was finally acquired by the legal streaming audio service Rhapsody in 2011. The version of the company and service that exists today is a legal subscription streaming audio service. Although the company's logo looks substantially the same as it did when Napster raised all sorts of legal controversies, gone is the sharing of music in infringement of copyright that was perhaps the biggest talk of college campuses for a few brief years and the focus of widespread media speculation about the possible death of the music industry during that same time.

Despite the fact that the original Napster was shut down so quickly after it had created a significant stir in popular culture, had apparently caused widespread damage to the music industry, and had taken college campuses (in particular) by storm, and despite the fact that the reincarnation of Napster functions quite differently than the original, only part of the story of the company was the access to free music that it enabled. The fact that Ashley Moreno's 2013 report on the Austin, Texas, Digital Revolution roundtable included Electronic Frontier Foundation's John Perry Barlow and Napster's Sean Parker, Shawn Fanning, and Alex Winter and that the discussion focused on the ongoing importance of peer-to-peer connectivity suggests that connectivity might well be the most important part of Napster's legacy. In that respect, Napster can be seen as perhaps the first widely used platform (albeit only for a short time) that anticipated social media platforms such as Facebook, which made its debut in 2004; Twitter, which made its debut in 2006; Instagram, which made its debut in 2010; and the numerous platforms that have helped people connect with friends, family members, like-minded strangers, and so on, ever since Napster's initial demise over two decades ago.

See Also: Desktop and Laptop Computers; iPods and Other Portable Digital Music Players; MP3s

Further Reading

ABC News. 2000. "Napster Has Hit a Sour Note in Court." *ABC News*, July 27. Republished January 7, 2006 as "Napster Shut Down." Accessed January 18, 2021. https://abcnews.go.com/Technology/story?id=119627&page=1.

Benkler, Yochai. 2006. *The Wealth of Networks: How Social Production Transforms Markets and Freedom*. New Haven, CT: Yale University Press.

King, Brad. 2002. "The Day the Napster Died." *Wired*, May 15. Accessed January 18, 2021. https://www.wired.com/2002/05/the-day-the-napster-died/.

McIntyre, Hugh. 2018. "The Piracy Sites That Nearly Destroyed the Music Industry: What Happened to Napster?" *Forbes*, March 21. Accessed January 18, 2021. https://www.forbes.com/sites/hughmcintyre/2018/03/21/what-happened-to-the-piracy-sites-that-nearly-destroyed-the-music-industry-part-1-napster.

Michel, Norbert J. 2006. "The Impact of Digital File Sharing on the Music Industry: An Empirical Analysis." *Topics in Economic Analysis & Policy* 6, no. 1. This article is available for download from the Recording Industry Association of America (RIAA) at https://www.riaa.com/wp-content/uploads/2004/01/art-the-impact-of-digital-file-sharing-on-the-music-industry-michel-2006.pdf.

Moreno, Ashley. 2013. "It's Still All about Peer-to-Peer Connectivity: Alex Winter Raps about Napster with Shawn Fanning and Sean Parker." *Austin Chronicle*, March 12. Accessed January 18, 2021. https://www.austinchronicle.com/daily/sxsw/2013-03-12/its-still-all-about-peer-to-peer-connectivity/.

Winter, Alex, director. 2013. *Downloaded*. Documentary film. New York: VH1.

Percussion Instruments

As we have seen in several entries in this volume, not all the connections between music and technology have involved or need to involve the use of sophisticated computer technology, or even noncomputerized electronic technology. Although percussion instruments may be the oldest musical instruments known to humankind, they have undergone technological advances over the years. Let us consider just a few of the examples of the kinds of technologies—some low-tech and some high-tech—that have shaped this diverse family of instruments over the last several hundred years of music history.

One of the more visible instruments at orchestral concerts of music from the 18th century through the present are the timpani, sometimes called the kettle drums. These tuned drums were used primarily during the late 18th century—the Classical period—to play the roots or sometimes the roots and fifths of the primary chords in the key of the composition. As music grew more complex and harmonic vocabulary expanded during the 19th century and beyond, it was necessary to be able to tune the drums to different pitches quickly and accurately.

Timpani are tuned by tightening (to raise the pitch) or loosening (to lower the pitch) the drum head. Various methods for accomplishing this have been used over the centuries, including a handle mechanism that was developed in the early 19th century. One of the standard methods that has been used since the 1870s is a pedal mechanism. Because the pedal mechanism for timpani tuning was developed in Dresden, Germany, it is sometimes called the Dresden mechanism or the Dresden method of tuning. Although this might seem to be a low-tech solution, it is quicker, easier, and more accurate than earlier methods.

On a completely different note, one of the most famous examples of creating a new instrument out of throwaways is the steel drum. In the 1930s, percussionists in Trinidad found that the bottom surface of discarded oil drums could be tuned by hammering the metal so that it would stretch different amounts from its original shape, thus essentially making different parts of the surface different thicknesses. They also found that they could use just part of the discarded oil drum to make similar but smaller instruments that could be tuned to higher pitches. These smaller drums, known as pans, form the basis of today's steel bands.

Another example of creating a percussion instrument out of a common, nonmusical materials is the cajon, an instrument often associated with folk music groups today but that started out in Peru, when enslaved people from Africa who were brought into the South American country were denied access to the drums that had been such a vital part of their culture in Africa. What probably started out as wooden boxes or drawers from discarded dressers has evolved into a wooden box instrument with snares inside and generally a sound hole in the back that enables better bass response.

Certainly, not all the developments in percussion technology have been of the low-tech variety. Because there is a separate entry for drum machines in this volume, the sole example of electronic technology applied to percussion that we will examine is the electronic drum set. The first commercially available electronic drums were offered by Pollard in 1976; however, they did not enjoy commercial success. The company most closely associated with electronic percussion, the Simmons Company, was formed in 1978. Simmons released its most successful and impactful electronic drum set, the SDS-5, in 1981. Today, a number of companies—electronic percussion specialists and more general musical instrument companies—produce electronic percussion. These electronic drum sets can be purchased for a wide range of prices and can be configured in a wide variety of ways.

One of the challenges posed by electronic drums is with the feel that they have for percussionists who are used to acoustic percussion instruments. Specifically, the rubber-like material used for electronic drums, ride cymbals, crash cymbals, hi-hats, and so on typically does not have exactly the same bounce responses when struck that drum heads and metal cymbals do. Despite that challenge, electronic drum sets are more portable, can produce both louder and quieter sounds than conventional drum sets, and generally are less expensive than acoustic drum sets. In addition, arguably, synthesized drum and cymbal sounds are closer in tone color and attack and decay characteristics than those of other synthesized instrument sounds (e.g., synthesized trumpet, flute, and clarinet).

As is the case with any MIDI instrument, a particular surface on an electronic drum set can be used as a trigger to the computer. Used this way, hitting a particular object on the set—say, the electronic large tom-tom—could trigger a synthesizer or a computer to play a particular sequence or to change some sound parameter. Thus, the electronic drum sets, or electronic percussion instruments in general, are far more than electronic versions of their acoustical counterparts; they are MIDI instruments with the same kinds of capabilities of MIDI keyboards and so on.

See Also: Drum Machines

Further Reading

Blades, James, Evelyn Glennie, and Neil Percy. 2020. *Percussion Instruments and Their History*. London: Kahn & Averill Publishers.

Boynton, Brad. 2017. "A Brief History of the Cajon." *Drum Magazine*, December 8. Accessed January 18, 2021. https://drummagazine.com/a-brief-history-of-the-cajon/.

Player Pianos

Although they are infrequently seen today—and then often in the context of a recreation of the feel of the late 19th century or early 20th century—player pianos played an important role in bringing pop hits, classical pieces, and ragtime and jazz to listeners during the last few years of the 19th century and the first couple decades of the 20th century. The basic principle of the player piano revolves around a roll of paper into which punches are made; the instrument's mechanism converts these punches into the notes that are played on the piano.

The Pianola, the first commercial player piano, was patented by Edwin S. Votey in 1897. Early player pianos did not have a way of reproducing dynamic changes or pedaling effects. They also suffered from the fact that the machine's operator kept the roll feeding through the mechanism using a treadle, similar to how early pre-electric sewing machines were operated. Still, even these early instruments provided live music even if a skilled pianist was not available. Eventually, more sophisticated systems allowed for a steadier mechanical feed of the encoded paper through the machine as well as for the performance of touch, sustain, and dynamic shading not possible on the earliest player pianos.

Melville Clark, one of the principals of Story & Clark pianos, left Story & Clark in 1900 to form the Melville Clark Piano Company, which was one of the leaders in early 20th-century development of the player piano. Other companies also produced player pianos, and companies emerged in the early 20th century that produced piano rolls, including QRS, Aeolian, Connorized, Vocalstyle, and others. One of the problems associated with the rapid expansion of the market for player pianos and piano rolls was that there were several systems and standards in use until an industrywide standard was established in 1908.

The classic player piano of the early 20th century played a significant role in the ragtime craze that began with the sheet music publication of Scott Joplin's *Maple Leaf Rag* in 1899 and that continued into the second decade of the new century. Compositions by Joplin, Eubie Blake, James Scott, and other ragtime musicians were published by piano roll companies, and the post-ragtime, pre-jazz novelty numbers, such as Zez Confrey's *Kitten on the Keys*, were also naturals for the player piano medium. It should be noted that the player piano arrangements of some of these pieces were different somewhat from the versions that were published in sheet music form. For example, a version for player piano might place the melody an octave higher on a repeat of a section. Sometimes, too, the piano rolls contained notes, octave doublings, reaches, and so on that would have been impossible for a human pianist to play. These were possible to place into player piano arrangements, however, because in the early days of the instrument, piano rolls were encoded by hand. A technician made the actual punches into the paper.

Although piano rolls could be encoded by an individual punching holes in the paper, the development of a reproducing piano made it possible for the paper to be encoded by a person playing the piano in real time. It should be noted, however, that these piano rolls could be embellished by punching additional holes in the paper roll. Some of the more famous examples of rolls produced using the reproducing piano include several rolls that Scott Joplin cut in the last year or so of his life. These provide the clearest examples of how Joplin intended his version of classic ragtime music to be played, although Joplin's last piano roll recordings exhibit the neurological decline due to syphilis that ultimately led to the musician's death in 1917. As the capabilities of the reproducing pianos grew to be more sophisticated, several other prominent pianists recorded piano rolls, including George Gershwin, Sergei Rachmaninoff, Ignacy Paderewski, and others. Among the early 20th-century jazz pianists, Fats Waller, James P. Johnson, Jelly Roll Morton, and others cut piano rolls. Composers such as Paul Hindemith and Igor Stravinsky also wrote pieces for player piano.

One particular type of player piano, the nickelodeon, allowed users to listen to a piece for—as the name suggests—a nickel. Nickelodeons could be found in public spaces in major cities, arcades, and in bars and other establishments. Years later, the 1949 Stephen Weiss and Bernie Baum song "Music! Music! Music! (Put

Another Nickel in)" seemed to reference these pay-to-play player pianos. Curiously, though, the late 1940s and early 1950s, during which time several hit recordings of "Music! Music! Music!" were made, was a time period in which the jukebox was the most popular pay-to-play mechanical music device.

The high point of sales of piano rolls was reached in 1927, the year in which the QRS Music Company sold 10 million piano rolls (https://www.qrsmusic.com/history.php), accessed November 26, 2021). A decline in popularity of player pianos soon followed, with contributing factors including the widespread popularity of radio, jazz music that centered on bands (as opposed to solo pianists), and the 1929 stock market crash and subsequent Great Depression, which greatly decreased disposal income.

Despite the downturn in popularity of the player piano in the 1930s and beyond, the instrument continued to pique the interest of art music composers. For example, between 1948 and the early 1990s, composer Conlon Nancarrow took to composing works specifically for the player piano, because the pieces he had written for live pianists proved to be too difficult to perform. Although Nancarrow's work was not widely known, even within avant-garde music circles, a series of recordings and live performances (using a player piano) in the 1970s through the 1990s brought his unusual use of the player piano into a more prominent place within the new music scene.

Aside from its popularity and the impact that the player piano had with the ragtime craze of the first decade or so of the 20th century, one of the important offshoots of the instrument's technology came decades later in the computer age. Specifically, the punch cards associated with computers of the 1950s and 1960s bear at least a passing resemblance to the theory behind how player pianos read the piano rolls of the early 20th century. Even as music moved into the digital age with MIDI and digital audio workstations, the idea of the piano roll and manual punching of rolls remained: one of the representations of encoded music used on computers mimics (and is named for) the old piano rolls.

With an increase in interest in restoring old player pianos that developed in the second half of the 20th century, the piano roll mainstay QRS continued to produce piano rolls, even as other former roll-producing companies turned away from the piano roll and player pianos as viable media. Although QRS is the only company that continues to produce piano rolls going into the third decade of the 21st century, the company also works with its affiliated Story & Clark Pianos to produce new player pianos that use digital encoding of music. QRS also offers a digital retrofit kit so that early 20th-century player pianos can make use of a technology that eliminates the problems associated with physical paper piano rolls.

Mechanical Rights

Perhaps one of the most misunderstood terms associated with music publishing, the term "mechanical rights" dates back to the days when distinctions were made between licensing rights to record music for commercial release, rights to publish sheet music, and licensing rights pertaining to public performance of musical works. The term "mechanical" itself refers to the late 19th- and early 20th-century player pianos and their *mechanical* reproduction of music. At the time of the advent of these popular instruments, disputes arose about who owned the rights to approve or disapprove the production of piano rolls of musical works, instrumental or vocal, for which sheet music was published. The same question of ownership rights also arose around the numerous other player piano–like mechanical instruments that proliferated at the time. The separate category of mechanical rights was created to take into account the manufacture of piano rolls and audio recordings, which eventually included records, cassettes, 8-tracks, CDs, digital recordings, and so on.

See Also: Mechanical Organs; MIDI; Yamaha Disklavier

Further Reading

The Editors of *Encyclopædia Britannica*. 2020b. "Player Piano." *Encyclopædia Britannica*. Accessed March 10, 2020. https://www.britannica.com/art/player-piano.

Quadraphonic and Surround Sound

Among audiophiles, quadraphonic sound enjoyed a fairly brief period of popularity, particularly in the 1970s and 1980s. Some quadraphonically encoded discs could be played on conventional stereo turntables, using conventional stereophonic sound systems. Some of the quadraphonic records released during the period were remastered from albums that originally had been mastered for stereophonic release. In short, some of these were not originally recorded with the idea that they would be released quadraphonically. Although quadraphonic in-home systems enjoyed only a short run of popularity among audiophiles, the concept of surrounding the listener with sound found a lasting home in the film industry.

When quadraphonic and surround-sound systems made their way into the consumer market for audiophiles, it was really not an entirely new technology. Karlheinz Stockhausen, Edgard Varèse, Iannis Xenakis, and some other experimental composers of electromagnetic tape–based compositions in the *musique concrete* genre created works in which sound was projected at the listener from more than two stereophonic channels. With the exception of Varèse's *Poème électronique*, which was part of the multimedia experience at the Philips Pavilion at the 1958 Brussels World's Fair, the 1950s' and 1960s' tape works in multichannel formats were experienced by relatively few people.

Quadraphonic recordings entered the commercial audiophile world at the end of the 1960s and enjoyed a run of popularity in the next couple of decades. Ultimately, several factors led to the technology's downfall in the consumer marketplace. One of the challenges is that there were several systems used for encoding and decoding quadraphonic records. As one might expect, systems that produced true, discrete four-channel sound were more expensive than those that essentially took part of the stereo signal and artificially made it sound like it was coming at the listener from behind. Quadraphonic releases were not limited to vinyl records, as quadraphonic reel-to-reel tapes and 8-track tapes also saw commercial release.

Quadraphonic systems required four speakers in order to produce the sound from the four discrete channels. When considering the additional cost to consumers, it is also important to keep in mind that special decoders and amplifiers were also required, and they were expensive. It is reasonable to assume that economic factors such as these probably played a significant role in the short life span of quadraphonic sound. However, there were not necessarily a lot of choices of quadraphonic discs available for purchase, especially recordings that were originally intended for quadraphonic systems. And, as one might expect, the quadraphonic discs tended to be more expensive than conventional stereophonic records. In addition, some of the commercial quadraphonic releases were of recordings that were never meant to be released in the format, for example, a 1973 greatest-hits collection of music by the Doors. So, a song such as "Light My Fire" could be remastered quadraphonically from the original master recordings, but the song was only originally mixed and released as a monophonic and stereophonic recording back in 1967.

Although it could well have been the cost of quadraphonic discs and the fact that one would need twice as many speakers compared with a stereo system, as well as the additional cost of special decoders and amplifiers that led to quadraphonic sound never fully getting off the ground commercially, it is also interesting to consider how different the surround sound and the quadraphonic experience are from the centuries-old live music experience. The centuries-old paradigm is that the performers face the audience. Although an audience member might hear some reflected sound from the back of a recital hall or auditorium, acoustic instruments and voices largely projected their sound waves toward the face of the listener, not the back of their head. Even in the age of amplification, the basic idea of sound projecting toward the audience from the stage remained. There were some exceptions, such as a quadraphonic performance by Pink Floyd during the 1960s' psychedelic period, and the experimental compositions by avant-garde composers such as John Cage, but those were rarities.

Some readers might think of the quadraphonic experience as being akin to sitting in the middle of an orchestra; however, that is not an accurate description of the experience. Because the instruments generally project their sound outward toward the audience, and because the loudest instruments—the brass and percussion instruments—are placed in the back of an orchestra, sitting in the middle of an orchestra is an entirely different sonic experience than quadraphonic sound in which all the performers essentially surround the listener and project at them from different spots around the room.

Although the quadraphonic records and tapes and the quadraphonic playback systems that were marketed failed to become the standard for in-home audio systems in the 1970s, surround sound found a home and took a firm hold in the movie industry. Some of the earliest surround sound experiences were part of the IMAX films of the decade, including *Man Belongs to the Earth*, which surrounded the audience visually and with its soundtrack. This 1974 film, directed by Graeme Ferguson, featured a complex score by jazz trumpeter and composer Don Ellis and is a very good example of how surround sound can work when music is recorded, mixed, and mastered with a full 360-degree sound experience in mind from the get-go.

Surround sound was not limited to the proprietary IMAX films of the 1970s and beyond. Dolby Laboratories, the company behind the famous eponymous noise reduction system most frequently associated with cassette tapes, developed Dolby Surround Sound, which became a standard format for commercial motion pictures. Another surround-sound system DTS, perhaps most closely associated with the 1993 Stephen Spielberg film *Jurassic Park*, is a chief competitor with Dolby Surround Sound and Dolby Digital. The other name frequently associated with high-fidelity theatrical surround sound is THX, originally associated with filmmaker George Lucas. Unlike encoding systems such as Dolby Surround

Sound and DTS, THX is a performance standard for surround-sound systems. After becoming popular in movie theaters, surround systems entered the home theater marketplace, and today, surround-sound systems that meet the THX standards can be purchased for the home. Although home theater sound systems might not have the advantage of the number of discrete speakers and channels that are utilized in some movie theaters, industry standards such as THX help to ensure that home theater audio systems come as close as they can to replicating the theatrical experience.

See Also: 8-Track Tapes; Motion Picture Soundtracks; Vinyl Albums

Further Reading

Calore, Michael. 2009. "May 12, 1967: Pink Floyd Astounds with 'Sound in the Round.'" *Wired*, May 12. Accessed January 18, 2021. https://www.wired.com/2009/05/dayintech-0512/.

Morgan, Robert P. 1972. Liner notes for *Edgard Varèse: Offrandes, Intégrales, Octandre, Ecuatorial*. Reissued on CD as Elektra/Nonesuch 9 71269–2.

R

Radio

When considering the connections of radio and music, one must keep in mind that radio began its life as a system for transmitting information and voice communication, not as a music transmission system. When radio and music came together as a commercial enterprise in the 1920s, the transmission of music over the airwaves played a significant role in 20th-century culture, particularly in popular culture.

The transmission of information—albeit not music—electronically and over long distances had been accomplished well before the invention of radio by various means such as the telegraph and through the use of underwater cables. The transmission of information via the sub-Atlantic cables dated back to the 1860s. And the basic principles of the telegraph were even put to musical usage in the mid-1870s, when American inventor Elisha Gray received a patent for his "Electric Telegraph for Transmitting Musical Tones," an oscillator-based device that anticipated oscillator-based musical instruments such as the theremin and the ondes Martenot by a half-century. However, unlike the electronic instruments of the 1920s—which generally are acknowledged as the first electronic musical instruments—Gray's invention was more focused on transmitting tones over a telegraph wire over long distances than on creating a practical musical instrument that was meant to be heard where it was played. Telegraph technology reached the point that it was possible for U.S. president Theodore Roosevelt to send a telegram to himself that literally traveled all the way around the world in 1903.

Radio itself came out of Scottish scientist James Clerk Maxwell's theories about electromagnetic waves, as published in the 1860s. In the 1880s, German scientist Heinrich Hertz proved Maxwell's theories about the existence of electromagnetic waves that traveled through the air. Then, Italian scientist Guglielmo Marconi demonstrated his expansion of Hertz's work in London in the 1890s. Throughout the rest of the 1890s, Marconi's electromagnetic wave–based wireless transmissions continued to break new records, even as competitors such as Nicola Tesla set their sights on grander schemes involving massive antennas to transmit electromagnetic waves to greater distances. Marconi's wireless ship-to-shore telegram transmissions and the work of other individuals and companies continued to demonstrate that electromagnetic signals traveling through the air could be an effective way to transmit information. To the extent that we view music as a form of electronic information, all of these innovations paved the way for the commercial radio industry—with its focus on music—that followed.

In-home radio took off in 1919, when RCA started marketing home radio receivers. The radio industry at this time was one in which each component of the entire commercial enterprise quickly fed off the others: more radios led to the incorporation of more radio stations, which led to more manufacturers marketing radios, and so on.

The earliest commercial radio broadcast in the United States is reputed to have been made by Pittsburgh station KDKA on November 2, 1920. The station was established by Pittsburgh area ham radio operator Frank Conrad, who collaborated with Westinghouse in setting up the station. A leading manufacturer of radios, Westinghouse was one of the companies eager to capitalize on the synergy of radio stations and home radio sales. Within four years of historic Pittsburgh broadcast, there would be approximately 600 commercial radio stations in the United States.

In addition to radio stations, in the 1920s and first half of the 1930s, the number of radios in U.S. homes also grew at a rapid pace. As Katy June-Friesen wrote in the magazine of the National Endowment for the Humanities, "In 1922, there were 60,000 radio sets in use in the United States; one year later, there were 1.5 million. By 1935 [despite the economic devastation of the Great Depression], two thirds of all homes in the country had one" (2015).

By 1924, RCA's David Sarnoff advocated for superstations that would produce signals that reached considerably farther than those transmitted by standard radio stations. By this time, too, hundreds of these small stations had been established around the country. Although it would take some time for Sarnoff's vision to be at least partially fulfilled, we can think of his vision as the genesis or jumping-off point for later superstations, radio networks, and even today's nationwide satellite radio and internet radio stations, which are available around the globe.

Beginning in 1934, the Cincinnati, Ohio-based WLW was the United States' lone superstation, with an output of 500 kW for five years (June-Friesen 2015). This was at a time period during which all other U.S. commercial radio stations were limited to 50 kW, as WLW had been prior to the approval of owner Powel Crosley Jr.'s application to substantially increase the station's output power and as it was after the experiment in allowing high-output clear channel radio stations ended. Interestingly, Crosley was involved in several ventures, including the Crosley automobiles; however, most pertinent to his desire to have a high-output radio station was that he also led a company that made home radios.

The 1930s also saw the development of high-powered radio stations in northern Mexico, just over the border from the United States. Perhaps the most notable of these were XER and XERA, across the border from Texas. It was said that powerful stations, which was owned by the American John R. Brinkley, could be heard throughout Mexico and as far north as Canada, as could the other so-called border blasters of the 1930s. By the 1940s, the U.S. experiment had ended, and the United States and Mexico had signed agreements limiting the power of radio stations in the two countries.

During the period in which WLW and XERA were operating at the 500 kW level (1934–1939), these two stations in particular were significant in the dissemination throughout much of North America of American country music. WLW presented a program of live country music called *Midwestern Hayride*, and XERA featured live performances by what many people think of as the first superstar group in country music, the Carter Family. Solo artists such as Red Foley, Jimmie Rodgers, and others also performed on XERA, which helped to establish them as international stars. Although apparently a major component of the programs on XERA were advertisements for various elixirs and tonics, as Mark Zwonitzer and Charles Hirshberg wrote in their biography of the Carter Family, "The beauty of the whole Texas arrangement was that the Carters didn't have to do the hard sell. They just made their music and let the announcers make the pitch. But they were still the draw. In fact, they were the biggest draw" (2002, 216–217).

The concept of the mega station that could reach millions of listeners expanded during World War II when Crosley Broadcasting Corporation collaborated with the U.S. government to build and operate Bethany Station, which became known during the war and during the Cold War as Voice of America. In comparison to Crosley's earlier efforts as WLW, Voice of America was mammoth, originally requiring 3.5 million watts to transmit its signal around the world (The Editors of *Ohio History Central* n.d.).

As well-known Cleveland, Ohio, television broadcaster and raconteur Neil Zurcher wrote, powerful radio stations were not without problems "especially for people who lived near what was later called Bethany Station. Some folks with metal bridgework in their mouths started receiving WLW's radio programs. Others said it was difficult to sleep at night, since their bedsprings started talking" (Zurcher 2005, 104).

In addition to its significant transmitting power, WLW was also one of the original four radio stations in the Mutual Broadcasting System. The others were Chicago's WGN, Detroit's WXYZ, and New York's WOR. This and other radio networks such as the National Broadcasting Company (NBC) and the Columbia Broadcasting System (CBS), brought programming of shows such as *The Lone Ranger*, *Amos & Andy*, *The Shadow*, and *Tarzan* to the entire country. Significantly, jazz clarinetist and bandleader Benny Goodman's *Let's Dance* was also part of the nationwide live broadcasts of the mid-1930s. Goodman's program is credited with playing a major role in bringing big-band swing music into popularity.

Goodman was not by any means the only star to have a radio program in the first half of the 20th century. Perhaps one of the other most famous examples of how a radio program affected American culture came in the form of singer Kate Smith's debut and subsequent radio performances of the Irving Berlin composition "God Bless America." In the late 1930s, Smith had both a radio talk show, *Kate Smith Speaks*, and a music program, *The Kate Smith Hour*. With World War II on the horizon in November 1938, on her talk show, Smith built up anticipation about this new song of Irving Berlin—which, incidentally, was really an earlier Berlin song that simply had never been released—and how it provided an important message for the times. Smith performed "God Bless America" on her evening music program, and the song was an instant hit, with Smith becoming and remaining the singer most closely associated with it for the rest of her career. Such was the power of the connection between music and radio in the 1930s. The examples of the impact on the popular culture of the time of musical radio stars such as Benny Goodman and Kate Smith illustrate the importance of radio on American music in the 1930s, and the radio's role in developing nationwide strands of popular culture continues till today.

The AM radio band on which programs such as *Let's Dance* and *The Kate Smith Hour* were broadcast was dominant well into the 1960s; however, it was not the only way in which music could be transmitted using electromagnetic waves in the atmosphere. During the 1930s, inventor Edwin Armstrong worked on developing FM radio, which had shown much promise for clearer sound, more depth, ability to use far less power than AM radio, and for being able to have its signals rebroadcast by various stations within a network without relying on wired connections, as was the case with AM radio. Although there is not sufficient space to detail the travails of Armstrong, suffice it to say that the major manufacturers of radios and the major radio networks were not particularly warm to Armstrong's invention of FM radio. However, as Scott Woolley wrote in *The Network: The Battle for the Airwaves*, "The Federal Communications Commission had been slow to approve many of Armstrong's experiments, but the technology's advantages were too overwhelming to ignore. In 1940, the commissioners agreed to allocate a range of wavelengths . . . for FM radios. If FM's takeover of the commercial radio market had seemed likely before the Second World War, its performance on the battlefield eliminated any remaining doubt" (Woolley 2016, 127). This was not, however, the end of the story. After this allocation had been made, FM radios that could pick up these signals had been built, sold, and purchased, the Federal Communications Commission (FCC) changed the band of frequencies allocated for FM in 1945. As Woolley notes, "Overnight, the FCC's decision made $75 million worth of FM radios obsolete" (2016, 136). Legal wrangling and lawsuits continued into the 1950s.

FM radio as a medium for music really came into its own in the 1960s. Public radio stations broadcast classical music that sounded significantly fuller, richer, and true-to-life than what was possible on AM radio. Particularly important was the development of stereophonic FM broadcasting, something that would not be done in AM broadcasting for years. Commercial FM stations of the 1960s tended not to have the same degree of commercial constraints as did their AM cousins, and disc

Many popular home radios of the first half of the 20th century were large pieces of furniture, such as this example. (Wisconsinart/Dreamstime.com)

jockeys could play lengthier material with fewer interruptions by advertisements. As a result, lengthy album cuts (e.g., the original album version of the Doors' "Light My Fire," as opposed to the truncated single version in which the electric guitar solo was exorcised) or even entire sides of albums could be played. An entire album-oriented rock style of radio programming developed around this possibility.

The technological development of radio is not limited to the story of commercial radio stations. Of equal importance were the developments that occurred over the years, particularly in the 1920s, that took radios from large battery-driven devices that were best listened to with earphones, because of the limitations of the speaker technology of the day, to all-electric, plug-in devices with high-quality built-in speakers. Although there is not sufficient space to detail these developments, they played a significant role in establishing radio as a principal form of in-home musical entertainment in the 20th century.

Perhaps just as important as the history and development of commercial radio stations and in-home radio technology is the development of car radios. In fact, one could reasonably argue that the development of car radios played a critical role in the development of the post–World War II youth subculture, the teenage culture, of the 1950s and 1960s. From a technological standpoint, car radios presented numerous challenges for Bill Lear and Paul Galvin of Galvin Manufacturing Corporation (later called Motorola, Inc., named for the first car radio the company made) and other companies that were intent on linking the popularity of radio and the popularity of automotive transportation in the United States. Not the least of these challenges was the interference posed by the automotive electric system, with its spark plugs, generators, and so on. However, by the early 1930s, the engineering challenges were largely met, and the automotive radio, which was at one time an option, is still very much with us today.

John R. Brinkley

One of the most colorful individuals involved in the early 20th-century radio industry, John R. Brinkley, owned several radio stations, both in the Midwest and in Mexico just across the U.S.-Mexico border. Brinkley's powerful border stations—as well as those owned by others—played a major role in bringing early country music stars, including Jimmie Rodgers, the Carter Family, and others, to listeners throughout a large part of the United States, Mexico, and all the way into Canada.

Brinkley was also famous as a self-professed doctor (whose questionable credentials were from what would later be described as "diploma mills"). He used his Mexican and Midwestern radio stations to advertise various patent medicines and to tout a procedure for men suffering from impotence in which he "sewed goat testicular glands into a patient's scrotum" (Lapin 2016).

Not surprisingly, medical boards, federal radio regulators, and others tried to shut down Brinkley's medical and radio operations. Surprisingly, Brinkley's use of radio was so effective in bringing him fame throughout much of the United States that despite the controversies surrounding his work in pseudo-medicine, he was nearly elected governor of Kansas. There are several fascinating biographies of this colorful character available, including Lee (2002) and Brock (2008).

> **Further Reading**
>
> Lapin, Andrew. 2016. "The Bizarre History of a Bogus Doctor Who Prescribed Goat Gonads." *National Geographic*, July 15. Accessed May 19, 2020. https://www.nationalgeographic.com/news/2016/07/documentary-interview-medicine-science/#close.
>
> Lee, R. Alton. 2002. *The Bizarre Careers of John R. Brinkley*. Lexington: University Press of Kentucky.

See Also: Internet Radio; Pandora and the Music Genome Project; Satellite Radio; Transistor Radios

Further Reading

Berkowitz, Justin. 2010. "The History of Car Radios. Car Tunes: Life before Satellite Radio." *Car and Driver*, October 25. Accessed January 18, 2021. https://www.caranddriver.com/features/a15128476/the-history-of-car-radios/.

Brock, Pope. 2008. *Charlatan: America's Most Dangerous Huckster, the Man Who Pursued Him, and the Age of Flimflam*. New York: Three Rivers Press.

DeMauro, Thomas A. 2021. "1967 Cadillac Fleetwood Eldorado." *Hemmings Classic Car* 17, no. 6 (March): 36–40, 42.

The Editors of *Ohio History Central*. "Voice of America." *Ohio History Central*. Accessed January 18, 2021. https://ohiohistorycentral.org/w/Voice_of_America.

Gray, Elisha. 1875. "Electric Telegraph for Transmitting Musical Tones." U.S. Patent 166095 granted July 27. Accessed November 30, 2020. https://patents.google.com/patent/US166095.

June-Friesen, Katy. 2015. "For a Brief Time in the 1930s, Radio Station WLW in Ohio Became America's One and Only 'Super Station.'" *Humanities: The Magazine of the National Endowment for the Humanities* 36, no. 3 (May/June). Accessed January 18, 2021. https://www.neh.gov/humanities/2015/mayjune/feature/in-the-1930s-radio-station-wlw-in-ohio-was-americas-one-and-only-sup.

Robles, Sonia. 2019. *Mexican Waves: Radio Broadcasting along Mexico's Northern Border, 1930–1950*. Tucson: University of Arizona Press.

Smith, Kathleen E. R. 2003. *God Bless America: Tin Pan Alley Goes to War*. Lexington: University of Kentucky Press.

Woolley, Scott. 2016. *The Network: The Battle for the Airwaves and the Birth of the Communications Age*. New York: Ecco Press.

Zurcher, Neil. 2005. *Strange Tales from Ohio*. Cleveland, OH: Gray & Company.

Zwonitzer, Mark, with Charles Hirschberg. 2002. *Will You Miss Me When I'm Gone: The Carter Family and Their Legacy in American Music*. New York: Simon & Schuster.

Reel-to-Reel Tape

Back in the 1950s and 1960s, particularly before the audio fidelity of cassette tapes was being improved by means of new magnetic tape formulations and the use of the several Dolby noise reduction systems, the principal audiophile home recording and playback format was the reel-to-reel tape. Although recording on electromagnetic media went back to the late 19th century, at least in terms of theories and noncommercial music applications, recording on a ribbon medium that had been coated with an electromagnetic compound did not enter the commercial music recording industry until after World War II. When it did, however, it helped to make possible multi-track recording and all the stereo and quadraphonic effects and delineation of listening space that continue (albeit rarely using electromagnetic tape) today.

In the commercial recording industry, before the introduction of magnetic reel-to-reel tape, recordings were done directly onto a disc, from which master copies would be made in order to press the copies that were sold to consumers. Before the widespread use of electromagnetic tape, metal-based wire was used as a recording medium; however, wire recording had serious limitations and made little impact except as a historical step in the audio recording process and as part of some of the legends of mid-20th-century live recordings. The use of magnetic tape provided significant enhancements to the recording process, including making the recording of alternate takes of pieces easier and less expensive than had been the case during the direct-to-disc era. Once multi-track recording was developed, magnetic tape, still moving from one reel to another, was divided into separate tracks, which could be recorded individually and in which two tracks could be combined into one open track, freeing up the original two for more overdubs, and so on. In short, the reel-to-reel medium was well suited to multi-track recording, particularly as specialized recorders were made that could use increasingly wider tape as recording studios moved from 4-track to 8-, 16-, 32-, and even more-track recording from the late 1950s on.

Although the fact might largely be forgotten today, during the 1950s, 1960s, and 1970s, reel-to-reel tape was a preferred format for audiophiles. Reel-to-reel tape was absent the crackles, pops, and other surface noise associated with vinyl records. Today, we often think of this period as being the heyday of the 45-rpm single and the 33-1/3-rpm vinyl album; however, some albums were commercially released on reel-to-reel tape. And with a high-quality reel-to-reel system, even with a portable reel-to-reel recorder, concerts, rehearsals, and so on could be recorded, offering more flexibility and opportunities than owning an audio system entirely built around discs. Although cassettes and 8-track tapes eventually entered the home recording and playback market in the 1960s, the reel-to-reel tape format offered some clear advantages. Whether in the recording studio or even in the home, reel-to-reel tape players typically recorded and played at a faster speed than cassettes (1–7/8 ips) or 8-track tapes (3–3/4 ips), which led to reel-to-reel recordings typically featuring better audio fidelity than any of the compact formats in the pre-digital age. Because the entire setup involved magnetic tape and two reels, reel-to-reel recording and playback were less susceptible to the mechanical problems that could plague cartridge-based systems such as cassettes and 8-tracks.

A 1969 article in *Popular Science* magazine compared the mechanics and positive and negative attributes of the then three leading magnetic tape formats that were in use at the time: reel-to-reel, cassette, and cartridge. The magazine's test results for the various formats showed that reel-to-reel had the best audio fidelity

Because of the duration of recordings that could be made, the possibility of multi-track recording, and the fact that the audio fidelity of reel-to-reel tape could surpass that of cassette and 8-track tapes, reel-to-reel tape was popular among home recordists and audiophiles in the second half of the 20th century. On a larger scale, wider reel-to-reel formats were staples of professional recording studios during the same period. (Viktorus/Dreamstime.com)

and frequency response. However, compared with the compact cassette tape, which ran at a speed of 1–7/8 inches per second, the magazine's tests suggested, "Eight-track cartridge tape, with 3 3/4-i.p.s. speed, is a happy compromise for good frequency response with reasonably long recording time" (Shatavsky 1969, 129). Interestingly, reel-to-reel and cassettes, particularly after the various tape formulations and noise reduction systems for cassettes were developed, ended up outliving the 8-track tape.

In addition to better audio fidelity when run at 7–1/2 inches per second, many reel-to-reel tape recorders offered one other easy-to-overlook advantage over the other tape formats. Many home tape decks could play and record at either 3–3/4 ips or 7–1/2 ips. Using this knowledge, musicians could make a tape recording of, say, an album cut of one of their favorite musicians at 7–1/2 ips and play it back at half speed. Because the tape was moving exactly at half the recorded speed, the tempo would be halved, and the pitch of all the voices and/or instruments would be exactly an octave lower. This way, a saxophone player could learn an intricate Charlie Parker improvised solo at half speed. The ability to record and play back at different speeds also was used as a studio track, along with playing tapes backward and other effects, all of which are covered in this volume's entry on Tape Manipulation.

One of the challenges faced by users of all tape-based formats was breakage of the electromagnetic tape. Because home reel-to-reel tape typically was 1/4-inch wide while cassette tape was 1/8-inch wide, and because reel-to-reel tape was not housed in a case like cassettes and 8-tracks, it was easier to fix breaks by splicing the tape back together. The cassette and 8-track tapes of the 1960s and beyond were treated more as disposables than were reel-to-reel tapes.

When quadraphonic recordings enjoyed a brief period of popularity in the late 1960s and through the 1970s, reel-to-reel tape was a natural medium for the technology because tapes commonly were already divided into two or four tracks. However, quadraphonic sound did not make a huge commercial impact, and as the recording industry and home recording entered the digital age, there became less and less of a need for reel-to-reel tape from the 1980s to the present.

See Also: Cassette Tapes; 8-Track Tapes; Multi-Track Recording; Quadraphonic and Surround Sound; Tape Manipulation; Wire Recording

Further Reading

Daniel, Eric D., C. Denis Mee, and Mark H. Clark, eds. 1999. *Magnetic Recording: The First 100 Years*. New York: John Wiley & Sons/IEEE Press.

Murnane, Kevin. 2017. "'Sgt. Pepper's' Was a Perfect Storm of Musical and Recording Creativity." *Forbes*, June 3. Accessed January 18, 2021. https://www.forbes.com/sites/kevinmurnane/2017/06/03/sgt-peppers-was-a-perfect-storm-of-musical-and-recording-creativity.

National Museums Liverpool. n.d. "Emergence of Multi-Track Recording." Accessed August 4, 2020. https://www.liverpoolmuseums.org.uk/emergence-of-multitrack-recording.

Shatavsky, Sam. 1969. "The Best Tape System for You: Reel, Cassette, or Cartridge." *Popular Science* 194, no. 2 (February): 126–129.

Resonator Guitars

In the early 20th century, one of the obstacles faced by vaudeville musicians was the fact that before electronic amplification became widespread, musical performances, vocal and instrumental, were acoustic. Some instruments, such as the guitar, were particularly difficult to hear in large theaters. Vaudeville performer George D. Beauchamp was eager to solve this problem and approached violin maker and repairman John Dopyera to create a guitar that would be audible in large venues.

The system that Dopyera created used metal resonators to amplify the sound without the use of electricity. In 1927, Dopyera and Beauchamp formed the

National String Instrument Corporation to produce resonator guitars. By 1929, Dopyera had left National to form Dobro with four of his brothers. Coincidentally, the name of their company, Dobro, was short for Dopyera Brothers and also a word that meant "good" in the Slovak language; the Dopyera family had immigrated to the United States from Slovakia in 1908. Eventually, in the 1930s, John Dopyera dropped the *y* from his name, so some instruments that he designed and built are credited to John Dopera.

By the early 1930s, there were three principal types of metal resonators being used: the single-cone resonator, which had been patented by George D. Beauchamp after John Dopyera left National; the tri-cone, a John Dopyera invention often used in metal-bodied resonator guitars used primarily by blues musicians; and the spider-bridge resonator, patented by Rudolph Dopyera as part of Dobro. The spider-bridge system is that which is primarily used today in instruments generically referred to as "dobros" and used primarily in country and folk music. Interestingly, the Beauchamp patent was for a type of resonator cone that John Dopyera claimed he had invented, experimented with, and ruled out because it did not make for an instrument that was guitar-like enough for him. Because Beauchamp held the resonator patent, the Dopyera Brothers were forced to develop the competing spider-bridge resonator system in order to own resonator intellectual property for their company.

Eventually, resonator guitars came to be used primarily in blues and country music. Some famous blues musicians, such as Son House, used metal-bodied resonator guitars, often (but not exclusively) tuned to an open major chord. At the time

A 1920s invention, the resonator guitar was a solution to the problem of achieving adequate guitar volume in large venues before the development and widespread use of the electric guitar. (Andrea Leone/Dreamstime.com)

of this writing, one can find several YouTube videos of Son House later in life performing his famous song "Death Letter Blues" using a metal-bodied resonator guitar and playing and singing in a style similar to his early 20th-century style (see, e.g., https://www.youtube.com/watch?v=NdgrQoZHnNY, accessed January 15, 2021). For a video by a more recent blues artist associated with the metal-bodied resonator guitar, a backstage video of Keb Mo preparing for a White House performance makes for excellent viewing (see, https://www.youtube.com/watch?v=NZyt6ogwbcw, accessed January 15, 2021). While blues musicians such as Son House played round-neck resonators that they held Spanish style (with the strings and the resonator facing out), country musicians as early as the 1930s by and large followed the Hawai'ian guitar style of playing with the resonator and strings facing up. The instruments that players who held their guitars in their laps played customarily had thicker square necks, which made them easier to hold in that position. Today, these square-neck resonator guitars are often generically referred to as dobros (lowercase), although the Gibson company currently owns the trademark of the name Dobro for its line of resonator guitars.

In country music, the square-neck resonator guitar received special attention as part of the bluegrass music of Lester Flatt and Earl Scruggs and the Foggy Mountain Boys in the person of dobro player Josh Graves. The visibility of the square-neck resonator guitar was helped by the fact that in the 1950s and early 1960s, Flatt and Scruggs and their band had a television program that featured their style of bluegrass music. In some of the songs and instrumental pieces, Josh Graves would play dobro solo breaks, and viewers had the opportunity to both see and hear his work on the instrument. In addition to Graves, one of the most prominent early and mid-20th-century dobro players was Bashful Cousin Oswald, who played for many years with singer-fiddler Roy Acuff.

Although the use of resonator guitars in acoustic blues music is not as prevalent as it was in the 20th century, the dobro has continued to be a significant part of country music, particularly through the late 20th and early 21st-century concert and extensive recording work of dobro virtuoso Jerry Douglas. At the time of this writing, several videos of Douglas can be found on YouTube (see, for example, https://www.youtube.com/watch?v=3dtPc3wNxOg, accessed January 15, 2021). Although the late 20th and the early 21st centuries have seen a number of successful dobro players (e.g., Mike Aldridge, Cindy Cashdollar, Phil Leadbetter, Rob Ickes, and others), Douglas generally is acknowledged as perhaps the most technically impressive player of the instrument, as well as perhaps the most recorded dobro player of his time.

The success of Douglas and other dobro players in the country genre and the success of Keb Mo in acoustic blues have been part of a general late 20th- and early 21st-century resurgence of interest in American roots music. The prominence of these artists in their respective genres has provided a continuing place for the resonator guitar, a technological development that was borne of necessity during the vaudeville era. It is particularly interesting that the instrument has far outlived its life as a technical necessity, particularly in light of the fact that the developments that Adolph Rickenbacker and the very same Dopyera Brothers who played a seminal role in the development of the resonator guitar independently created versions of the electric guitar within just a handful of years after the development of the resonator-based acoustic instrument. The use of amplifiers and the magnetic coil pickups in the new electric guitars basically eliminated the need for the resonator guitar in many venues; however, arguably, its unique tone color and its association with bottleneck and steel bar playing—as well as the fact that it did not require electricity—kept the resonator guitar alive in rural settings where country music and acoustic blues remained popular genres beyond the 1920s. So, an acoustic-based technology born out of necessity found its lasting strength not in its volume—the principal reason for which it was developed—but for the unique tonal qualities that it possesses.

See Also: Pedal Steel Guitars

Further Reading

Beauchamp, George. 1929. Patent granted 1931. Patent application for "Stringed Musical Instrument." Accessed January 18, 2021. https://patents.google.com/patent/US1808756.

Dopyera, Rudolph. 1933. Patent application for "Musical Instrument." Accessed January 18, 2021. https://patents.google.com/patent/US1896484.

Reverb

What today is known as reverb is related to, but not exactly the same as what one might think of when thinking about reverberation in, say, an auditorium or other large acoustical space. This electronic effect made its way into guitar amplifiers and became one of the defining features of electric guitar tone in several popular genres, particularly in surf music. Artificial reverberation is also frequently used in recording studios and live performance settings to warm up the tone and improve the blend of the human voice.

During the 1950s, some recordings that incorporated artificial echo or reverberation usually revolved around the room in which music was recorded or was accomplished through a particular recording studio's contrivance. For example, the reverberation heard on some recordings by the Crickets and Buddy Holly and the Crickets (a variety of versions of the band's name were used) was a result of the signal from the band's microphones being fed into another room in which amplifiers in that room were recorded with additional microphones. Similarly, the so-called slap-back echo associated with rockabilly musicians recorded in the 1950s at Sun Studios in Memphis by the studio's owner Sam Phillips was accomplished by using two tape recorders, one of which received the microphone signals directly and the other of which fed its signal into the master tape recorder. The master recorder, then, receives the two signals with a slight delay in the second, creating the echo effect. At the time of this writing, there is a brief YouTube video recorded at Sun Records in which the procedure is demonstrated (see, https://www.youtube.com/watch?v=FuStmPbG528, accessed December 7, 2020).

As one can imagine, these echo/reverb-related effects were difficult to take out of the confines of the studio in no small part because of the almost Rube Goldberg manner in which they had been developed. At least that was true until an electronic way of creating the same effects was developed. In the world of the electric guitar, Leo Fender ranks as perhaps the most important proponent of electronic reverb. The Fender Reverb Unit of 1961 helped to bring this effect into the hands of guitarist who could use it inside and outside the recording studio. The stand-alone reverb unit was followed in 1963 by the Fender Twin Reverb and Deluxe Reverb amplifiers. As implied by their names, these amplifiers featured electronic reverberation, which made it an easy tool for guitarists to incorporate into their bags of tricks.

Fender's reverb unit and his new amplifiers quickly made an impact on popular music, particularly in surf music of the first half of the 1960s. Famed surf guitarist—and arguably the founder of the surf genre—Dick Dale made reverb an integral part of his sound, as did groups such as the Ventures, the Beach Boys, and countless others. Although interpreting the implications of musical tone colors can be tricky business, and probably varies considerably from person to person and culture to culture; however, the openness of the electric guitar's sound when treated with reverb and perhaps a touch of tremolo can be interpreted as reflective of the expanse of the ocean and the shimmering waves upon which one would surf.

One of the important things to note about reverb as used in genres such as surf music and in other electric-guitar-based genres is that it often defines the sound of the lead guitar. To put it another way, it is used as part of a particular instrument's overall sound, regardless of whether or not other instruments, such as the electric bass, drums, and rhythm guitar, have a dry sound. This stands in sharp contrast to our understanding of reverberation as a natural part of the sound of, say, a cathedral or a large auditorium. So, generally the intent of a reverb setting on a guitar amplifier is not to mimic or emulate the natural acoustics of a large performance space. If it were, then the drums, rhythm guitar, keyboards, bass, and so on would be treated with the same type and degree of reverberation.

Artificial reverb applied to the human voice also mellows the sound of the voice. The addition of reverb to vocal tracks in the recording studio is quite common, particularly because it mimics the natural resonance of the voice in a performance space such as auditorium, church, and recital hall. With the coming of home studios, cassette-tape-based multi-track recording equipment, and now studio programs, studio apps, and audio-editing software, reverb is available for professional and amateur musicians alike. And the effect is available even with free multi-track recording software, such as *Audacity*. One of the things that many online tutorials for using vocal reverb rightly warn about is the amount of the effect that the engineer uses. Too much reverb can make the clarity of the vocals suffer, particularly with regard to the crispness of enunciation. Reverb is clearly an electronic effect, but it is considerably less invasive than, say, such vocal manipulations as running a singer's voice through a revolving Leslie speaker, or a more recent electronic effect such as *Auto-Tune*.

See Also: Amplification; Effects Pedals; Virtual Studio Software

Further Reading
Teagle, John, and John Sprung. 1995. *Fender Amps: The First Fifty Years*. Milwaukee, WI: Hal Leonard.

S

Sampling

The earliest electronic synthesizers were developed around the premise of manipulating electronic oscillators to change wave forms, which in turn would change the timbre—or tone color—of the sound. One could analyze the tonal spectrum and overtone structure of an instrument—say, the clarinet—and program an analog synthesizer so that the same wave forms—and thereby the same tone color—would be produced. One of the challenges of imitating the sound of acoustic instruments in this way is the fact that the overtone structure of an instrument such as the clarinet is not exactly the same on every pitch or in every register. As a result, one might take a representative note or range of notes, analyze the wave forms and overtones and so on, and capture at least part of the feel of the acoustic instrument but not the subtle tonal changes between registers and so on. Similarly, sampling an instrument that customarily plays with vibrato (e.g., flute and member of the bowed orchestral string family) could result in an unnatural sameness to the vibrato produced on the sampling keyboard—the same on every note, as opposed to an expressive, changing part of the acoustic instrument's tone.

Sampling, whether analog or digital, made it easier to imitate the sound of acoustic instruments; however, unless a variety of pitches on, say, a clarinet, were sampled, the subtle changes of overtone balance and pitch tendencies of all the individual notes would still be absent from the sampled sounds on the synthesizer. Computer-based sampling came into the mainstream in the music industry with the release of the Fairlight CMI (computer musical instrument) in the late 1970s. Over the years, the Fairlight's inventor, Peter Vogel, participated in several television programs, interviews, and demonstrations not only of the sampling capabilities of the Fairlight CMI but also of the graphic display–based way in which the user could shape, manipulate, and design wave forms (see, e.g., https://www.youtube.com/watch?v=yldcZ65PmVU, accessed January 15, 2021; https://www.youtube.com/watch?v=afzw6B5apxk, accessed January 15, 2021; and https://www.youtube.com/watch?v=iOlPCpSmhRM, accessed January 15, 2021). Perhaps one of the best-known recordings to make use of the Fairlight CMI's sampling capabilities is Thomas Dolby's song "She Blinded Me with Science." In this song, Dolby sampled the well-known British scientist Magnus Pyke exclaiming the word "Science!" and then inserted Pyke's voice at numerous places in the recording.

Other electronics firms followed suit with sampling keyboards and keyboards that had built-in digital samples. One of the downsides of the emphasis that came to be placed on sampling and on the presets that dominated digital synthesizers is that some of the creativity that was available—in fact, required—in programming, mixing and modifying wave forms, and so on in analog additive and subtractive synthesis was lost.

Sampling keyboards allowed one essentially to take a digital sonic snapshot of a sound and use it. However, it was not just acoustic instruments that were and remain part of the sampling world. In the hip-hop genre, the DJ techniques of pioneers such as DJ Kool Herc, in which instrumental groove sections of existing recordings were looped and eventually used to create new songs, evolved into sampling. The sampling of bass lines, percussion breaks, horn section figures, and so on quickly became widespread. In fact, the first highly commercially success rap single, the 1979 release "Rapper's Delight," by the Sugarhill Gang, finds the three rappers performing over an instrumental groove interpolated (e.g., re-recorded; not sampled) Chic's song "Good Times." In large part, interpolation would give way to sampling, with rare exceptions such as in the work of the Roots, a band that specializes in live performances that incorporate the effects and tone colors of sampled music but that is played by members of the group. The prominence of sampling and using the samples to create new hip-hop arrangements and songs is evidenced by the large number of songwriter credits that one can find attached to numerous recordings in the genre. Interestingly, the fact that so many recordings today incorporate samples is confirmed by the fact that the American Society of Composers, Authors and Publishers (ASCAP) has placed on the fillable web form for its members to register their compositions a question about what, if any, samples are included in the piece a member is registering, as the writers of any sampled recordings receive at least partial writer's credit for the composite recording.

Samples can be captured in any number of ways, including using sampling keyboards/synthesizers, using professional and even relatively inexpensive studio/virtual studio software. Some of the common samples include snippets of a distinctive bass line or drum break from an already-existing recording. Sometimes sampling is combined with other techniques, such as looping. So, a brief drum part from a well-known song might be sampled and then looped to form the basis of the percussion part in the new composition. Horn section chords, guitar parts, a backing vocal ensemble chord, and other pieces from pre-existing materials have been used, sometimes as a clear homage to the original and sometimes more stealthily such that it is difficult to detect the source of the material or even that some of the material is derived from a pre-existing recording.

Throughout the hip-hop era, sampling has led to some unlikely pairings and later collaborations between artists whose work would not ordinarily be thought of as compatible. For example, Coolio's 1995 hit song "Gangsta's Paradise" samples Stevie Wonder's 1976 Songs in the *Key of Life* track "Pastime Paradise." Although there are lyrical connections between the two songs, it is mostly the musical hook that connects the two songs. It is interesting to note that this is one of the rare Coolio raps that does not contain profanity. According to Coolio, when Wonder was approached about allowing "Pastime Paradise" to be sampled for a song about contemporary urban gangsters he was reticent but eventually approved the project, "his only stipulation was that I had to take the curse words out"

(Epstein 2015). So, in this case, the writer and performer of the source material were able to exercise a degree of control over how their material was used.

Another intriguing connection of sampled and new material can be found in Puff Daddy's 1997 sampling of Sting's 1983 composition "Every Breath You Take" for the rap hit "I'll Be Missing You." Particularly interesting in this case is how the Sting song, with which Sting's group, the Police, enjoyed a no. 1 single in the United States, is turned from a song that comes from the viewpoint of a stalker at the end of a relationship into a rap about a departed colleague rapper and friend. Interestingly, in the case of both "Gangsta's Paradise" and "I'll Be Missing You," the artists associated with the source material, Stevie Wonder and Sting, later made live appearances with the rappers who incorporated the source material.

Sylvia Robinson

When one thinks of leading figures in music technology, one might not immediately think of Sylvia Robinson. Robinson first came to fame in the music industry as Sylvia Vanterpool as half of the 1950s' duo Mickey & Sylvia. Mickey & Sylvia's 1956 recording of "Love Is Strange" was a hit on the R&B and pop charts and had appeared in a variety of movies, television shows, and television commercials over the ensuing years. Robinson also enjoyed a solo hit with "Pillow Talk" in 1973. Robinson later founded and was CEO of Sugarhill Records, and it is in this capacity that she played a significant role in music technology. She is credited with being the driving force between two of the earliest significant hits in the rap genre: the Sugarhill Gang's "Rapper's Delight" and Grandmaster Flash and the Furious Five's "The Message." These records helped to bring the DJ turntable techniques, competitive rapping, sampling, and loops of hip-hop, a street genre born in the South Bronx, to a national audience as a commercially appealing popular music form.

See Also: Digital Synthesizers; Loops

Further Reading

Epstein, Dan. 2015. "Coolio's 'Gangsta's Paradise': The Oral History of 1995's Pop-Rap Smash." *Rolling Stone*, August 7. Accessed December 14, 2020. https://www.rollingstone.com/music/music-news/coolios-gangstas-paradise-the-oral-history-of-1995s-pop-rap-smash-50357/.

Schloss, Joseph. 2012. "Sampling Ethics." In Murray Forman and Mark A. Neal, eds., *That's the Joint!: The Hip-Hop Studies Reader*, 2nd ed., 610–630. New York: Routledge.

Vogel, Peter, and James Gardner. 2012. "Interview: Peter Vogel." Radio New Zealand (RNZ), May. Accessed January 18, 2021. https://www.rnz.co.nz/concert/programmes/hopefulmachines/20131119. This is an update and correction of an interview transcript originally published on May 8, 2010. The exact May 2012 date of republication is not given.

Satellite Radio

A 21st-century technology, satellite radio, began with the launch of Sirius and XM satellites in 2000 and 2001. XM began satellite radio broadcasts in 2001, and

Sirius followed a year later. Today, services such as Sirius and XM are virtually inescapable, so dominant has satellite radio become, and the two services that started the satellite radio phenomenon as business rivals are now joined together as SiriusXM.

One of the basic premises of satellite radio that at least in part hearkens back to the commercial radio networks of the early 20th century is that one can be anywhere in the country and can have access to radio broadcasting of a known quality. However, with their emphasis on very specific programming and with a large number of choices of focused channels (e.g., a blues channel, a comedy channel, a 1980s' new wave rock channel, and a 1960s' soul channel), satellite radio services such as Sirius and XM provided more consistency of programming than did any of the broadcast radio networks of the past.

The fact that the signals for Sirius and XM came from several satellites in orbit above Earth also meant that—perhaps short of traveling through mountain tunnels—one could drive in their car and listen to the same type of programming on the same channel without ever having to change the tuner in their radio, for literally hundreds and hundreds and hundreds of miles. Satellite radio services also offer their subscribers access to internet-based radio, with the same channel offerings that are available with the satellites. So, although the principal benefit of satellite radio might seem to be in cars, trucks, and other similar transportation devices, users can access the services at home or at work.

Satellite radio services such as Sirius and XM use a subscription model. Given the focus of these services on transportation, it has been common throughout the 21st century for companies such as General Motors, Ford, Hyundai, Toyota, MINI, and so on to offer free three-month trial subscriptions for purchasers of vehicles that are outfitted with an XM or Sirius satellite radio. To get a sense of how many of these trial subscriptions are active, consider that according to SiriusXM's 2019 earnings report, at the time the report was published, the company had approximately 30 million self-paying subscribers, "plus another 4.9 million not paid by the listener (such as the familiar three-months-for-free SiriusXM deal for car buyers)" (Peoples 2020). The combined Sirius and XM company grossed approximately $7.8 billion in revenue for 2019 and added just over 1 million new self-paying subscribers during the year (Peoples 2020). When considering the impact of the company—aside from the satellite radio business—it should also be noted that SiriusXM also owns the Pandora streaming audio service. And although "it is no secret that Pandora has been losing listeners due to rising industry competition over recent years" (Zarczynski 2020), Pandora still has a larger market share than Spotify, its closest competitor in the streaming audio field.

Some commentators, including Peoples (2020), have noted that satellite radio presents something of an anomaly or paradox in the subscription service sphere. Although this technology competes against free (but with commercial advertisements) streaming audio services such as Pandora, satellite radio continues to increase its listenership and generate significant revenue. And this is not because of how prevalent satellite radio is in the automotive, SUV, and truck markets, as many of today's new vehicles offer *Apple CarPlay* or *Android Auto*, which provide access to streaming audio services. Anecdotally, it seems that it is also because the convenience factor of receiving a reliable and consistent style of programming wherever one might travel and the fact that the specialized channels on satellite radio stations have their own brand of DJs, thereby linking the satellite radio experience to the local AM and FM experience.

See Also: Internet Radio; Radio; Streaming Audio Services

Further Reading

Peoples, Glenn. 2020. "How SiriusXM Is Beating the Subscriber Paradox & Scoring a Better Return Than Streaming Services." *Billboard*, February 7. Accessed December 14, 2020. https://www.billboard.com/articles/business/8550404/siriusxm-listener-subscriber-growth-data-analysis-streaming-radio/.

Zarczynski, Andrea. 2020. "Record High 34.9 Million Paid Subscribers Marks SiriusXM Milestone Year." *Forbes*, January 7. Accessed December 14, 2020. https://www.forbes.com/sites/andreazarczynski/2020/01/07/record-high-349-million-paid-subscribers-marks-siriusxm-milestone-year.

Sequencers

The basic definition of a sequencer is a device with which the user can record, play back, and/or edit digital music. The early sequencers were capable of an extremely limited number of notes that could be programmed and played back. However, all that changed in 1977 with the introduction of the Roland MC-8 MicroComposer. This synthesizer was capable of storing a sequence of 5,200 notes.

With the increased power of the Roland MC-8 MicroComposer and later sequencers, entire sections of pieces could be pre-programmed into synthesizers and activated at the touch of a button or key. This was especially helpful for musicians who performed live concerts and wanted to re-create the sound of their studio recordings but absent the presence of string sections, brass sections, or other studio musicians. In considering how sequencers could be and were used in the late 20th century—and how sequencers and sequencing software are used today—it is important to keep in mind that the stored sequence of notes—5,200 in the case of the Roland MicroComposer—could use any of the synthesizer's sounds or voices. In the case of polyphonic synthesizers with sequencing capabilities, this means that the sequence—the recorded, stored, and played back music—might represent an entire brass section, with different tone colors for the synthesized trumpets, trombones, and so on.

Because a synthesizer player could activate the sequence with a touch of a key or button, they could do so and then go about the business of playing live, nonsequenced figures perhaps on a second keyboard, or activating addition sequences on another keyboard as needed for the piece. Throughout the 1980s, sequencers became increasingly common but not without some controversy. The ability to pre-program synthesized passages note-by-note, even by a musician with limited technical skills, made the musicianship of artists who relied heavily on synthesizers suspect for some members of the public. Hence, the synthesizer-based group the Human League included the sentence, "There are no sequencers on this record," in the liner notes of their 1986 album *Crash* (Jam and Lewis 1986).

The virtual studio, digital audio workstation (DAW) software that began to appear in the late 1990s and early 2000s, such as Sonic Foundry's *Acid* and *Ableton Live*, provided users the ability to create, sequence, and play back musical passages and loops without having to use a sequencing keyboard. Software such as this led to new virtual studio software, including such well-known names as *GarageBand*, but these early titles were also important irrespective of the software that followed. Early 21st-century DJs, in particular, made extensive use of the looping and sequencing capabilities of *Ableton Live*. Some of the early sequencing software—perhaps most notably *Acid*—have continued to evolve over the years, both in terms of corporate ownership and in terms of technical capabilities.

It is important to note that what at one time in sound synthesis were discrete operations, for example, playing tones on a keyboard, manipulating wave forms, sequencing, sampling, and so on, increasingly came together in the keyboard-based synthesizers of the last couple decades of the 20th century through the present, as well as in the digital audio workstations that began to appear around the start of the new millennium. As a result, today, we might not think of sequencing as the kind of separate operation that it once was—it is fully integrated into the everyday operations of keyboard-based synthesizers and virtual studio software.

See Also: Digital Synthesizers; DJ-ing; Virtual Studio Software

Further Reading

Chang, Jeff. 2005. *Can't Stop Won't Stop: A History of the Hip-Hop Culture*. New York: Picador.

DJ Booma. 2017. *How to Be a DJ in 10 Easy Lessons: Learn to Spin, Scratch and Produce Your Own Mixes*. London: QED Publishing.

Jam, Jimmy, and Terry Lewis, producers. 1986. Liner notes for *Crash* by the Human League, LP, A&M Records SP-5129.

McKee, Ruth, and Jamie Grierson. 2017. "Roland Founder and Music Pioneer Ikutaro Kakehashi Dies Aged 87." *Guardian*, April 2. Accessed January 18, 2021. https://www.theguardian.com/music/2017/apr/02/roland-founder-and-music-pioneer-ikutaro-kakehashi-dies-aged-87.

Rushent, Martin, and the Human League, producers. 1981. Liner notes for *Dare* by the Human League, LP, A&M Records SP-4892.

78-rpm Records

Although the battle between Thomas Edison's wax cylinders and the flat shellac discs of Emile Berliner was resolved in favor of the flat disc in the early 20th century, there were relatively few standards at the time. In fact, some of the corporations that manufactured both record players and records seemed to go out of their way in developing proprietary standards for their own companies. For example, Edison, once the company began manufacturing flat discs and flat-disc record players, made their discs significantly thicker than those of other companies. The way in which the grooves were cut in Edison and Victor discs also fundamentally differed, so much so that one could ruin an Edison disc by playing it on a Victor machine. Various other companies made records that spun at various speeds for their proprietary record players.

Eventually, like the Sony Corporation's Betamax would do in the world of the video cassette several generations later, the various propriety speeds, thicknesses, and so on fell by the wayside. Until approximately 1950, the standard record would be 10 inches in diameter and would be played at 78 revolutions per minute. The discs would be made of shellac and, with rare exceptions, would be black in color.

The 78-rpm discs typically had one selection on one side and another on the other side, with the exception of releases of some classical works in which lengthier pieces might be split between sides or even between multiple discs. The album as we know it was born during the era of the 78s when multiple singles would be collected in a photo album–like package. In fact, the photo album–like packaging led to the collection being known as an "album."

The 78-rpm shellac discs were meant to be played using steel needles. One of the drawbacks to the steel needle stylus is that each one was only good for a very limited number of players until it was too dull to work properly. Interestingly, the stylus that Thomas Edison had used for his wax cylinder players could last for a lifetime, because the softness of the wax with which the cylinders were made caused very little wear to the stylus.

Early sound recordings were made using an acoustical process. Music was played into a device with a large horn. The sound waves were transferred to a needle that moved back and forth as the disc (or cylinder, as Edison's cylinders were recording similarly) rotated. Through this process, a master disc was created. The company would then create negative copies of the master that were used to stamp out the copies that were sold at retail. In order to boost production, record companies would sometimes set up multiple disc-recording machines for a single performance so that multiple master discs could be made. The acoustical process had a number of drawbacks. For one thing, subtle dynamic changes and quiet volume levels did not reproduce well. As a result, louder instruments tended to record better. A detailed explanation of the entire acoustical recording process can be found in several sources, including Steffen (2005). After the development of microphones, a new electrical process was developed in the middle of the 1920s and became the norm, thus allowing instruments such as the double bass to be heard in jazz recording. In the past, recorded bass lines would have had to be played by a tuba or sousaphone. The electrical process not only made it easier to record quieter instruments but it also made it possible to produce discs that featured a wider dynamic (volume) range. Despite this change in the recording process, it should be noted that until recording on magnetic tape became the norm after World War II, recordings continued to be onto disc, with master discs being made to stamp out the records that consumers purchased.

One of the challenges presented by World War II was a shortage of material for making records. The record industry was also hampered in the early 1940s by a recording ban instituted by the American Federation of Musicians to protest the increasing use of phonograph records—as opposed to live studio musicians—on radio. These two factors meant that what records were produced tended to be from recordings made before the recording ban. Interestingly, because of these factors, some of the innovative new music that was making an impact on the live music scene in the first half of the 1940s—most notably bebop jazz—was not released commercially on disc until several years after the music first created a sensation in clubs.

Interestingly, as suggested in the title of one of the chapters in Kathleen E. R. Smith's study of Tin Pan Alley popular music during the World War II era, *God Bless America: Tin Pan Alley Goes to War*, "Even Stale Music Sells Like Nylons," record companies issuing music that had already been recorded before the recording ban did not curb Americans' thirst for records (Smith 2003).

One notable exception to the black 78-rpm singles came in the form of picture discs. There were examples of picture discs produced during the 1930s; however, these were not by any means common, reputedly were of inferior sound quality, nor did they sell well because of the economic hardships of the Great Depression. After World War II, the most famous of the 78-rpm picture discs came in the form of the elaborate picture discs released by Vogue Records. Vogue's discs used an aluminum core covered with a new, nonshellac material that resembled the vinyl that would finally become the new standard for records in the 1950s. Vogue only stayed in business until 1947; however, the company's use of new materials, sound quality that far exceeded that of any previous picture-disc 78-rpm records, and the elaborate photographs and drawings that graced the Vogue discs marked them as the culmination of the development of the 78-rpm disc.

The release of the first vinyl 33–1/3-rpm albums and 45-rpm singles right around 1950 signaled a new age in the record industry. However, the infrastructure for 78-rpm discs was such that 78s would continue to be produced into the 1950s. Record players and turntables designed to play the new discs would continue to also be able to spin at 78 rpm in order to play the shellac discs. Some of the new record players of the 1950s and 1960s had two separate styluses—one for vinyl/plastic records and one for shellac 78-rpm records—that could be accessed by turning the end of the tonearm. This is particularly important to note because damage to stylus and record can occur from using the steel needle intended for use on 78s with vinyl discs and from using a needle designed for use on vinyl on the shellac discs. In addition, 78-rpm discs and spring-wound record players would continue to be used in locations around the world in which electricity was scarce or not used (e.g., remote locations in India) for years and years after it appeared in the United States that the era of the 78-rpm single had ended.

See Also: Electrical Recording

Further Reading

The Association of Vogue Picture Disc Collectors. n.d. "What Is a Vogue Picture Record?" Accessed June 9, 2021. http://www.voguepicturerecords.org/records.html.

Coleman, Mark. 2004. *Playback: From the Victrola to MP3, 100 Years of Music, Machines, and Money*. Cambridge, MA: Da Capo Press.

The Library of Congress. n.d. "Emile Berliner and the Birth of the Recording Industry." The Library of Congress. Accessed March 30, 2020. https://www.loc.gov/collections/emile-berliner/about-this-collection/.

Smith, Kathleen E. R. 2003. *God Bless America: Tin Pan Alley Goes to War*. Lexington: University of Kentucky Press.

Steffen, David J. 2005. *From Edison to Marconi: The First Thirty Years of Recorded Music*. Jefferson, NC: McFarland.

Sheet Music

Although one might not ordinarily think of printed sheet music as an example of technology, the printing of music ties in directly with how music was used by professional and amateur musicians, the genres that made impact on society at different points in history, and much more. Because we deal with the printing of music using movable type in another entry, our focus in this section is on sheet music and the publishing industry, regardless of whether the music was printed from handwritten original, typeset, photocopied, or printed using computer-generated notation.

Musicologists believe that musical notation began its development in Europe primarily to play a descriptive role. That is, early European music notation—the root of modern Western notation—developed in religious communities apparently as a way of reminding individuals who already had learned the monophonic chants by rote of how the tunes went, something that was important as the chant repertoire grew. As additional detail was added over many years, notation could play a prescriptive role: as long as one understood what the notation symbolized, one could sing a melody from the notation having never actually heard the melody before.

This added detail and the prescriptive nature of the evolving notation led to the development of a publishing industry, especially as learned individuals who were not part of a clergy order learned to read notation. So, by the Renaissance, collections of madrigals were published for home musicmaking. As detailed in this volume's entry on Music Printing with Movable Type, some of the earliest secular publications that used this new technology were madrigal collections of this sort.

Publishing played a role in the dissemination of the European art music of the Baroque, Classical, and Romantic periods; however, the publishing industry was in a state of flux, and there essentially was no such thing as copyright laws until the end of the 18th century. The sale of sheet music in the late 18th and early 19th centuries was lucrative for composers and publishers. In fact, sales of sheet music were so lucrative that famed composer Ludwig van Beethoven was known to have sold some of his works to multiple unknowing publishers as exclusives. However, by the 1840s, the publishing industry had changed. As economic historian Stafan Albinsson writes, "The paradigm shift in politics, economics and music which occurred at the end of the eighteenth century made it possible for [Robert] Schumann to make a living predominantly from what he earned from publishers' fees" (2012, 265).

If the art music composer of the mid-19th century could make a living through music sales, it was nothing like what would happen with the development of a distinct popular culture in countries such as the United States in the second half of the 19th century. In his book *Doo-Dah! Stephen Foster and the Rise of American Popular Culture*, Foster's biographer Ken Emerson makes a compelling argument that the September 11, 1847, premiere of Foster's song "Oh! Susanna," in Pittsburgh, Pennsylvania, "is a firm date for the birth of pop music as we still recognize it today" (Emerson 1998, 127). Although minstrel songs such as this and the parlor ballads of Foster and his contemporaries were heard by audiences in shows, concerts, and the like, they also sold by the thousands in sheet music publications.

Despite some of the headway that had been gained in songwriters' rights under copyright law, some musicians of the mid-19th century, including Foster himself, sold their songs outright for far less money than what they might have collected in royalties. Still, songs such as "Oh! Susanna" ushered in a dramatic increase in the size and scope of the publishing industry in the United States.

Arguably, the high point of the sheet music publishing industry came during the Tin Pan Alley era, roughly the 1880s through the 1940s. As an example of the impact of sheet music publication and distribution, one of the most noteworthy Tin Pan Alley successes was Charles K. Harris's 1893 song "After the Ball," which quickly sold a million copies and eventually was reputed to have sold over 5 million copies in sheet music form.

The Tin Pan Alley publishing industry was centered in New York City, where publishers had the opportunity to work with musical theater producers to have new songs plugged into shows. This was an important way to turn a song into a hit, particularly before commercial radio and the recording industry getting into full swing by the 1920s. Tin Pan Alley built an established corporate structure/technology that both served it well and helped to prepare a later generation of successful songwriters. Specifically, a typical Tin Pan Alley publishing company employed staff songwriters who penned many of the hits of the day. Some lower-tier employees of a publisher—individuals who had musical talent but generally were just breaking into the industry—worked as song pluggers. The song pluggers might play the piano and sing some of the publisher's latest songs at music stores, for theatrical producers, for well-known Broadway singing stars, and so on. Some successful song pluggers who also demonstrated a talent for writing lyrics or music might then become staff songwriters. Well-known star songwriters such as George Gershwin and Irving Berlin got their start as song pluggers at New York

From the Tin Pan Alley era of the late 19th century to the middle of the 20th century, sheet music sales were a particularly important part of the popular music industry. Many sheet music covers used artwork to suggest the mood of songs and to attract buyers. "Hilo Bay" was published by Forster Music Publisher in 1935. (David Pillow/Dreamstime.com)

publishing companies. The song plugger-to-staff writer paradigm established in the early years of the Tin Pan Alley era continued in the publishing and recording industry of the 1950s and early 1960s, as song pluggers morphed into musicians who recorded demo records that were produced to interest recording executives, popular recording artists, and so on. Later well-known singer-songwriters such as Carole King and Paul Simon got their professional starts as teenagers making demo records for a New York publisher.

Despite the prominence of New York's Tin Pan Alley in the sheet music industry of the late 19th century until approximately the middle of the 20th century, there were other noteworthy music publishing centers around the United States. However, even expanding our scope to include cities such as Los Angeles, Chicago, and Cincinnati does not really provide a full sense of the importance of the simple technology of publishing, printing, distributing, and selling sheet music to the music industry of that time period. Particularly during the first couple of decades of the 20th century, the city directories of even some small cities and towns might include one or more music publishers.

The widespread adoption of radio, records, and television led to a decrease in the need for sheet music for home musicmaking; however, publishers continued to adopt the technologies of changing times, including using new types of duplicating machines, computer-generated notation, and turning to publication of pdf files and even MIDI-based files. So, although the sheet music industry is not as large or pervasive as it once was, it lives on.

> *The Lightning Arranger*
>
> Not all the technological developments over the years have been earth shattering, nor have all caught on to the same extent. An example of a minor development that it is likely to have flown under the radar of the general public, performing musicians, and even composers, orchestrators, and arrangers is the Lightning Arranger. Patented by Samuel M. "Skip" Rapaport of the Allentown, Pennsylvania, area in 1931, the Lightning Arranger and its accompanying booklet, *The Lightning Instructor*, were advertised by the Lightning Arranger Company as "music's most perfect slide rule" (The Lightning Arranger Company 1931). The device itself consists of two wheels that, turned to various positions, show the spellings of all 12 major triads, all 12 minor triads, all fully diminished seventh chords, all major-minor seventh chords, all minor-minor seventh chords, and all dominant ninth chords. In addition, the device shows the spellings for all these chords for E-flat and B-flat transposing instruments.
>
> The accompanying booklet explains musical arranging principles (e.g., which instruments typically play the melody), as well as how to use the "musical slide rule" to determine which instruments should play which chord tones as part of the accompaniment. Although it is possible to use the device to create a workable musical arrangement, and its sales of "between 100,000 and 200,000 arrangers" between 1931 and 1984 (Schaffer 1984) suggest that it had some appeal to orchestrators and arrangers, the device has some limitations. Among these are the fact that it does not account for the French horn, a commonly used instrument in concert bands and orchestras, which is a transposing instrument in the key of F; the instructions in the booklet are focused on arranging for small ensembles that include instruments such as the banjo, an instrument commonly used in Dixieland bands of

the time period in which the Lightning Arranger was patented, but certainly not in widespread use outside of Bluegrass and folk music by the 1940s. The booklet also suffers some credibility from the fact that saxophone is misspelled throughout as "saxaphone." Such common arranging techniques as the creation of countermelodies and even how to deal with rhythmic figures in the accompaniment are not addressed. However, the Lightning Arranger is an example of an under-the-radar music-related technology that enjoyed continuous availability for decades.

Further Reading

The Lightning Arranger Company. 1931. *The Lightning Instructor for The Lightning Arranger.* Allentown, PA: Lightning Arranger Company.

Schaffer, Stan. 1984. "Math Music Add up for Whitehall Inventor." *Morning Call*, March 22. Accessed January 18, 2021. https://www.mcall.com/news/mc-xpm-1984-03-22-2400593-story.html.

See Also: MIDI; Music Notation Software; Music Printing with Movable Type

Further Reading

Albinsson, Staffan. 2012. "Early Music Copyrights: Did They Matter for Beethoven and Schumann?" *International Review of the Aesthetics and Sociology of Music* 43, no. 2 (December): 265–302.

Chang, Jeff. 2005. *Can't Stop Won't Stop: A History of the Hip-Hop Culture*. New York: Picador.

DJ Booma. 2017. *How to Be a DJ in 10 Easy Lessons: Learn to Spin, Scratch and Produce Your Own Mixes*. London: QED Publishing.

Emerson, Ken. 1998. *Doo-Dah! Stephen Foster and the Rise of American Popular Culture*. New York: Da Capo.

Jam, Jimmy, and Terry Lewis, producers. 1986. Liner notes for *Crash* by the Human League, LP, A&M Records SP-5129.

McKee, Ruth, and Jamie Grierson. 2017. "Roland Founder and Music Pioneer Ikutaro Kakehashi Dies Aged 87." *Guardian*, April 2. Accessed January 18, 2021. https://www.theguardian.com/music/2017/apr/02/roland-founder-and-music-pioneer-ikutaro-kakehashi-dies-aged-87.

Rushent, Martin, and the Human League, producers. 1981. Liner notes for *Dare* by the Human League, LP, A&M Records SP-4892.

Smartphones

In the spoken introductions to the bonus tracks on the "D.L.X.," or deluxe, edition of the 2014 album *1989*, Taylor Swift discussed—among other things—her use of the *Voice Memos* app on her smartphone to record song ideas that she had (Swift 2014). The use of this technology expanded shortly thereafter with the development and publication of apps such as *Music Recorder*, *Voice Record*, and *Music Memos* for Apple's iPhone, to name just three, and not to mention additional Android-based apps. Some of the apps that appeared on the market shortly after Taylor Swift's *1989* included the ability for the app to automatically add a drum track and bass line to the song memo, as some of the apps were able to determine the key of the music and basic rhythmic feel. The increasing sophistication of apps was noted in an article in *USA Today* (Graham 2016). Although Swift's notes on her use of a smartphone as part of the songwriting process suggest the power of these 21st-century devices, smartphones have proven to offer numerous other opportunities for musicians and music fans, including in the areas of listening, recording, rehearsing, and so on.

It probably goes without saying that the principal importance of the abundance of useful smartphone music apps is portability. Of course, some music-related applications, such as notation programs and virtual studio applications that are available for both smartphones and tablets, are easier to use on tablets because of the larger display and the opportunity for more precise screen manipulation. However, tuners, decibel meters, voice and music memo, metronome, and other such applications that are focused on musicians having them available in the studio, in the practice room, or that can be used at a moment's notice are better suited to a device that is relatively small, portable, and likely to be with a potential user virtually all the time. And there are free and/or relatively inexpensive apps for all these musical uses.

Interestingly, some of these smartphone apps are actually considerably more powerful than the physical devices that they replace. For example, some tuner apps, such as *Cleartune*, offer Equal Temperament and other more unusual tunings. Specifically, *Cleartune* offers stretch-tuned guitar, Pythagorean, standard Just Intonation, Pythagorean Just Intone, several versions of Mean-Tone intonation, several versions of tempered tunings, some of the French intonation systems of the Renaissance, and others. These options either are not available at all on stand-alone tuners or are only available on electronic tuners that cost multiples of the price of apps such as *Cleartune*. Similarly, some of the apps designed for music students and performing musicians combine several virtual devices into one. For example, *MusiciansKit* includes a tuner, recorder, and a metronome. The metronome area of the app includes a tap function. This allows the user to tap a pulse and see what the numerical value of the pulse is in beats per minute; many stand-alone metronomes, mechanical or electronic, do not include a tap feature. And, smartphone apps such as *MusiciansKit* cost a fraction of the price of physical devices that include metronome, tuning, and recording features.

Although mp3 players enjoyed a strong run of popularity at the end of the 20th century and into the early 21st century, the widespread adoption of smartphones soon rendered them if not obsolete, then significantly less necessary as they had been just shortly earlier. Of course, different levels and costs of smartphones offer varying amounts of memory; however, today's smartphones generally can store significantly more digital music than the mp3 players of just a couple of decades ago, in addition to all the other applications that smartphones offer. In addition, Pandora, iHeartRadio, American Roots, and other streaming audio services—free and subscription based—can easily be used on smartphones.

Perhaps the best demonstration of the computing power and the high quality of today's smartphone microphones and cameras come with the huge amount of usage that they received during the COVID-19 pandemic, both through virtual concerts, music lessons, tutorials, and so on and through the numerous Facebook Live, YouTube Live, and other streaming and downloadable virtual concerts. Numerous artists who did not necessarily have access to expensive professional video and audio recording equipment made videos and did live streaming concerts using nothing more than a smartphone, the video setting on the phone's photo application, and a good WI-FI connection. Despite the minute microphones on smartphones, and despite the fact that these devices are considered little more than enhanced phones by some users, the frequency range and clarity provided by the microphones and the smartphone's electronics produced numerous high-quality videos during the pandemic.

In addition to virtual concerts, smartphones have been used to record how-to-play-type tutorials. YouTube hosts numerous videos of this type, some of which are stand-alone videos—recorded either by amateur musicians or by professionals—and some of which are samples that are intended to drive the viewer to a subscription or paid site. Regardless, in some cases, the person who recorded and posted the video mentions doing so using their smartphone, and in other cases, one can see the instructor move toward a device to turn it off in between sections of the video, suggesting the use of the video component of the camera feature of either a smartphone or a tablet.

Smartphones have also been at the center of the use of various musical collaboration apps, such as *Acapella*. The fact is that apps such as these allow users to share their audio and video recordings with their collaborators—and with a virtual worldwide collaboration community, if they choose to do so—via links through email, SMS texts, Line, WhatsApp, Facebook Messenger, Instagram, and so on. The fact that several of these methods—most notably SMS text—are more smartphone focused than laptop or tablet focused suggests the importance of the smartphone for these collaborations.

Ringtones

Although perhaps easy to overlook because a couple of decades through the 21st century they are so pervasive, the cellular phone ringtone is another interesting intersection of music and technology. Ringtones on the early cellular phones were limited, and they were monophonic. Some of the early ringtones, particularly Nokia's so-called Nokia Waltz, were iconic. This famous ringtone dated from 1992 and was based on a phrase-ending figure from the first section of Francisco Tárrega's *Gran Vals* for solo guitar. Incidentally, several performances of Tárrega's Gran Vals can be found on YouTube (see, e.g., https://www.youtube.com/watch?v=-JaHnOppeLk, accessed June 1, 2020).

In his memoir, musician Thomas Dolby writes of receiving a query from Nokia for a synthesizer that could be built into a later generation of the company's cellular phones to make them competitive with the several Japanese phones that by the late 1990s included a Yamaha MIDI chip. The Beatnik Audio Engine that Dolby's company had developed for quickly loading synthesized sound files for web pages was licensed for use in the Nokia phones. Dolby also writes that when he heard a young woman's phone playing a synthesizer figure from Eminem's "The Real Slim Shady" and realized that young people would actually pay for ringtones, "that was the moment I recognized that the U.S. ringtone market was about to go ballistic" (Dolby 2016, 256). The global ringtone market became a multi-billion-dollar part of the recorded music industry, and the Beatnik Audio Engine was a major part of the industry. Because Beatnik's licensing deal with Nokia was not exclusive, the company soon secured arrangements for its polyphonic cellular phone synthesizer technology to be licensed to additional companies such as Motorola, Sony Ericsson, Siemens, and Samsung. By 2005, Beatnik's technology was found in over 250 million cellular phones (Business Wire 2005).

The speed at which the cellular phone music market grew—from the advent of the "Nokia Waltz" in 1992 to polyphonic sound synthesis developed by just one company finding itself in a quarter-billion phones approximately a dozen years later—illustrates the rapid early growth of the cellular phone industry, even before the availability of smartphones. It also suggests the importance of ringtones as a means of personalizing one's phone and phone experience, almost as a form of self-branding.

Further Reading

Business Wire. 2005. "Beatnik's Audio Engine Reaches a Quarter of a Billion Mobile Phones; One of the Fastest Adoptions of Embedded Software Technology in the Handset Industry." *Business Wire*, February 14. Accessed June 1, 2020. https://www.tmcnet.com/usubmit/2005/Feb/1116860.htm.

Dolby, Thomas. 2016. The Speed of Sound: Breaking the Barrier between Music and Technology. New York: Flatiron Books.

See Also: Boomboxes; Collaboration Software; Electronic Tuners; Internet Radio; Metronomes; Streaming Audio Services

Further Reading

Graham, Jefferson. 2016. "App Makes It a Cinch to Add Backup Band." *USA Today*, January 21, Money section: 3b.

Swift, Taylor. 2014. Spoken introductions to bonus tracks. *1989*. CD. Big Machine Records BMRBD0550A.

Social Media

Since its inception, social media has played a significant role in the development and maintenance of the careers of numerous musicians. Although others have enjoyed strong social media presences in the 21st century, Taylor Swift is perhaps one of the best examples of how one particular popular artist has used various platforms. Even a brief study of Swift's use of social media illustrates how platforms and the usage of social media have changed over the years.

In 2006, Swift used MySpace to promote her first single: "Tim McGraw." At the time, MySpace was reputed to be the largest social media platform in the world. Significantly, too, MySpace was favored by teens, and Swift was approximately 17 at that time during which she was using the platform to promote the song. The radio and print media made note of Swift's use of social media early on. For example, a 2007 article in the music trade magazine *Billboard* quoted Becky Brenner, the program director of Seattle county radio station KMPS-FM, as saying that Swift's use of MySpace as a promotional tool not only generated interest among music fans but also helped to generate interest in "Tim McGraw" among radio disc jockeys (Hasty 2007, 9).

As Swift matured, she turned to Facebook as her principal means of staying in touch with her fans. Then, as social media platforms continued to expand and the demographics of their user bases continued to evolve, Swift increasingly turned to Twitter, Instagram, and Tumblr, while she maintained ties with the older platforms with which she had been associated. By maintaining a presence on multiple platforms, as Taylor Swift has continued to do, she has managed to cover multiple audience demographic groups. Swift and other major artists have used the various social media platforms to (in the current vernacular) drop new tracks, preview new albums, and generally stay in contact with fans. In this way, social media has replaced and greatly expanded the possibilities of the music fan magazines of the past.

The connection between social media and star performers has not always been pleasant. For example, Lizzo announced in early 2020 that she was taking a break from Twitter because of the large number of trolls using the platform. And social media provided music fans center stage for the 2016 tiff between Taylor Swift and Kanye West and Kim Kardashian. Similarly, the legal troubles of pop stars such as Justin Bieber and others have resulted in large amounts of commentary on social media platforms such as Twitter.

On a more positive note, the COVID-19 pandemic led to several major artists, including Lizzo, Keith Urban, and Chris Martin, mounting Facebook Live concerts in the last couple of weeks of March 2020, as several states across the United States shut down concerts and concert venues (CNN Business 2020). And Garth Brooks and Trisha Yearwood's March 23, 2020, Facebook Live concert, which they called "Inside Studio G," had a viewership of 3.4 million, which apparently exceeded the platform's capacity, making it necessary for some viewers to have to refresh the page multiple times (Garvey 2020). Similarly, YouTube Live concerts also became common during the pandemic. Unlike Facebook Live, in which the individuals or groups producing the performance are not required to have a certain number of followers, YouTube artists must have at least 1,000 subscribers to livestream on the platform through a mobile device. However, "creators who have less than 1,000 subscribers can still livestream through a computer and webcam" (https://support.google.com/youtube/answer/2853834?hl=en, accessed January 15, 2021). Arguably, trying to livestream a musical event using a desktop or laptop computer poses far more challenges (e.g., finding a tripod adapter for a laptop), so Facebook Live was generally the platform of choice for lesser-known artists.

During the period of high use of Facebook Live for virtual concert streaming, it has not just been major stars who have used the technology. Although in the context of the Justin Biebers, Taylor Swifts, and so on it might not be particularly impressive, within a month of the live event, the Americana music group in which I play had over 490 views of a March 2020 Facebook Live virtual concert that we performed after the venue at which we were scheduled to perform was closed because of the pandemic. Ultimately, the concert had over 850 views. Based on past performances at the venue, we might have expected 50 or so people in attendance, so the Facebook Live event allowed a significantly larger number of people to hear the performance. A November 2020 Facebook Live concert that we performed under the auspices of the county arts agency had over 1,200 views before the end of the calendar year. This is just one example of the various chamber musicians, virtual choirs, orchestras, folk groups, bluegrass bands, hip-hop artists, rock bands, and so on who have reached multiples of their usual live audience size through social media. Of course, the technology was already there; the pandemic made it necessary for musicians to take advantage of it. With the options for including donation links and selling virtual tickets for these streaming events on social media, it will be interesting to see if artists continue to make such high use of the technology as concert venues reopen. Possibly, a new type of concert may emerge, in which there is an audience in attendance with additional virtual seats offered through social media.

See Also: Smartphones

Further Reading

Benkler, Yochai. 2006. *The Wealth of Networks: How Social Production Transforms Markets and Freedom*. New Haven, CT: Yale University Press.

CNN Business. 2020. "Music's Biggest Stars Are Performing Online during the Pandemic." *CNN Business*, March 17. Accessed March 18, 2020. https://www.cnn.com/videos/business/2020/03/17/coronavirus-musicians-social-media-orig.cnn-business.

Garvey, Marianne. 2020. "Garth Brooks and Trisha Yearwood's Emotional Home Concert Crashes Facebook Live." *CNN*, March 24. Accessed March 24, 2020. https://www.cnn.com/2020/03/24/entertainment/garth-brooks-trisha-yearwood-concert-trnd/index.html.

Hasty, Katie. 2007. "Swift's Un-Swift Climb." *Billboard* 119 (August 4): 9.

Perone, James E. 2017. *The Words and Music of Taylor Swift*. Santa Barbara, CA: Praeger Publishers.

Sony Walkman and Discman

The Sony Corporation brought a new dimension to portability of personal audio to the world with the advent of the Walkman in 1979. The Walkman was a portable cassette player to which the user could listen using headphones. The connection to headphones, the light weight of the Walkman, as well as the fact that it was designed with motion in mind meant that Sony's Walkman was well suited for walkers and joggers. The personal nature of the Walkman—unlike portable cassette players and boomboxes that had built-in speakers, the Walkman allowed the user to keep their music to themselves—became a standard. Other manufacturers followed Sony's lead, and the concept of an intrinsic connection between music player and headphones found its way into Sony's Discman, as well as the mp3 players and devices such as the iPod that began to appear at the end of the 20th century and early in the 21st century.

The original cassette-based Walkman weighed 14 ounces and, in contrast to cassette-playing predecessors, was only used for playback through headphones.

As *Time*'s Meaghan Haire wrote on the 30th anniversary of Sony's introduction of the player, "the Walkman's unprecedented combination of portability (it ran on two AA batteries) and privacy (it featured a headphone jack but no external speaker) made it the ideal product for thousands of consumers looking for a compact portable stereo that they could take with them anywhere" (Haire 2009).

In addition to the Walkman proper, the personal cassette player, Sony has also produced portable music players for compact discs (the Discman), minidiscs, mp3s, and streaming audio. The original Walkman, which the company offered until 2010, was the most popular, however, and, as of 2014, accounted for half of the 400 million personal music players that Sony had produced (Franzen 2014).

Of Sony's post-Walkman portable music devices, perhaps the other most notable was the Discman, which made its debut in 1984, shortly after the first consumer compact discs were introduced. Although the Discman was not as immediately portable as the Walkman—because of the diameter of the physical CDs—it did make it possible to listen to CDs while on the move. Perhaps more importantly to some users, Sony's Car Discman served as a transitional device during an era in which CDs were commanding more and more of the market from vinyl albums and cassette tapes, even as automobile manufacturers continued to focus on offering factory-installed cassette players in cars. Using a cassette-shaped adapter that sent an electrical signal directly from the adapter to the cassette player's play head, the Car Discman allowed users to listen to CDs in cars that were fitted with factory-installed or aftermarket cassette players but no CD player. A power adapter for the Car Discman plugged into the automobile's cigarette

The Sony Walkman brought portability and personal listening to the world of the cassette tape through its small size and use of headphones. (Philip Kinsey/Dreamstime.com)

lighter opening (today, generally referred to as the auxiliary or 9-volt power output). The fact is that the factory-installed CD player and the widespread availability of affordable, aftermarket, automotive CD players in the late 20th and early 21st centuries made devices such as Sony's Car Discman less necessary. Interestingly, given the fact that in the second and third decades of the 21st century, the new standards in automotive audio center around Bluetooth connectivity, USB drives, and applications such as *Apple CarPlay* and *Android Auto*, fewer and fewer new cars can be outfitted with factory-installed CD players. In fact, anecdotally, it seems that the transition away from factory-installed CD players really started in the 2010s. Ironically, this means that drivers who feel that they must play compact discs in their cars can, if they still have their old Sony Car Discman around, return to compact disc technology.

Clearly, the original Walkman from 1979 was not the first personal listening device that provided privacy; many highly portable transistor radios well over a decade before had earphone jacks. Given the fact that it did not contain a speaker and that it was designed for stereophonic listening using headphones, though, Sony's Walkman increased the level of privacy and the quality of the music to which one could listen on the move. And, unlike the portable transistor radios of the 1960s and 1970s, the Walkman was designed around the idea of having the music that the consumer truly wanted to hear—without a DJ playing a role in selecting it and without commercial interruption—at their disposal practically anywhere. That control aspect cannot be overlooked, as it continued to be a necessary part of future personal listening device design, from Sony's products, such as the Discman, to the mp3 players that made a commercial smash at the very end of the 20th century and early in the 21st century, to Apple's subsequent iPods, and today's smartphones.

Sony Corporation

Founded by Masaru Ibuka in 1946, Tokyo Tsushin Kogyo K.K. was the immediate predecessor of Sony Corporation (the corporate name was changed in 1958). The company quickly became a major player in consumer electronics. Although Tokyo Tsushin Kogyo and Sony developed and sold many products over the years, let us focus primarily on the company's contributions that relate to music. In the 1950s, the company produced Japan's first magnetite-coated, paper-based recording tape and tape recorder, Japan's first transistor and first transistor radio. In the 1960s, Sony continued to move in transistorized electronics, including radios, televisions, video tape recorders, and amplifiers. In 1968, Sony joined forces with CBS to form CBS/Sony Records, Inc. The 1970s found Sony doing further work in video, including introducing the ill-fated Betamax VCR in 1975. In 1979, Sony released its first Walkman, a portable cassette tape player that used headphones and, in 1982, the world's first CD player. That was followed by the Discman, a portable, headphone-based CD player. In the late 1980s, Sony acquired CBS Records, with the record company taking the name Sony Music Entertainment in 1991.

Over the past 30 years, Sony has continued to build a presence in the music and film industries and has also moved into robotics and computers and, in 2010, introduced its digital music service, "Music Unlimited."

See Also: Cassette Tapes; Compact Discs (CDs); Headphones; iPods and Other Portable Digital Music Players; MP3s; Stereophonic Sound

Further Reading

Du Gay, Paul, Stuart Hall, Linda Janes, Anders Koed Madsen, Hugh MacKay, and Keith Negus. 2013. *Doing Cultural Studies: The Story of the Sony Walkman*. 2nd ed. London: Sage Publications.

Franzen, Carl. 2014. "The History of the Walkman: 35 Years of Iconic Music Players." *Verge*, July 1. Accessed June 10, 2020. https://www.theverge.com/2014/7/1/5861062/sony-walkman-at-35.

Haire, Meaghan. 2009. "A Brief History of the Walkman." *Time*, July 1. Accessed December 8, 2020. http://content.time.com/time/nation/article/0,8599,1907884,00.html.

Stereophonic Sound

The stereo soundscape has been a regular part of the recording industry for so long—particularly since it was exploited for effect so dramatically during the psychedelic era of the 1960s (e.g., the Jimi Hendrix Experience's iconic recording of "Purple Haze") that today's music fans might assume that stereo simply has always been around. In fact, the encoding stereophonic music for playing on 33–1/3-rpm disc was first done by the British firm E.M.I. in 1957.

When considering stereophonic sound and its path to becoming the norm, it must be considered that in the late 1950s through the 1960s, popular music, particularly pop, rock, soul, and country, was closely tied to AM radio. FM stations of the time focused on jazz and classical music and, eventually in the late 1960s, with album-oriented rock. Because AM radio was monophonic, the norm until the late 1960s was for record companies to release monophonic singles. In the world of popular music, particularly that enjoyed by young people, monophonic albums made sense in the record industry because less expensive record players of the late 1950s and early 1960s were designed for monophonic reproduction of sound, both in the styluses with which they were outfitted and with the single speaker that some had. When single-groove stereophonic records were standardized, a new groove structure in the records (in which vertical and horizontal movements were encoded) and stereo-specific styluses that could "read" those grooves were necessary, in addition to stereo decoding by the phonograph/turntable, and a pair of speakers. As one might expect, the stereo equipment was more expensive than cheap monophonic record players, so stereo and mono coexisted longer in pop music marketed to young people, who presumably had less disposal income than adults.

In fact, the monophonic and stereophonic mixes in the world of popular music were sometimes done entirely separately and can provide fundamentally different aural experiences. To put it another way, the monophonic mixes sometimes were not simply the stereo mix redistributed equally across the aural space. Some recording engineers, producers, and fans of music of the first half or so of the 1960s still debate the relative merits of stereophonic versus monophonic mixes of such classic albums as Beatles' *Revolver* and others, based on the greater rhythmic impact that monophonic mixes often had and based on the balances of instruments and voices on the different mixes. Similarly, some of the soul recordings of the early 1960s, perhaps best represented by the singles of Wilson Pickett, arguably exhibit more impact in their monophonic mixes than in their stereo mixes.

More likely to be enjoyed by nonteens and audiophiles, genres such as jazz, easy listening, and classical music better lent themselves to the stereophonic soundscape. In the world of orchestral music, the aim was to produce the most life-like concert hall experience possible. As a result, and provided that one's turntable, amplifier, and speakers were wired correctly, one would typically hear the violins on the left side of the stereo spectrum, the double basses and cellos on the right, percussion on the left and in the middle, and the woodwinds and brass generally in the middle with some of the brass skewed toward the right, much like an orchestra's players traditionally are placed on the stage.

Interestingly, even in the upscale automotive market of the second half of the 1960s, stereophonic sound was not a universal standard by any means. For example, the 1967 Cadillac Fleetwood Eldorado, described by *Hemmings Classic Car*'s Thomas A. DeMauro as an "icon of the upwardly mobile" of the era (DeMauro 2021, 36), came with three possible radio configurations: AM, AM/FM, or AM/FM stereo (DeMauro 2021, 38).

When stereophonic albums entered the sphere of rock and roll music in the early 1960s, the results could be hit or miss. As mentioned earlier, for some music fans, monophonic mixes pack more sonic impact. Some of the early stereo mixes, too, could be somewhat artificial sounding, for example, when the instruments were recorded on one channel and the voices on the other. In some cases, stereophonic releases of this sort were never intended to be heard by the public by their producers. As producer George Martin put it, he used a stereo machine to make recordings of groups such as the Beatles because he "found it convenient to use twin-track [as opposed to recording directly to mono], putting all the instruments on one track and putting all the vocals and any lead solos (like guitars or piano) on that track as well and compressing the two together.... But that backfired because, many years later, some idiot at EMI took these and assumed they were stereo and put them out" (Giuliano and Devi 1999, 263).

By approximately 1966 and 1967, however, popular music producers were making fuller use of the stereophonic soundscape. This was made possible by the fact that studios increasingly had 4-track and 8-track tape machines with which the recordings could be made. Multi-track recording—beyond the 2-track recording Martin described earlier—gave producers the very real possibility of making viable stereo mixes of the recordings, as well as the opportunity to use the stereophonic space in creative ways. As mentioned earlier, Jimi Hendrix's song "Purple Haze" is a prime example in which instruments are placed in specific parts of the stereo soundscape, while others pan from side to side, and the various drums and cymbals of the drum set have their individual spaces across the sound field from the left to the right. Other psychedelic era recordings, such as the Beatles' song "Strawberry Fields Forever" and the album *Sgt. Pepper's Lonely Hearts Club Band*, also showed the extent to which producers, such as George Martin and others, played significant roles in the creation of music that was the product of the studio and not something that could reproduced live.

It is within the milieu of where the use of the stereophonic soundscape had progressed through the creative mixing down of multi-track recordings to which singer-songwriter-guitarist-composer Frank Zappa provides context:

> On a record, the overall timbre of the piece (determined by equalization of individual parts and their proportions in the mix) tells you, in a subtle way, *WHAT* the song is about. The orchestration provides *important information* about what the composition *IS* and, in some instances, assumes a greater importance than *the composition itself.* (Zappa with Occhiogrosso 1989, 188; italics and capitalizations in the original)

Later writers such as Gracyk (1996) and Zak (2001) concur with Zappa and, in fact, are even more emphatic about the point that the best-known recording of a piece of music essentially equals the piece for listeners of rock music. Full use of the stereophonic soundscape (e.g., placement of the voices and instruments and panning from left to right), then, joins timbral effects, tape manipulation, balance, tone color, and so on, as part of what defines a composition, going well beyond melody, rhythm, harmony, instrumentation, and lyrics. Despite the development of quadraphonic sound and various formats of surround sound, stereophonic sound remains the norm, well over 60 years after it was standardized.

See Also: Multi-Track Recording; Quadraphonic and Surround Sound

Further Reading

Daniel, Eric D., C. Denis Mee, and Mark H. Clark, eds. 1999. *Magnetic Recording: The First 100 Years*. New York: John Wiley & Sons/IEEE Press.

DeMauro, Thomas A. 2021. "1967 Cadillac Fleetwood Eldorado." *Hemmings Classic Car* 17, no. 6 (March): 36–40, 42.

Giuliano, Geoffrey, and Vrnda Devi. 1999. *Glass Onion: The Beatles in Their Own Words*. Cambridge, MA: Da Capo Press.

Gracyk, Theodore. 1996. *Rhythm and Noise: An Aesthetics of Rock*. Durham, NC: Duke University Press.

Wegman, Jesse. 2000. "The Story Behind 'Purple Haze.'" National Public Radio (NPR), September 18. Accessed January 18, 2021. https://www.npr.org/2000/09/18/1088122/jimi-hendrix-purple-haze.

Zak, Albin J. III. 2001. *The Poetics of Rock: Cutting Tracks, Making Records*. Berkeley: University of California Press.

Zappa, Frank, with Peter Occhiogrosso. 1989. *The Real Frank Zappa Book*. New York: Poseidon Press.

Streaming Audio Services

Two of the prominent ways in which consumers obtained and listened to music in the 20th century—that is, music that the consumer did not experience at a live performance—included what we might think of as polar opposites. Either one purchased music that was encoded on some sort of physical medium (e.g., a shellac disc, a polyvinylchloride disc, a cassette tape, an 8-track tape, and a compact disc), or one listened to commercial or public radio. In the former case, the consumer had control over the music they listened to; in the latter case, the consumer might have been able to select a radio station that focused on a format or genre, but the actual musical pieces most likely were selected by a program manager of disc jockey or based on an industry chart (e.g., one of *Billboard* magazine's ranking charts). Streaming audio services offer 21st-century music consumers other, but related, kinds of listening experiences.

Although Pandora is detailed in this volume's entry Pandora and the Music Genome Project, it is important to note that at the time of this writing, Pandora remains perhaps the major player in this music industry category. Pandora's closest rival, Spotify, also maintains a strong share of the streaming audio market. Various, seemingly ever-changing, online reviews of these services suggest that each has advantages, considering factors such as the number of musical works offered by the service, ease of use, quality of the offerings, the quality of the algorithms that are used to create essentially an artificial intelligence–curated playlist for the user, and so on. However, Pandora and Spotify are joined by several other streaming audio services. One, Napster, retains much of the branding—such as the logo—of the original Napster; however, the reincarnation of Napster is a legitimate streaming audio service. In fact, the current Napster was once called Rhapsody, a streaming audio service that has been affiliated with some publishers of high school–level and higher education–level music appreciation books. Some of these books are packaged with online access to the musical examples as streamed through the Rhapsody/Napster service.

Tidal, YouTube Music, and other streaming audio services also have strong adherents that are drawn to the unique features that each offers. Increasingly, some of the major names in purchased digital music, including Apple and Amazon, also have turned to streaming audio while retaining their digital download sales presence. Despite retaining their sales presence, however, some users might sense that Apple and Amazon both are increasingly placing emphasis on subscription-based streaming services, based on the high visibility of the streaming services on their apps and websites.

Some of the current streaming audio services use strictly a subscription model. For a monthly fee, the user can stream as much music as they wish. Others operate on a free-to-the-user with commercials model, while others offer a free version—which has commercial advertising—and a paid subscription version that does not interrupt the programming with advertisements. Again, because these streaming services are a fairly recent phenomenon and because they are evolving, it is difficult to specify which service offers exactly what, as any given company's business model might change at any time.

Another aspect of the evolution of streaming audio is that some of the services are expanding and diversifying their content, with Spotify being a notable example. According to stock trader Todd Gordon, quoted by CNBC, in late 2020, "they're [Spotify] making a big move into content, streaming content, diversifying away from music, which I think they did a really good job fending off all the streaming music services, Apple Music included" (Lahiff 2020).

Streaming audio services are not, however, limited to well-known, big-name corporations. For example, American Roots offers roots music, traditional country, bluegrass, and blues programming that straddle the line between a streaming audio service and internet radio service. And the blurring of lines between these digital categories is ongoing with other services, with satellite radio, internet radio, and streaming audio services providing fairly similar experiences to the end user, with the exception of such things as disc jockey commentary, nonmusical programming (e.g., interviews with artists), and commercial advertising aside. And some umbrella companies offer multiple brands that cover these categories. For example, Pandora and SiriusXM are distinct brands under one corporate umbrella.

Many streaming audio services can be accessed by means of desktop or laptop computer, Apple iOS and Android smartphones, iPads, and other tablet computers. As a result of this saturation of technological platforms, one can access streaming audio while driving using Google's *Android Auto* or *Apple CarPlay*, which again blurs the lines between these services, satellite radio, and musical selections that might reside on one's smartphone.

See Also: Internet Radio; Pandora and the Music Genome Project; Satellite Radio

Further Reading

Lahiff, Keris. 2020. "Two 'Surprising' Stocks to Sell—and One to Buy—Heading into the New Year, according to Trader." CNBC, December 24. Accessed December 24, 2020. https://www.cnbc.com/2020/12/24/sell-zoom-beyond-shares-into-the-new-year-according-to-trader.html.

McIntyre, Hugh. 2018. "The Piracy Sites That Nearly Destroyed the Music Industry: What Happened to Napster?" *Forbes*, March 21. Accessed January 18, 2021. https://www.forbes.com/sites/hughmcintyre/2018/03/21/what-happened-to-the-piracy-sites-that-nearly-destroyed-the-music-industry-part-1-napster.

T

Tablature

Instead of indicating musical pitches as traditional musical notation does, tablature indicates fingerings for a particular music instrument. Keyboard tablature systems were first developed in the middle of the 15th century; however, traditional musical notation remained the norm for harpsichords, organs, and subsequent keyboard instruments. By approximately 1500, lute tablature systems were in use in Europe. These were the ancestors of the guitar tablature systems that are still in use more than 500 years later.

A typical guitar tab (short form of tablature) will have six lines, each of which corresponds to a string on the typical 6-string guitar (electric bass tabs typically have four lines, although some are geared toward 5-string basses). The fret position of the fingers for each note is indicated. One of the odd quirks about tablature that some particularly spatially oriented readers might notice is that the lines are designated E, A, D, G, B, E, assuming traditional tuning, from lowest to highest pitch; however, because of the way in which the guitar is typically held, the strings are in the opposite order from low to high in terms of their placement on the instrument. In other words, the open A string is indicated by the second line from the *bottom* in the tablature; however, viewed by the player as they hold the guitar, the open A string is the second string from the top in relationship to the floor.

The one area in which some string tablature system tends not to be nearly as prescriptive to the player as traditional music notation is in the notation of rhythm. In present usage, however, guitar tablature tends to be used for music that the player has heard before. Not all published tabs include rhythmic notation; however, in some cases, tablature is presented with rhythmic notation based on that of traditional music notation included but under the notation of the strings, frets, and so on. In these cases, because the rhythmic notation is not entirely integrated into the fingering notation in some versions of tablature, it arguably presents the player who is not familiar with how the music sounds from having heard it before more of a challenge to sight-read than traditional, integrated music notation.

The area in which tablature perhaps has the greatest value comes in the fact that it provides the player who has choices with how to finger a given passage a clear indication of the fingering suggested by the composer, arranger, professional player, or amateur guitarist who has published the tablature. For example, there are two principal ways in which acoustic guitar players typically play Paul McCartney's guitar accompaniment for the 1968 Beatles song "Blackbird." Some tabs have the two contrapuntally moving melodic lines primarily limited to two strings, while other tabs of the guitar part have the upper line of the figure divided roughly equally between two adjacent strings, bringing a total of three strings into play. Both fingering patterns sound roughly the same, because the pitches are the same; however, each has an advantage. The tablature that uses two strings for the moving lines requires bigger hand stretches but doesn't involve switching strings for the upper line, while the other tab requires little in the way of stretches.

Another area in which tablature is particularly helpful to players is when a capo is used to raise the pitch of the instrument's open strings. A player who learns a piece from tablature in a given key who then has to transpose it to a different key needs only to use a capo and follow the same tablature with the location of the capo taking the place of the open strings. Using traditional music notation, transposition would result in the music looking entirely different.

Tablature across the fretted string instruments is fairly uniform, with the lines that represent the strings being in the arrangement from low to high as per how the instruments that are played Spanish style (strings facing out, away from the player). As mentioned earlier, for guitar, this means that the highest-pitched strings are on the bottom of the tablature. Once the player gets used to this, it does not create a significant problem. A complication arises, however, for instruments such as the ukulele and five-string banjo in which the strings are not all in which the strings are not arranged in consecutive order from low to high in pitch. Some readers of ukulele or banjo tablature might find the aural sensation of the lowest line on the tablature to be sounding a note that seems to be in a counterintuitive range, particularly if the player is more used to standard guitar tablature.

Musical passages presented in traditional notation and guitar tablature. (Rui G. Santos/ Dreamstime.com)

In the late 20th century and early 21st century, guitar tablature has been published in some sheet music editions along with a version of the same piece written using traditional musical notation so that players who read one but not the other can find the publication useful. It should be noted, however, that although tablature in and of itself might seem like one of the lowest-tech "technologies" profiled in this volume, the expansion of the internet and the proliferation of websites and smartphone apps have been tied to the expansion of the availability of tablature. Some guitar tablature websites carry almost a social media feel, with subscribers to the site being able to comment on published tabs and being able to publish their own versions. Guitar tablature websites, in particular, have raised issues with regard to copyright protection, particularly in so far as traditional music notation traditionally has been used in copyrighting music. An interesting article about the legal wrangling over tablature and copyright, particularly in relationship to tablature websites, can be found in Kempema (2008). At the time of this writing, there are several guitar tablature websites that can be accessed for free. In recent years, some of these websites have added embedded YouTube-style instructional videos and, in some cases, links to audio recordings of the pieces. Some of the tablature sites have expanded to include smartphone apps. In addition, computer, tablet, and smartphone applications such as *TEFview* can play the tablature using MIDI technology, as well as showing where the notes are fretted on the fingerboard.

See Also: Musical Notation

Further Reading

Kempema, Jocelyn. 2008. "Imitation Is the Sincerest Form of . . . Infringement? Guitar Tabs, Fair Use, and the Internet." *William & Mary Law Review* 49, no. 6: 2265–2307. Accessed online January 18, 2021. https://scholarship.law.wm.edu/cgi/viewcontent.cgi?referer=https://www.google.com/&httpsredir=1&article=1166&context=wmlr.

Weiss, Piero, and Richard Taruskin. 1984. *Music in the Western World*. New York: Schirmer Books.

Tape Manipulation

The widespread availability of electromagnetic recording tape after World War II ushered in a new era in audio recording possibilities. Eventually, multi-tracking, overdubs, and a variety of sonic manipulations were possible—all things that were impossible with the direct-to-disc, wire recording, and other technologies of the earlier part of the sound-recording era.

The earliest famous use of magnetic recording tape as a medium for assembling musical compositions came in the form of *musique concrète*. Theories about the construction of musical compositions assembled from prerecorded sources were developed by Pierre Schaeffer in the 1940s; however, it was not until the 1950s and 1960s that tape pieces came into their own. Some of the famous composers associated with *musique concrète* include Schaeffer, Karlheinz Stockhausen, Iannis Xenakis, Pierre Boulez, and Edgard Varèse.

While it might appear that experimental tape compositions were probably not likely to be heard by masses of listeners, the 8-minute electronic music tape piece *Poème électronique*, by composer Edgard Varèse, was part of the multimedia Philips Pavilion at the 1958 Brussel's World's Fair. During the exhibition's run, it is estimated that over 2 million people heard this tape piece as they moved through the pavilion.

Some composers during the 1950s created randomly—and sometimes not randomly—assemblages of electronic tape to create tape compositions, including John Cage in whose music of the time randomness and accident played a significant role. Composers such as Yoko Ono and the Beatles' John Lennon were influenced by *musique concrete* and Cage's assemblages. In the case of Lennon, the 1968 track "Revolution No. 9," which includes snippets from various BBC programs, sound effects and voices by members of the Beatles, and so on, represents a *musique concrète* heard by millions of people on the album *The Beatles*, widely known as the "white album." Although "Revolution No. 9" is the best-known example of *musique concrete* produced by a rock band, other groups put together *musique concrete* pieces, such as Creedence Clearwater Revival's "45 Revolutions per Minute," which was released as a promotional single, and which at the time of this writing can be found on YouTube (see, e.g., https://www.youtube.com/watch?v=p5pntMNOPAg, accessed January 16, 2021), and which was included as a bonus track on the CD release of the group's album *Pendulum*.

Because recordings and playback on electromagnetic tape could be made at various speeds, in the 1950s several musicians took advantage of that what we might call mild examples of tape manipulation. One of the more adventurous musicians in the world of popular music in that decade was jazz guitarist Les Paul, designer of what might have been the first solid-bodied electric guitar but for the fact that it was not put into production until years after Paul came up with the idea; by that point, Fender had beaten Paul and Gibson to the punch. In recordings that Paul made with singer Mary Ford, several studio techniques were used that were by no means mainstream at the time, including overdubbing and speed changing. In some recordings, Paul recorded his guitar solos at a slower speed. When these were sped up on playback and for mastering, the solos sounded impossibly fast.

Songwriter and record producer Ross Bagdasarian Sr., who used the professional moniker Alan Seville, used speed and its accompanying pitch change (when a recording was sped up using electromagnetic tape, pitch rose; conversely, pitch went down when tape playback was slower than the recording speed) in a very different way. Seville used the effect for humor, as first demonstrated with his 1958 recording "Witch Doctor." However, Seville is probably best remembered for his subsequent work as the real-life person behind the animated group, the Chipmunks. In the still-often-heard holiday song "The Chipmunk Song (Christmas Don't Be Late)," Seville recorded the Chipmunks voices at a slow speed and at normal pitch. As a result, when the recording was sped up to what we might call normal speed, the voices were much higher pitched, vibrato was correspondingly faster, and so on, giving the voices of the animated characters their perceived chipmunk-like quality.

There are other more subtle uses of tape speed manipulation that might be easy to miss—certainly, some are not as obvious as Alan Seville's work. For example, in his memoir about producing the recordings of the Beatles, George Martin discusses the genesis of the Beatles' well-known song "Strawberry Fields Forever." According to Martin, songwriter John Lennon had the group record two different versions of the song. Ultimately, Lennon wanted to use the beginning of one recording with the rest of the finished product coming from the other recording. Problematically for producer Martin, the two recordings were in different keys and were different speeds. As Martin put it, "I thought: If I can speed up the one and slow down the other, I can get the pitches the same. And with any luck, the tempos will be sufficiently closer not to be noticeable. I did just that, on a variable-control tape machine, selecting precisely the right spot to make the cut, to join them as nearly perfectly as possible. That is how 'Strawberry Fields' was issued, and that is how it remains today—two recordings" (Martin and Hornsby 1979, 201). Martin also reported that tape speed doubling and the random assemblage of randomly arranged snippets of recordings of Victorian-era steam organs were used for the instrumental break section of the Beatles' song "Being for the Benefit of Mr. Kite."

One of the other manipulation techniques used by George Martin and Beatles, as well as by other rock musicians primarily in the 1960s, involved playing a taped recording backward. For example, John Lennon sings part of the Beatles' song "Rain" backward near the end of the track. What he actually sings is the first verse, but the tape was played backward. In the song "I'm Only Sleeping," the Beatles' lead guitarist George Harrison plays a backward electric guitar solo. In order to fit with the song's harmonic structure, the solo had to be constructed so that it would fit when the tape direction was reversed. Backward effects were also used in the Beatles' "Strawberry Fields Forever." After the breakup of the group, John Lennon, in particular, continued to use the occasional backward snippet. For example, in the single release of the song "Meat City," the B side of the "Mind Games" single, recording engineer Roy Cicala says, "Check the album," which the listener hears as an incomprehensible backward and sped-up snippet. In the corresponding place in the mix used for "Meat City" on the *Mind Games* album, Cicala says, "F*** a pig," which, again, is turned backward and sped up.

Interestingly, the move to digital technology made it possible to do all these sorts of operations and more on a laptop or iPad/tablet. So, even a free program such as *Audacity* allows the user to do multi-track, change speed with changing pitch (as could be done on tape), change speed without changing pitch (which could not be done with tape), reverse musical passages, splice passages in whatever order the user wants, add reverb, add echo, add wah-wah effects, and on and on. Programs such as *Audition* and others for which there is an outright purchase or a subscription fee offer even more effects that replicate tape manipulation procedures, and some offer more ease of use, higher quality, and more features than some of the free programs.

> ### Artificial Double Tracking (ADT)
>
> Although it might be one of the lesser-known recording technologies of the 1960s, nevertheless the results of Artificial Double Tracking (ADT) were heard by millions of pop music fans in the middle of the decade, particularly because the Beatles, their record producer George Martin, and Martin's technicians made so much use of it in the studio. Martin explained the procedure as "taking an image of the sound and delaying it slightly, or advancing it slightly, so that it forms double.... If you take the image farther out [than a few milliseconds], to about twenty-seven milliseconds, you get what we call Artificial Double Tracking—two definite voices" (The Beatles 2000, 211). By using this technique, which was pioneered by engineer Ken Townsend and at one time was unique to Abbey Road Studios, a singer could sound like they were dubbed over themselves while only having to sing their part once. Townsend stated that the impetus for the technology was a recording session in which Paul McCartney recorded take after take trying to accurately double track his lead vocal. According to Townsend, "As I was driving home from that session, I began to think that there music be an easier way of double-tracking, if you simply want to reinforce your voice" (Babiuk 2002, 184). Townsend's technique drew on the experimentation of fellow Abbey Road engineer George Barnes to vary the speed at which tape recorders ran to create special effects. According to Mark Lewisohn, author of *The Beatles Recording Sessions: The Official Abbey Road Studio Session Notes 1962–1970*, "almost every song on [the album] *Revolver* was treated to ADT" (Lewisohn 1988, 70). The Abbey Road Studios and Waves Audio made a brief documentary film in which Ken Townsend explains the history of ADT and the technical side of how it works (see, e.g., https://www.youtube.com/watch?v=TgnSVdjfSwk, accessed May 28, 2020).
>
> **Further Reading**
>
> The Beatles. 2000. *The Beatles Anthology*. San Francisco: Chronicle Books.
>
> Lewisohn, Mark. 1988. *The Beatles Recording Sessions: The Official Abbey Road Studio Session Notes 1962–1970*. New York: Harmony Books.

See Also: Digital Recording; Reel-to-Reel Tape

Further Reading

Babiuk, Andy. 2002. *Beatles Gear: All the Fab Four's Instruments, from Stage to Studio*. Revised ed. San Francisco, CA: Backbeat Books.

Martin, George, with Jeremy Hornsby. 1979. *All You Need Is Ears*. New York: St. Martin's Griffin.

Morgan, Robert P. 1972. Liner notes for *Edgard Varèse: Offrandes, Intégrales, Octandre, Ecuatorial*. Reissued on CD as Elektra/Nonesuch 9 71269–2.

Theater Organs

Although this volume contains an entry for Organs that includes pipe organs and electronic organs, as well as an entry for Mechanical Organs that includes a wide range of sizes and styles of instruments from street organs to carousel band organs and dance hall organs, the massive theater organs that were part of American life predominantly in the first half of the 20th century deserve special consideration as instruments that the movie-going public saw as technological wonders.

In considering the theater organs of the early 20th century, it is important to keep in mind that their home was in the large theaters that housed vaudeville shows and that showed movies. And in considering the specifics of their role, it is important to keep in mind that prior to the 1927 release of *The Jazz Singer*, movies, with extremely rare exception, did not include any kind of pre-recorded soundtrack. The incidental music that accompanied various scenes and set the mood of those scenes required live music, often provided by the piano in small theaters or by the significantly louder theater organs in large venues.

During the first half of the 20th century—although primarily during the 1910s, 1920s, and into the early 1930s—several organ-building companies manufactured and installed theater organs across the United States. The best-known name in theater organs, however, was Wurlitzer. The company's founder, Rudolph Wurlitzer, was a German immigrant who started his business in Cincinnati, Ohio, in 1853. Initially known as an importer of musical instruments, Wurlitzer supplied military units with instruments during the U.S. Civil War. Wurlitzer later began manufacturing pianos and moved its headquarters and manufacturing facilities to North Tonawanda, New York. The company's involvement with theater organs lasted from 1915 to the early 1930s, essentially marking the heyday of the theater organ.

Because of their size, the large number of manuals on the instruments' consoles, and the variety of stops and the instruments' volume, the Rudolph Wurlitzer Company's instruments earned the name "the Mighty Wurlitzer." The creation of the theater organ as a separate type of organ generally is credited to the English organ builder Robert Hope-Jones. After coming to the United States and experiencing the failure of his own company, Hope-Jones began working with Wurlitzer on their instruments. Theater organs such as the "Mighty Wurlitzer" frequently included stops not as frequently found on church organs of the day. And, in fact, the most impressive of these instruments included such atypical sounds as drums, mallet percussion, cymbals, and so on. In this respect, they can be understood as human-played extensions of the carousel band organs of the period. In addition, the theater organs of the early 20th century included effects such as a tremolo that, while unlike that of the church organs of the past or even of the day, is suggestive of a large orchestral string section playing with expressive vibrato.

Perhaps one of the most impressive Wurlitzer theater organs still in existence, and probably one of the most impressive ever built by the company, is the instrument that was installed in the Riviera Theatre in North Tonawanda, New York, in 1926. Because this instrument was located in a large theater in the company's hometown, it was used by Wurlitzer as a demonstration organ to show off the capabilities of the company's theater organs to potential customers. As such, the organ features all the requisite stops, as well as several percussion instruments that are controlled by the organist. The percussion instruments are prominently displayed near the organ console just in front of the Theatre's stage. To fully appreciate how numerous theater organs were in the first third of the 20th century, consider that longtime organist and Buffalo, New York, theater organ expert Harvey Elsaesser was quoted as telling the *Buffalo News* in 1991 that "there was a time when any theater with 800 seats or more had an organ. I'd estimate there

The console of the "Mighty Mo" organ at the historic Fox Theatre in Atlanta, Georgia, suggests the complexity and the range of sonic possibilities offered by early 20th-century theater organs. (Library of Congress)

were at least 50 in Buffalo alone." Elsaesser continued by estimating that if all of Erie and Niagara Counties in New York were counted, the total would be "between 60 and 70" and that "about two-thirds of these were either Wurlitzers or Marr & Coltons" (The Staff of the *Buffalo News* 1991).

As Elsaesser stated, Wurlitzer was not the only company that constructed large, impressive organs for theaters in the first half of the 20th century. For example, Kilgen, which had its roots in early 19th-century Germany, became a major organ maker in the United States by the end of the 19th century. Although the company was best known as a maker of church organs, including that built for Saint Patrick's Cathedral in New York City, Kilgen also made theater organs. Over the years, virtually all of Kilgen's theater organs have been disassembled, destroyed, or moved to different locations. The Kilgen Wonder Organ, still functioning and played on a regular basis before films at the Canton, Ohio, Palace Theatre, has been located there since its installation in 1926. The Theatre claims that this is the only Kilgen theater organ still in its original location.

Ironically, the two organs described earlier date from just a year before a significant event that would virtually eliminate the need for theater organs: the release of *The Jazz Singer*. Although technically *The Jazz Singer* was not the first film with pre-recorded sound, the prominence of the soundtrack and its tight synchronization with the visual element in *The Jazz Singer* signified a major change for the movie industry. Within a few years of the 1927 release of the film, movie theaters were flooded with more "talkies," including numerous musicals, both those that were written specifically for film productions and those that were film adaptations of live shows. However, from the 1910s well into the 1930s, theater organs were numerous, provided musical accompaniment for silent films, provided pre-film concerts and sing-alongs, and were an important and sonically impressive part of popular culture.

See Also: Mechanical Organs; Motion Picture Soundtracks; Organs

Further Reading
Bush, Douglas E., and Richard Kassel, eds. 2006. *The Organ: An Encyclopedia*. New York: Routledge.
The Canton Palace Theatre. n.d. "A Brief History of the Kilgen Organ." Accessed January 18, 2021. https://cantonpalacetheatre.org/about-us/brief-history-of-the-kilgen-organ/.
Miller, Mary K. 2002. "It's a Wurlitzer: The Giant of the Musical Instrument Collection Makes Tunes Rootin 'Tootin' or Romantic." *Smithsonian Magazine* (April). Accessed January 18, 2021. https://www.smithsonianmag.com/history/its-a-wurlitzer-61398212/.
The Staff of the *Buffalo News*. 1991. "Pipes & Bells & Whistles for a Feast of Pull-out-the-Stops Music, There's Nothing Like Listening to Buffalo's Classic Theater Organs." *Buffalo News*, September 20. Accessed December 17, 2020. https://buffalonews.com/news/pipes-bells-whistles-for-a-feast-of-pull-out-the-stops-music-theres-nothing-like/article_90d400bd-ade1-5e96-865d-b49037ff7fc0.html.

Theremins

The Russian inventor and musician Leon Theremin developed the electronic instrument that bears his name around 1920 but first patented the instrument in 1928. Although widely acknowledged as the first electronic musical instrument, the theremin remains popular 100 years after it was invented. At time of this writing, several companies manufacture professional-level theremins that, unlike their predecessors from the 1920s and 1930s, must be connected to an amplifier in order to be heard. In addition, vendors on internet marketplaces such as Amazon.com, eBay.com, and others offer less-expensive theremins as well as theremin kits.

The theremin was by all accounts an instrument that developed out of an accident. Leon Theremin was working on the wiring of a radio when he inadvertently created a circuit that produced a howling sound that changed frequency (pitch) when he moved closer or further away from the antenna. Theremin, being a musician in addition to an electronics genius and inventor, figured out what was happening in the circuit and designed the instrument that bears his name based on the electronics principles he uncovered that were at work in the radio.

Basically, a theremin has two pairs of oscillators; one pair affects the pitch, and the other pair affects volume. Within each pair, one oscillator produces electronic vibrations of a constant frequency. The other member of the pair changes its speed of oscillation based on how close one gets to either the pitch or the volume antenna on the device. The human body, which consists in large part of water, acts as a capacitor, and when it enters the electromagnetic field around the antenna, it absorbs increasing amounts of electricity as one gets closer to the antenna. In the case of the pitch antenna, the difference between the frequency of the

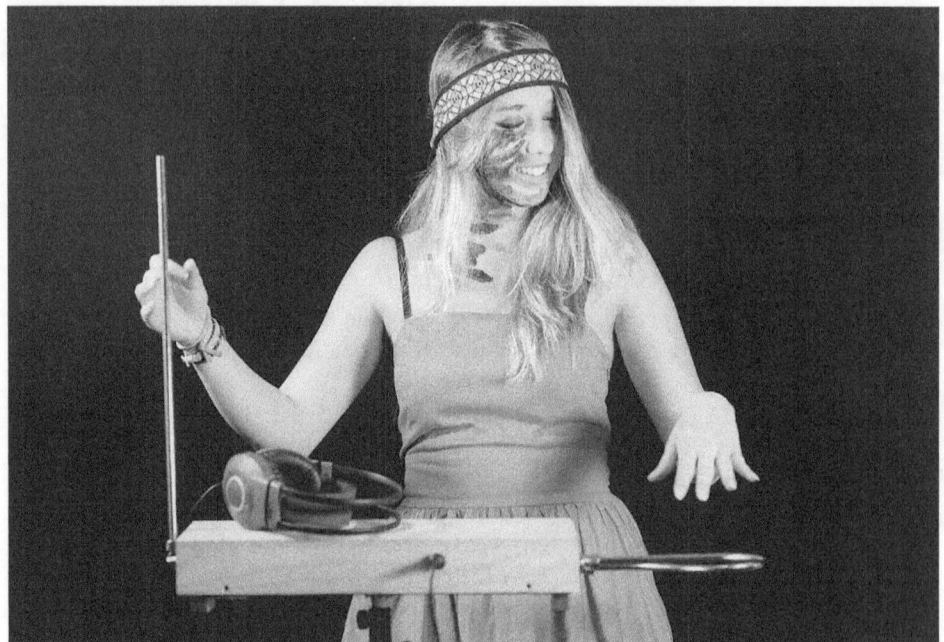

A modern theremin which uses the same principals as the instruments of the 1920s: the player's right hand controls pitch, and their left hand controls volume, both without actually touching the instrument. (Michelangelo Oprandi/Dreamstime.com)

never-changing oscillator and that of the circuit that includes changes caused by the body's capacitance becomes the pitch that one hears. In the case of the volume antenna, the sound becomes quieter as one gets closer to the antenna until the theremin becomes silent.

Around the time that Theremin patented the instrument, RCA began marketing theremins that included a built-in amplifier and speaker for home use. Although touted as the easiest instrument in the world to play in RCA's advertising copy of the day, it is actually one of—if not the—most difficult to play in tune, because of the fact that the hand controlling pitch has no tactile reference point. To put it more colloquially, one has to literally pull the pitches out of the air on the theremin. The RCA theremin failed to catch on and was only offered for a short time, possibly because of the difficulty in controlling pitch on the instrument but also certainly because of the 1929 stock market crash and subsequent Great Depression, during which time people generally did not have sufficient disposable income to purchase something as fundamentally unnecessary as a theremin.

In Hollywood, the 1940s and 1950s represented something of a heyday for the theremin, particularly in science fiction films. However, the instrument's use was not limited to this genre. As popular movie critic Roger Ebert wrote in his review of the 1993 documentary film *Theremin: An Electronic Odyssey*, the theremin's "otherworldly sound was used by Hitchcock to suggest mental illness in

Spellbound by the composer Bernard Herrmann, to accompany alien life forms in *The Day the Earth Stood Still*, by Billy Wilder to suggest alcoholic disorientation in *The Lost Weekend* and by Jerry Lewis to suggest looniness in *The Delicate Delinquent*" (Ebert 1995).

A ribbon controller that used the basic principles of the theremin was the source of the theremin-like melodic effects in the Beach Boys' song "Good Vibrations." The ribbon controller—which was also available on a contemporary electronic instrument of the theremin, the ondes Martenot—allowed the player to have tactile points as pitch references, akin to the fingerboard on a violin. This meant that the player could still slide between pitches, but it also meant that one could accurately find discrete pitches much more easily. More recently, Smash Mouth has used the theremin, and Jimmy Page dramatically incorporated the theremin in the concert tours Page did with his fellow former Led Zeppelin member Robert Plant in the mid-1990s. The theremin continues to be heard at the time of this writing as th e principal melodic instrument in the theme music, as well as in some of the incidental music for the long-running British detective television show *Midsomer Murders*, which has been in production since 1997 and is still in production at the time of this writing.

Although the theremin or close relatives of the instrument have had a place in popular culture since the mid-20th century, the instrument is also important as one of the influences on the synthesizer of the second half of the 20th century through the present. In fact, analog synthesizer pioneer Robert Moog built theremins—and published an article on transistorized theremin construction (see, Moog 1961)—several years before he released the first Moog synthesizer.

Ribbon Controller

Since Leon Theremin's invention of the electronic instrument that bears his name in the 1920s, the theremin has posed one overwhelming challenge to players: because the theremin is played without touching the instrument, it is extremely difficult to accurately find pitches. Musicians who wish to create the science fiction film-like effects of melodic playing, wide vibrato, and sliding between discrete pitches, then, have sometimes used a ribbon controller. By sliding one's fingers along a metal ribbon to change capacitance and change pitches—and therefore giving the player a tactile interaction with the inner electronics of a theremin—one can achieve the sonic effects of the theremin without having to rely on the hand's proximity to an antenna to literally pull a pitch out of thin air. Perhaps the most famous use of a ribbon controller was in the Beach Boys' 1966 song "Good Vibrations" (although the black-and-white video quality is not great, see https://www.youtube.com/watch?v=mdt0SOqPJcg, for a Beach Boys lip-synched performance of "Good Vibrations" in which Mike Love is shown with the ribbon controller, accessed May 19, 2020). Similar ribbon-like pitch controllers were used on the early 20th-century ondes Martenot, an electronic instrument that made its debut at about the same time as the theremin, as well as on late 20th-century synthesizers.

See Also: Moog Synthesizers; Ondes Martenots

Further Reading

Ebert, Roger. 1995. "Theremin: An Electronic Odyssey." Rogerebert.com, December 15. Accessed January 18, 2021. https://www.rogerebert.com/reviews/theremin-an-electronic-odyssey-1995.

Garner, Louis E., Jr. 1967. "For That Different Sound, Music a la Theremin." *Popular Electronics* 27, no. 5 (November): 29–33.

Glinsky, Albert. 2000. *Theremin: Ether Music and Espionage*. Urbana and Chicago: University of Illinois Press.

Martin, Steven M, director. 1993. *Theremin: An Electronic Odyssey*. VHS Video. Orion Classics. Released on DVD MGM/Fox Video, 2003.

McNairn, Bonnie, with James Wilson. 1993. "A Simple Theremin from Schematic to Performance." *Experimental Musical Instruments* 8, no. 3 (March): 26–27.

Moog, Robert A. 1961. "A Transistorized Theremin." *Electronics World* 65 (January): 29–32, 125.

Transistor Radios

The development of the transistor dates back to 1947, although these semiconductors did not fully replace vacuum tubes in electronic devices until years later. Ultimately, the use of transistors to amplify and modify electronic signals meant that televisions, radios, amplifiers, and other devices could be significantly smaller and lighter weight and consume far less energy than their vacuum-tube predecessors.

Although Bell Laboratories rolled out the first transistors in 1947, the connection of these replacements for the vacuum tube became a significant part of the musical experience in the beginning in the mid-1950s with the 1954 introduction of the Regency TR-1 and the 1957 introduction of the Sony TR-63. And it was in radios that members of the public first experienced fully transistorized electronics; televisions continued to use vacuum tubes and cathode ray tubes for years, as did many guitar amplifiers, PA system amplifiers, and so on. The transistor radio became an important part of youth culture during the 1960s and into the 1970s; thus, its inclusion as a separate entry in this volume. Another way to think of the transistor radio's importance is to consider that these devices were so plentiful from the late 1950s through the 1960s that they clearly showed the promise that semiconductors held.

In the "Inventing the Teenager" episode of the 2007 program *James May's 20th Century*, a series sponsored the U.K.'s Open University, program host May examines several new technologies that appeared after World War II that helped to establish the teenager as a sociological phenomenon, including the invention of the transistor and, in particular, the transistor radio. The premise of the program's discussion of the transistor revolves around the portability that marked transistor radios of the 1950s and 1960s. This, and their relatively low cost, made transistor radios particularly attractive to teenagers at the same time that commercial radio stations increasingly embraced the teen and rock and roll market. As a result, the transistor radio held a special place for teens and, as a result, played a role as part of the generation gap that marked the turbulent 1960s.

To be sure, there were earlier radios that were favorites of teens. The crystal radio sets of the 1920s, for example, were very inexpensive to make and required no power source other than the radio waves themselves. With long trailing wire antennas and very little audio output to an earphone, crystal radios were far less portable than the transistor radios that would start to come along starting in the mid-1950s.

Typical transistor radios of the 1960s and early 1970s offered a small built-in monophonic speaker for AM radio broadcasts as well as an output jack for an earphone. Typical headphones of this period used 1/4-inch jacks, so the single earphone was the norm for private radio listening. The fact that these teen-oriented radios typically offered only the AM radio band is significant, as AM was still the band most closely associated with Top 40 popular music.

The transistor radios of the 1960s and 1970s began a trend toward portable listening devices that has continued into the 21st century. (Funniefarm5/Dreamstime.com)

The fact that transistor radios required far less power and were considerably smaller than their vacuum tube-based predecessors also meant that manufacturers could design their radios in a variety of shapes and sizes. There were larger units that also offered FM broadcasts and larger speaker, and these can be seen as antecedents of the boomboxes that would appear in the 1970s; however, there was a considerable degree of variety in small radios. In fact, the Japanese company Panasonic turned the AM transistor radio into a fashion accessory and fashion statement in the early 1970s with its Panapet and Toot-a-Loop radios. The round Panapet, more formally Panasonic's model R-70, used a single 9-volt battery, typical of the transistor radios of the 1960s and early 1970s. Panasonic offered Panapets in several colors, including blue, white, red, and green, and each radio had a loop and chain attached to aid in carrying the radio, which was 4-3/8 inches in diameter. The Toot-a-Loop, Panasonic's model R-72, also used a single 9-volt battery to power its small built-in amplification circuit and .25w speaker. As the name suggests, the Toot-a-Loop was built as a loop that could pivot open so that it could be worn around the wrist. The outer diameter of the loop was just under 6 inches, which seems to make it a somewhat clumsy bauble, but it was for a time a popular fashion accessory for teenaged girls. Like the Panapet, the Toot-a-Loop was available in a variety of colors, including red, orange, blue, and yellow. Interestingly, but perhaps not surprisingly given the market for mid-century modern and slightly post-mid-century modern furniture, accessories, clocks, radios, and so on, there is an active collectors' market for these Panasonic designs today. More so than the collectors' market, the iconic nature of these Panasonic transistor radios is confirmed by the Museum of Modern Art in New York, which includes a Toot-a-Loop as part of its collection (for a description of the Museum's radio, see, https://www.moma.org/collection/works/3443, accessed December 1, 2020). Likewise, the Indianapolis Museum of Art at Newfields includes a Panasonic Panapet in its collection (for information on the museum's radio, see, https://www.artsy.net/artwork/panasonic-corporation-panapet-70-r-70-radio, accessed January 16, 2021).

See Also: Boomboxes; Radio

Further Reading
Riordan, Michael, and Lillian Hoddeson. 1998. *Crystal Fire: The Invention of the Transistor and Birth of the Information Age.* New York: W.W. Norton.

Thomas, Helen, Claudine King-Dabbs, and Dan Walker, producers. 2007. *James May's 20th Century.* London: BBC Productions. Issued on DVD in 2012 as Athena Learning AMP-8788.

Video Games

Video games, originally available primarily on arcade-type machines, have become standard forms of entertainment in many households and, on the highest level, a rapidly expanding and lucrative competitive phenomenon known as eSports. The relationship of music with video games has an interesting history and, at the beginning of the third decade of the 21st century, appears to be on the cusp of yet more important changes.

For all intents and purposes, the video game phenomenon of the 1970s can be viewed as an extension of the pinball games that had been arcade favorites for years. Generally acknowledged as the first mechanical pinball-type game, *Pachinko*, which was invented and still is popular in Japan, dates from the 1920s. In the United States, the countertop mechanical game *Bally Hoo*, dating from the early 1930s, is considered the earliest ancestor of the later freestanding pinball games. The Bally Manufacturing Corporation would expand into numerous different pinball games and play a major role in supplying arcade-type video games to arcades, restaurants, bars, and so on during the first golden age of video games of the 1970s and 1980s.

Created in Japan by Tomohiro Nishikado and first published in Japan in 1978 by Taito, *Space Invaders* quickly became a popular hit, both in Japan and in the United States. The extent of the use of music in *Space Invaders* was a brief monophonic bass line figure that repeated throughout the game. In early video games, such as *Space Invaders*, simple and brief monophonic material such as this could be programmed onto a chip that was part of the overall workings of the game. Incidentally, the concept of having a rudimentary chip that contained a limited amount of pre-programmed musical material inspired the musical genre known as chiptune, which has influenced some artists in the electronic dance music (EDM) field.

Two other early arcade favorites, *Donkey Kong*, created by Shigeru Miyamoto and published by Nintendo in 1981, and *Frogger* (1981), created by Konami and published by Sega in 1981, each includes lengthier synthesized musical sequences that accompany the various sound effects than did *Space Invaders*. Although *Frogger* also has a more complex texture—melody-plus-accompaniment—neither of these games goes very far beyond *Space Invaders* in terms of tying together action, the player's moves, and the music.

Eventually, in-home game machines allowed the user to play arcade favorites using their television as the monitor; however, the rudimentary chipped music remained, as they were established as an integral part of the games. In fact, even in the early 1990s, there remained little in the way of fully coherent transitions between musical themes, high-quality sound, or interaction between the progress of the game, the player's choices, the player's pace of play, and the musical underscoring—despite the fact that the computer technology of the games themselves and their graphics had advanced considerably over *Space Invaders* and the like and despite the fact that synthesized music outside of the world of video games had progressed well beyond what was possible in the 1970s.

Beginning in the late 1980s, computer sound cards, such as the popular Sound Blaster cards used on Windows machines, brought more polyphonic FM synthesizer power to computers. This was exploited by video game developers and composers. The growing use of MIDI and other video game-specific technologies started to change the connections between games and music in the 1990s. Among other things, LucasArts' *Interactive Music Streaming Engine* (*iMUSE*) technology, developed by Michael Land and Peter McConnell, sought to smooth out the transitions between music cues that accompanied the various scenes. As Michael Sweet explains in his book *Writing Interactive Music for Video Games*, *iMUSE* "allowed the music system to wait for a musical phrase to end before transitioning to a new cue . . . This system ensured a seamless transition by waiting for the current musical phrase or melody to end instead of interrupting it" (Sweet 2015, 45). Sweet continues by explaining that the importance of this development was that it provided a better sound continuum for the player, thus keeping them more fully engaged in the entire game experience. Interestingly, in a 2016 interview, McConnell explained that contrary to the reputation for interactivity that *iMUSE* had back in the 1990s, he and Land had actually created the perception of interactivity simply by composing and recording multiple versions of sound cues of different durations. The game would play the appropriate one based on what the gamer was doing, and the phrase-to-phrase, cue-to-cue connection described previously would make it appear that the game was composing the music based on what was happening in the game. Despite the impact that *iMUSE* made in the video game world, in the same interview, McConnell lamented the quality of sound provided by the Sound Blaster audio cards with which many computers used by gamers were outfitted in the late 1980s and early 1990s; it was as if the structural work in matching music to action was steps ahead of the quality of the polyphonic FM synthesis of the time (Mackey 2016).

At approximately the same time as Land and McConnell were developing *iMUSE*, singer-composer and synthesist Thomas Dolby was taking a different approach to addressing the same technological issues. In his memoir, Dolby discusses his dissatisfaction with some of the video game music both in the quality of the synthesized sounds and in how it integrated with the action and the visuals (or failed to do so) of the games. Dolby describes his work in the early 1990s on his *Audio Virtual Reality Engine* (*AVRe*). According to Dolby, "In a computer game, the musical changes have to happen *instantaneously*—in real time, without unpleasant glitches or overlaps or gaps [italics in the original]" (Dolby 2016, 162). Ultimately, the video game music system that Dolby and collaborator Steve

Ellison developed (*AVRe*) reacted to the moves that the gamer made, as opposed to maintaining a static cue in, say, a particular room, as was the case with some of the games of the time. In this respect, not only did *AVRe* match visuals, situations, and music but it also had this interactive attribute that helped to bring the programming of music without video games closer to artificial intelligence, as the game and its music became more intertwined with the gamer's choices, speed of motion, and the potential situations in which their character might find themselves.

The entire issue of fully integrating music and game progression and gamer choices is still being pursued by programmers and composers. In fact, between 2016 and 2020, several articles appeared in magazines and online about new developments that were either being promised or made with regard to the development of fully real-time artificial intelligence–composed video game music systems that would be able to accomplish what early 1990s' technology such as *iMUSE* and *AVRe* began. These include *Melodrive* and *WolframTones*, both of which can be used to provide music that is composed and changes based on the progression of the game.

It should be noted that even as programmers and composers try to even more fully integrate music with all the variables that can occur in video games, music's place within video games has grown exponentially since the late 1970s and *Space Invaders*. Composers have used the various sound synthesis and computer-based real-time music composition tools available to get the level of sophistication of the music up to the level of the visuals, the stories, and so on. As for the connection between music and the gaming experience, it is interesting to note the observations of *Popular Science* writer Sara Chodosh, who wrote, "This is, by far, the best Life Pro Tip I've ever gotten or given: Listen to music from video games when you need to focus. It's a whole genre designed to simultaneously stimulate your senses and blend into the background of your brain, because that's the point of the soundtrack. It has to engage you, the player, in a task without distracting from it. In fact, the best music would actually direct the listener to the task" (Chodosh 2018).

One other specific video game is also important to consider because of its musical impact on popular culture: *Guitar Hero*. This game, introduced in 2005, allows players to simulate the motions that they would use in playing lead guitar, rhythm guitar, and electric bass. Although the game might help players develop accurate and solid rhythm skills, perhaps the most important benefit of *Guitar Hero* was that it introduced a new generation to the rock music that was an important part of popular culture decades before they were born.

See Also: Artificial Intelligence; Computer-Composed Music; Computer-Generated Music; Digital Synthesizers; MIDI

Further Reading

Chodosh, Sara. 2018. "You Should Be Listening to Video Game Soundtracks at Work." *Popular Science*, January 26. Accessed January 18, 2021. https://www.popsci.com/work-productivity-listening-music/.

Collins, Karen. 2008. *Game Sound: An Introduction to the History, Theory, and Practice of Video Game Music and Sound Design.* Cambridge, MA: The MIT Press.

Dolby, Thomas. 2016. *The Speed of Sound: Breaking the Barrier Between Music and Technology.* New York: Flatiron Books.

Mackey, Bob. 2016. "Day of the Tentacle Composer Peter McConnell on Communicating Cartooniness." *USGamer*, March 8. Accessed January 18, 2021. https://www.usgamer.net/articles/day-of-the-tentacle-composer-peter-mcconnell-on-communicating-cartooniness.

Rogers, Scott. 2014. *Level Up! The Guide to Great Video Game Design.* 2nd ed. Chichester, U.K.: John Wiley & Sons.

Sweet, Michael. 2015. *Writing Interactive Music for Video Games.* New York: Addison-Wesley.

Vinyl Albums

Compared with the shellac records that they eventually replaced, vinyl records had less surface noise, could fit a significantly higher number of grooves per inch, were not as susceptible to groove collapse if the grooves were cut close together, were less susceptible to breakage, and spun at a significantly slower speed than the 78-rpm shellac records, which gave them the ability to hold significantly longer playing time. The new polyvinylchloride plastic that was available after World War II was used to make singles and albums. In this volume, we are primarily concerned with vinyl albums, which spun at 33–1/3 rpm; the plastic 45-rpm singles have a separate entry in this volume.

Even early in the age of the 33–1/3 rpm vinyl album, the new format and material provided easier-to-use and more musically coherent than the old multi-disc shellac albums, which one could think of as usually a disparate collection of several singles dating from approximately the same time period. With vinyl albums, the instrumental or vocal tracks need not have ever been on a single. In some cases, particularly with pop music, this led to somewhat mixed results. Some of the pop records by rock and roll groups of the early 1960s, for example, consisted of tracks that were part of hit singles with other inferior material that music fans often refer to as "filler." However, the added playing time of vinyl albums also made it easier for record companies to produce theme-based albums. In the 1950s, these theme-based albums typically came from jazz and pop vocalists; in the late 1960s, these would become the concept albums of the rock age.

One of the major enhancements to albums occurred with the introduction of stereophonic records in 1957. Because most radio airplay of popular music in the late 1950s through most of the 1960s was on AM radio, which was monophonic, because of the fact that some turntables and portable record players that pre-dated the almost exclusive use of stereo, and other factors, monophonic records continued to be manufactured well into the 1960s. Fans of classical music, who tended to be older and better financially established than pop music fans, were treated to stereophonic orchestral, instrumental solo, vocal ensemble, opera, and other recordings, as new stereophonic equipment was easier for these music fans to purchase than it was for the teens of the day.

Despite the high-quality sound that could be achieved on vinyl albums, the recording industry continued to strive for even greater quality. Even aside from such recording studio technologies as 8-, 16-, and 32-track recording equipment—which made extensive multi-tracking and overdubs significantly easier to achieve than on earlier tape recorders—attempts to improve audio fidelity were made throughout the 1960s and beyond. One of the more intriguing record producers at the center of experimentation who is not as well remembered today as others was Enoch Light. A violinist and light, pop music/easy listening bandleader, Light led Command Records for a brief period. While with Command, Light produced albums by several pop singers and instrumentalists (e.g., *Tonight Show* bandleader and trumpeter Doc Severinsen) recording onto 35mm movie film instead of multi-track tape recorder. For one thing, because the film was wound using sprockets, the sound was absent the wow and flutter that sometimes marred the sound of tape recordings. The film also allowed Light additional tracks to use in multi-track recordings that what was available on magnetic tape. Enoch Light's productions were also notable for ping-pong-like effects of sounds moving around the stereophonic soundscape, years, in fact, before those effects were taken up by rock bands during the psychedelic era of 1966–1968. Although perhaps not as innovative as his use of 35mm film at Command, Light's pop productions with Project 3 Records also are notable for the disc mastering process, selection of microphones, and so on that the liner notes for Project 3 releases claimed produced "distortion-free sound" at any volume level (Light 1967).

Another easy listening bandleader, Brad Hill of Mystic Moods Orchestra fame, also made a notable impact on the quality of vinyl albums in the 1970s. Hill's Mobile Fidelity Sound Lab produced half-speed mastered versions of popular classical, jazz, pop, rock, and country albums on special high-quality vinyl. Another 1970s' trend was to record music direct to a master disc. By eliminating magnetic tape virtually altogether (I say "virtually," because typically these performances were also made on tape as a backup), wow and flutter and tape hiss were also eliminated. The challenge is that direct-to-disc recording also eliminated the possibility of overdubbing and made the possibility of recording multiple takes—in case mistakes were made—economically disadvantageous for the record company. Direct-to-disc and half-speed mastered albums were aimed at audiophiles, and the vinyl albums produced by the manufacturers of these products were more expensive for the consumer.

Given the nature of the project, the fact that the playing technology is similar, and the fact that the introduction of the compact disc in the early 1980s led to a decline in sales of vinyl records, one might reasonably consider the Voyager Golden Record as part of the legacy—and perhaps as the culmination—of pre-digital recording and stylus-based music reproduction technology. In addition to welcome messages in various languages from world leaders, environmental and other representative sounds, the gold record that was carried on the *Voyager I* and *Voyager II* spacecrafts when they were launched in 1977, included examples of music from around the globe such as various genres of ethnic folk music, Western art music, jazz, 1950s rock and roll, and so on. Although the records on each of the spacecraft were metal, they played using a stylus, so the technology chosen to possibly introduce the sounds of human beings and their music to some alien civilization was part of the continuum used here on the earth since the late 19th century.

As mentioned earlier, the coming of the CD in the 1980s ushered in a new era in the consumer's experience of purchasing and listening to pre-recorded music. A 2019 article in *Pitchfork* magazine, however, reported that sales of vinyl records had reached a 30-year high in 2018. Paul McCartney's 2020 album, *McCartney III*, was noted as another driving force in the upsurge of interest in vinyl. As *Billboard*'s Keith Caulfield noted, "The album was available in more than 10 vinyl variants, which combined to sell nearly 32,000 copies in its first week—the third-largest sales week for a vinyl album since Nielsen Music/MRC Data began electronically tracking music sales in 1991. Only the debut weeks of Jack White's *Lazaretto* (40,000) and Pearl Jam's *Vitalogy* (34,000) were larger" (Caulfield 2020).

With the increased interest in what at one time had appeared to be a dying format came new environmental concerns, issues that had not garnered nearly as many headlines or interest back in the heyday of vinyl records. As Michelle Kim writes in the *Pitchfork* article, PVC "comes from refined oil and can take up to 1,000 years to decompose in a landfill. Traditional [record] pressing machines are powered by steam boilers that require fossil fuels to generate heat and pressure; the water used is treated with anti-corrosive chemicals in order to prevent rusting, thus creating more wastewater. And that's just the pressing procedure" (Kim 2019). Add to that the cardboard used for packaging, plastic shrink wrap, and so on, and the vinyl album might appear to be far less environmentally benign than digital music for streaming or download. As the recent upturn in sales suggests, though, the vinyl record is a technology that is still relevant well into the 21st century.

The Album

In the 21st-century download and streaming audio society in which we live, it is easy to lose sight of just what albums were in the past. Today, they may seem to be the equivalent of a playlist in a music application such as *iTunes*: a collection of audio tracks that somehow go together or were released at the same time in the same batch.

During the age of the compact disc (CD), from the mid-1980s into the early 21st century, the album was a collection of music released on the same disc or discs. Often, they had elaborate booklets that presented the consumer with song lyrics, background on the performers and/or the selections, lists of performers, visual imagery that was tied to the music, and so on. During the age of the long-playing vinyl, 33–1/3-rpm disc, from approximately 1950 until the CD became dominant, albums most frequently were single two-sided discs that perhaps had liner notes about the performers and selections on the back of the cardboard sleeve, or in the case of some rock and pop albums (e.g., the Beatles' *Sgt. Pepper's Lonely Hearts Club Band*), printed lyrics of the songs.

Although some jazz albums as early as the 1950s and rock, soul, pop, and other albums of the late 1960s and beyond were built around clear themes, these were known as concept albums. Other less tightly constructed discs contained perhaps a few A sides and B sides from hit singles with the rest of the material essentially being filler.

The term "album" for a collection of music, however, dates back to the pre-LP days of 78-rpm shellac records that held one song per side. Albums were actually collections of these single discs, with each disc placed in a paper sleeve. The sleeves were bound together like the pages of a photo album, and the entire package had a hard protective front and back cover, also like a photo album; hence the term "album."

See Also: 45-rpm Records; Stereophonic Sound

Further Reading

Caulfield, Keith. 2020. "Paul McCartney's *McCartney III* Debuts at No. 1 on *Billboard*'s Top Album Sales Chart." *Billboard*, December 28. Accessed January 7, 2021. https://www.billboard.com/articles/business/chart-beat/9504857/mccartney-iii-debuts-top-album-sales-chart/.

Kim, Michelle. 2019. "How the Record Industry Is Trying to Make Vinyl More Environmentally Friendly." *Pitchfork*, May 9. Accessed January 18, 2021. https://pitchfork.com/thepitch/how-the-record-industry-is-trying-to-make-vinyl-more-environmentally-friendly/.

Light, Enoch, producer. 1967. Liner notes for *Kites Are Fun* by the Free Design. LP, Project 3 PR5019SD.

Mobile Fidelity Sound Lab. n.d.a. "History." MoFi.com. Accessed May 29, 2020. https://mofi.com/.

Mobile Fidelity Sound Lab. n.d.b. "Limited Editions Form the Original Masters." Advertising liner note supplement.

Scott, Jonathan. 2019. *The Vinyl Frontier: The Story of the Voyager Golden Record*. London: Bloomsbury Sigma.

Virtual Accompanists

One of the challenges faced by music students and professional musicians is that scheduling time with collaborative pianists, fellow chamber musicians, jazz bands, rock bands, and symphony orchestras is difficult, expensive, or completely impossible. Over the years, a number of evolving technologies have emerged that help aspiring jazz improvisers, students of classical repertoire, and professionals who are preparing for orchestral auditions or even a performance with a symphony orchestra build their skills to a performance level and become used to the sound of the instruments or full ensembles with which they might later perform.

One of the famous such technologies from decades past was Music Minus One. Founded by Irv Kratka in 1950, Music Minus One produced LPs (vinyl albums) that featured well-known repertoire, with each disc including two versions: one of a well-known performer playing or singing the music with accompaniment and one with the principal part not recorded. The concept was that someone working on the music could play or sing along with the accompaniment. Music Minus One was popular with music teachers, particularly those with students preparing for recitals, state solo and ensemble contests and festivals, and so on, because practicing with the recording would at least provide a simulation of what they would experience when performing with a live collaborative pianist. The fact that a Music Minus One disc included a full performance by a recognized expert professional musician enabled students to hear subtleties of phrasing and other musical points not necessarily fully expressed in the musical notation alone.

Similar to the Music Minus One approach were the accompanying recordings—first on LP and later on cassette and CD—produced by jazz saxophonist, educator, and publisher Jamey Aebersold. Beginning with his first volume in 1967, Aebersold produced dozens of recordings with accompanying books that allowed jazz musicians who played E-flat (e.g., alto and baritone saxophone), B-flat (e.g., tenor saxophone, clarinet, and trumpet), and concert-pitched instruments (e.g., trombone, piano, and guitar) to play and practice improvisation on well-known jazz standards. One of the unique aspects of Aebersold's recordings was that they placed the piano part entirely on one of the stereo channels and the bass part on the other. The drum part was audible on both the stereo channels. Because of this approach, jazz pianists could play along with the piano part of the ensemble essentially muted and practice their melodic improvisation and jazz accompanying (often called "comping" by jazz musicians) skills.

As one might imagine, playing along with an analog audio recording posed some challenges. For one thing, although speed could be adjusted on turntables, changing the speed of the recording would also change the pitch. This issue finally could be addressed during the digital age. One of the notable examples is the Hal Leonard Corporation's *Amazing Slow Downer*, which has been available on enhanced CDs that could be played on Windows and Apple Mac OS computers. If the user wanted to do so, they could slow down the play-along track without affecting pitch. So, a purchaser of the company's play-along CD and book of, say, surf guitar favorites associated with the Ventures could work their way up to the tempo of the famous versions of the pieces. Alfred Music Publishing offered its version of enhanced CDs with some of their popular music publications. The *Tone 'N' Tempo Changer* software that accompanied titles such as the *Acoustic Rock Riffs* book and CD functioned similarly to the *Amazing Slow Downer* and allowed the user to make tempo and other changes to the practice tracks. In fact, the current version of the *Tone 'N' Tempo Changer* opens a virtual mixing board that allows the various instruments to be muted or sound as a solo track.

At the time of this writing, the *Amazing Slow Downer* is also available as a download for Windows, Mac OS, iPhone, iPad, and Android operating systems from the software company Roni Music. The downloads range in price from $9.95 to $49.95 (see, https://www.ronimusic.com/, accessed January 16, 2021). Hal Leonard offers a free downloadable version of the *Amazing Slow Downer Lite* for Windows and Mac OS machines that is only usable with the company's enhanced CDs (see, https://www.halleonard.com/ASD/, accessed January 16, 2021).

Another approach to providing accompaniment tracks came in the form of *Band-in-a-Box*, a software package that originally allowed users to put together MIDI recordings that could be looped for practice in a wide variety of genre styles. *Band-in-a-Box* has also been popular among composers/songwriters, as one can create a chord progression, or use one that is standard in popular music, loop it, create verse and chorus sections, and so on, and then create, say, a song's melody to go over the top of the accompaniment. *Band-in-a-Box*, which was created by Peter Gannon of PG Music in 1990, has continued to evolve into the 21st century and has been used in the creation of karaoke backing tracks as well as in making practice tracks for instructional downloads and streaming subscription series. As PG Music's website describes the software, "Band-in-a-Box® is so easy to use! Just type in the chords for any song using standard chord symbols (like C, Fm7, or C13b9), choose the style you'd like, and Band-in-a-Box® does the rest . . . Band-in-a-Box® automatically generates a complete professional-quality arrangement of piano, bass, drums, guitar, and strings or horns" (https://www.pgmusic.com/, accessed August 19, 2020). PG Music's *RealBand* takes the virtual accompaniment idea one step further by incorporating live tracks that can be worked into one's accompaniment arrangement using artificial intelligence. At the time of this writing, several internet-based music instructors use PG Music's products to produce their accompaniment tracks. *Band-in-a-Box* is available for Windows and Mac OS platforms, but *RealBand* is a Windows-based program.

Although the use of digital recordings and MIDI-based recordings allows for tempo to be changed without changing pitch using software such as the *Amazing Slow Downer*, one of the issues that performers encounter particularly in performing the classical repertoire is with coordinating rhythm nuances—known as rubato—between performers. In practicing along with an early Music Minus One or even its digital successor, the soloist is locked into the recording's rhythmic and metrical approach. To put it another way, these recordings do not coordinate with, or follow, a soloist like a good collaborative pianist does.

As computing power grew and artificial intelligence became more sophisticated, this challenge was met by *SmartMusic*. Although *SmartMusic* can also be categorized as music education software, and thus is also included in the Instructional Software entry in this volume, the software's piano accompanying feature allows the virtual pianist to follow the soloist's rhythmic nuances (rubato). The degree to which the program does so can be set and reset. So, if someone wants to practice and learn a piece at a steady pace, the degree to which the virtual accompanist follows the soloist can be minimized. Conversely, if a performer is working on a composition that demands high levels of rhythmic flexibility (or if the soloist is still struggling with a piece technically), the degree to which the virtual pianist follows the soloist can be maximized. In the late 2010s, *New SmartMusic* was launched, and its producer, MakeMusic, Inc., rolled out plans for the replacement of the original *SmartMusic* software.

As has been the case with the proliferations of apps for iPhones, iPads, and Android devices in the 21st century are various aspects of computer technology, the state of the virtual accompanist is quite different at the time of this writing than it was even just a few years ago. Today, there are numerous applications available that are geared toward specific repertoire and are fully interactive so that the user virtually becomes part of the ensemble (e.g., jazz trumpeter Christian Scott aTunde Adjuah's *Stretch Music*) or geared toward and designed for specific instruments (e.g., the musical theater–focused vocal accompanying app *Appcompanist*). In fact, even a cursory study of an applications vendor such as Apple's App Store reveals titles such as aforementioned *Appcompanist*; *Cadenza Live Accompanist*, which focuses on famous solos from the orchestral repertoire; *MyPianist*, a virtual piano accompanist that uses artificial intelligence to follow the soloist, in the manner of *SmartMusic*; *MTI RehearScore*, a virtual musical theater accompanying app published by Music Theatre International (MTI), the licensing agency that owns the rights to many of the best-known and most popular musical shows of the past and present; and several others.

In addition to the changes in technology for virtual accompanists, the business model has shifted from user purchases of individual commodities (e.g., LPs or CDs with accompanying sheet music) to subscriptions that allow access either to an entire library of repertoire for one subscription price or to individual downloads that are individually priced for that particular application.

Perhaps one of the most exciting aspects of the developments in virtual accompanists who work collaboratively with the soloist is that these applications can be used in conjunction with the smart practice rooms and virtual performance venues that are also available to simulate real-life performance in a real-life space. For example, a college music student could use the Wenger Corporation's V-Room or Virtual Acoustical Environment technology (the latter was jointly developed with Lexicon Pro) interfaced with virtual accompaniments in a relatively small practice room and prepare for a senior recital that is scheduled to take place in a large recital hall. The student's sound, the accompaniment, and the virtual acoustical environment (e.g., reverberation and echo from the back of the virtual recital hall) would resemble that which the student will experience for real in the recital hall on the day of their recital.

Smart Practice Rooms and Virtual Auditoriums

One of the challenges faced particularly in university music units revolves around practice rooms. Often practice rooms are small and represent an acoustical environment that bears little resemblance to what students will encounter when they perform, say, their junior recital in a 150-seat hall, or when they perform a concerto with the college orchestra in a 1,000-seat auditorium. One of the challenges of many practice rooms is that their small size tends to discourage students from projecting their sound like they would need in order to adequately fill a performance space. An attendant issue is that typically students do not have enough access to university recital halls and auditoriums in order to become fully accustomed to their acoustics before they perform.

Wenger Corporation rolled out its V-Room (*v* for virtual) in the early 21st century. This was followed in 2006 by the Virtual Acoustical Environment (VAE) technology jointly developed by Wenger and Lexicon Pro. These technologies allow an equipped small-footprint room to simulate nine different acoustical environments: practice room; baroque room; medium and large recital hall; small, medium, and large auditorium; cathedral; and arena (Wenger Corporation n.d.). Anecdotally, this has continued to be one of the more impressive demonstrations at the annual Ohio Music Education Association annual conventions and appears to address many of the space and acoustical environment issues associated with university music units.

See Also: Karaoke

Further Reading

Beuttler, Bill. 2020. "Christian Scott aTunde Adjuah: Don't Stream; Stretch." *JazzTimes*, March 24. Accessed January 18, 2021. https://jazztimes.com/features/profiles/christian-scott-atunde-adjuah-dont-stream-stretch/.

National Association of Music Merchandisers (NAMM). 2005. "Irv Kratka." NAMM Oral History Program, September 13. Accessed January 18, 2021. https://www.namm.org/library/oral-history/irv-kratka.

Odell, Jennifer. 2019. "The History of Play-A-Long (and More)." *JazzTimes*, April 25. Accessed January 18, 2021. https://jazztimes.com/features/profiles/history-of-play-a-long/.

Wenger Corporation. n.d. "What Is VAE® Technology?" Wenger Corporation. Accessed January 18, 2021. https://www.wengercorp.com/Lit/Wenger_VAE_BRO_LT0373.pdf.

Virtual Studio Software

Perhaps one of the most accessible and economical pieces of virtual studio software—as well as one of the best-known today—is Apple Computer's *GarageBand* (see separate entry in this volume), which the company introduced in 2004. Part of the beauty of *GarageBand* is its near ubiquitous nature, as Apple produces versions for Macintosh computers, iPads, and even iPhones, and it is a free program/app for users of these Apple devices. The history and the present-day offerings in virtual studio software, however, do not begin and end with *GarageBand*.

One might reasonably consider the precursor of the virtual studio programs and apps that are available in the third decade of the 21st century to be Sonic Foundry's *Acid* software package of the 1990s. Although in retrospect *Acid* was more akin to sequencing software than what we might think of today as virtual studio software, composers, arrangers, and DJs used this software package to create instrumental riff and beat patterns, song accompaniments, all the way up to elaborate full compositions. *Acid* allowed users to highlight sections of an arrangement and copy or drag and drop to another location within the piece. Sony Digital Films purchased all of Sonic Foundry's software library in 2003, and MAGIX purchased the Sonic Foundry library from Sony in 2016, but *Acid* in various forms has continued to be modified and made more powerful through the present, and the program's drag-and-drop feature remains at the core of today's virtual studio software, from free applications such as *GarageBand* and *Audacity* to expensive sophisticated software.

Sonic Foundry's *Sound Forge* was closer to today's virtual software than its labelmate *Acid*. Like *Acid*, *Sound Forge* became part of the Sony Digital Films library and is now published in its 14th generation by MAGIX. Other current professional virtual studio software includes Apple's *Logic Pro X* professional studio software, which at the time of this writing carries a retail list price of just a tick under $200. Avid's *Pro Tools* has been widely used in university and commercial studios and is currently available on a subscription basis for $199 per year. The "Ultimate" version of *Pro Tools* is available at the time of this writing for just under $600. PreSonus, which also manufactures interfaces to connect microphones, electronic instruments, electric guitars, and so on to computers using USB, offers the virtual studio software package *Studio One Professional*, which is widely available for $399.95, with less expensive upgrade versions available for owners of PreSonus's less expensive and less powerful studio software.

The differences between software and programs such as *Audacity* and *GarageBand* include ease of use, power, the number of tracks that one can record, bit rate at which material can be saved, and other factors. However, even a program such as *Audacity* offers speed adjustment, track syncing capabilities, multi-track recording, a wide variety of electronic effects, echo, reverb, flipping a segment of a recording backward, the ability to drag and drop or copy and drop, the ability to zoom in on very small chunks of the audio so that a single stray note can be exorcised, and so on. In short, even some of the free and low-cost programs can replicate what could be done in an electromagnetic tape-based recording studio as well as some features that have only been possible in digital studios (e.g., changing a recording's speed without changing its pitch, or changing a recording's pitch without changing its speed).

Although they might not offer as many features as even an *Audacity* or a *GarageBand*, in recent years there has been a proliferation of studio/song construction applications for tablets and smartphones. Because there are so many apps currently available and because new apps seem to come to app stores virtually each week, it is difficult to generalize about trends in the world of inexpensive and free virtual studio apps. Suffice it to say, however, that it does appear that quite a few of these applications, many of which seem to bill themselves as virtual recording studios, focus on constructing backing tracks from pre-recorded loops. In other words, many of them lack the ability to allow the user to record multiple tracks of their own original material, in contrast to earlier virtual studio software from *GarageBand* to *Studio One* and *Pro Tools*. Although this may be a function of the low cost of the applications, the prepackaging of loops, drum beats, and so on resembles the prepackaging of sequences and patches that came into the world of sound synthesis, as post-Moog analog and then digital synthesizers were introduced in the 1970s and 1980s. To a certain extent, particularly in these apps and in some of the virtual studio software such as the latest iterations of *GarageBand*, convenience seems to be taking the place of allowing for full creativity.

See Also: Multi-Track Recording

Further Reading

Collins, Mike. 2015. *In the Box Music Production: Advanced Tools and Techniques for Pro Tools*. Burlington, MA: Focal Press.

Huber, David M. 2018. *Modern Recording Techniques*. 9th ed. New York: Routledge.

Lendino, Jamie. 2019. "Apple *GarageBand* (for Mac) Review." *PC Magazine*, September 26. Accessed October 29, 2020. https://www.pcmag.com/reviews/apple-garageband-for-mac.

Virtual Synthesizers

The first synthesizers available to performing musicians and general members of the public were stand-alone units, such as the Moog synthesizers of the late 1960s famously used by musicians as diverse as the Monkees, Wendy Carlos, the Byrds, and the Beatles. Later analog and some of the digital synthesizers that followed interfaced with computers but still had a base unit that almost always contained a piano-style keyboard. Eventually, the need for the keyboard itself to include extensive synthesis capabilities diminished to the point that today, virtual synthesizer software eliminates the need for an external keyboard.

Interestingly, one of the first moves in the direction of using a computer outside the confines of room-sized mainframe machines involved integrating the synthesizer and computer essentially within the synthesizer itself. The Fairlight CMI (computer musical instrument) synthesizer, which was introduced at the end of the 1970s, was the first digital sampling keyboard. The instrument allowed the user to digitally record sound samples and then digitally edit them on the instrument's display screen. The Fairlight CMI was quite expensive, with prices going into five figures. Soon thereafter, digital synthesizers such as Yamaha's DX7 and others brought the cost of digital synthesis down considerably, but they functioned as stand-alone units and tended to favor preset synthesized sounds over sampling or user-controlled additive or subtractive synthesis.

The release of Roland's Jupiter-6 and the Prophet 600 (both MIDI compatible) in 1982 and the publication of the MIDI standard the following year dramatically changed the respective roles of synthesizers and computers. MIDI made it possible for interaction between the two types of machines that, although not entirely seamless, was absent the kinds of battles between standards that have complicated other technological advances (e.g., home video tapes, cassette noise reduction systems, and early sound recording media). Fully computer-based synthesis, however, was marred by such factors as the limitations of early computer sound cards.

Beginning in the late 1980s, computer sound cards, such as the popular Sound Blaster cards used on Windows machines, brought more polyphonic FM synthesizer power to computers. The quality of these sound cards was a matter of concern to composers, video game music programmers, and so on. For example, in 2016 interview, Peter McConnell, one of the developers of LucasArts' *Interactive Music Streaming Engine* (*iMUSE*) technology, lamented the quality of sound provided by the Sound Blaster audio cards with which many computers used by gamers were outfitted in the late 1980s and early 1990s. McConnell suggested that the work that he and other programmers were doing at the time was steps ahead of the quality of the polyphonic FM synthesis provided by the sound cards of the time (Mackey 2016).

As the power and speed and sound capabilities of desktop laptop computers increased and improved, it became increasingly possible for musicians to use less-expensive and often smaller (far fewer than the 88 keys of a piano keyboard) keyboards as controllers of computer-based sound synthesis programs, music notation programs, and so on. To put it another way, improvements to consumers' computers and synthesizer programs shifted sound synthesis away from stand-alone keyboard-containing units.

The next step in the development of fully computer-based consumer synthesis came with the development of virtual keyboards, perhaps most conveniently found on iPads and other tablet computers. The touchscreen nature of these machines made them more intuitive for playing a screen-based keyboard than computers with point-and-click mice or touchpads. They are a practical improvement, too, over some of the earlier software with which the user could select pitches using the alphanumeric keyboard of their desktop or laptop machine. Now, the computer itself contained the keyboard (virtual), sound synthesis capabilities, playback capabilities, and so on. A musician could use a music notation program such as *Finale* or *Sibelius*; enter the pitches, rhythms, and so on from the device's screen; select voices (instrumental sounds); save and play back the entire project; and print out a score and parts. It should be noted, however, that for the type of project described here, a keyboard controller is still preferable because more complex music can be played in real time, and such things as touch/velocity sensitive keys make it possible to more easily control dynamics of passages and individual notes.

See Also: Desktop and Laptop Computers; Digital Synthesizers; Virtual Studio Software

Further Reading

Mackey, Bob. 2016. "Day of the Tentacle Composer Peter McConnell on Communicating Cartooniness." *USGamer*, March 8. Accessed January 18, 2021. https://www.usgamer.net/articles/day-of-the-tentacle-composer-peter-mcconnell-on-communicating-cartooniness.

Sweetwater Sound. 2004. "Virtual Analog Synths vs. Analog Synths." Sweetwater.com, April 22. Accessed January 18, 2021. https://www.sweetwater.com/insync/virtual-analog-synths-vs-analog-synths/.

Vocoders and Talk Boxes

Various effects, including reverb, fuzz tone, and others, have been part of the arsenal of electric guitar players for decades. Electronic keyboards, too, offered various effects before there were commercially viable and accepted ways of modifying the sound of the human voice electronically. Perhaps not as well known is the fact that electronic production of the virtual human voice was also developed between the late 1920s and the late 1930s. However, until the 1970s, electronic modification of the human voice in commercial music generally was limited to effects that could be generated in the recording studio or by using effects that were originally intended to be part of amplification systems for instruments. That changed in the 1970s with Bob Heil's development of the Heil Talk Box and the use of the Vocoder—the device that dates back to the late 1930s—as means of changing vocal tone color in popular music.

Working at Bell Labs, Homer Dudley invented the vocoder in 1938, after approximately a decade of work had been done on developing speech synthesis. Although music fans think of the vocoder as part of the electronic arsenal of tools available for modifying the voice in popular music, film soundtracks, and so on, the original use was tied to encoding speech in such a way that it could be transmitted in smaller chunks to preserve bandwidth in telecommunications. The encoded speech also ensured the security of transmissions as electronically encoded information (as compared, say, with talking over a telephone line). On the receiving end, the synthesized speech was played through a loudspeaker.

The speech synthesis capabilities of the vocoder and the robot-like sound of the speech that it can produce lend themselves to use in science fiction films. The modifications that it can make to human speech and singing also lend themselves to musical compositions and productions that are meant to have a machine-like quality. So, the vocoder provided a fitting sound for Laurie Anderson's singing and narration in her technology-focused piece "O Superman." Other examples of the use of the vocoder include Imogen Heap's song "Hide and Seek," Earth, Wind & Fire's "Let's Groove Tonight," and Lipps, Inc.'s "Funkytown," to name just a few of the better-known examples.

Perhaps the best-known use of the vocoder in popular music, however, was Jeff Lynne's use of the device in the 1977 Electric Light Orchestra (ELO) song "Mr. Blue Sky." In fact, Lynne's voice is treated so heavily by the vocoder technology that it is difficult to discern some of his words. One of the more interesting results of this is that listeners often mistake the last sung phrase of the song as being the song's title. Lynne confirmed in a 2012 interview on the British television program *The One Show* (see, e.g., https://www.youtube.com/watch?v=ZFexdd_-tL8, accessed January 16, 2021) that he actually sang the words "Please turn me over"—this, because "Mr. Blue Sky" originally appeared at the end of the third side of the Electric Light Orchestra's double album *A New World Record*.

Bob Heil's 1973 invention of the Heil Talk Box was taken up by musicians immediately. For example, Joe Walsh's 1973 song "Rocky Mountain Way" made prominent use of the Heil Talk Box and remains one of the classic examples of the use of the device (for Walsh's original recording of the song, see, https://www.youtube.com/watch?v=4Fz-mHGXgzs, accessed January 16, 2021). Three years later, Peter Frampton's album *Frampton Comes Alive!* contained "Do You Feel Like We Do" and "Show Me the Way," both of which featured Frampton's use of Heil's device. Frampton's album remains one of the best-selling live rock albums of all time, and the popularity of the recording, as well as the unusual sound of Frampton's voice on the songs, instantly made him the best-known user of the Talk Box.

Heil's device allows the performers instrument to integrate with their voice to essentially make it sound like their electronic instrument was producing vowel sounds and so on. Although performers such as Joe Walsh and Peter Frampton used the Talk Box to interface their voices with their electric guitars, some other performers used the device to interface their voices with other electronic instruments. A video of Stevie Wonder performing on the *David Frost Show* using a Talk Box is particularly interesting because it allows the viewer to see and hear how Wonder uses the device (see, https://www.youtube.com/watch?v=PnR19lNlXV8, accessed January 16, 2021).

One of the challenges posed by the Heil Talk Box and some later similarly designed devices by other companies is that the user has to keep a tube in their mouth or very near their open mouth. This was how the user would produce the vowel and diphthong shapes that would be integrated into the tone color of their instrument. More recently, a number of pedals that produce the same effects as the earlier vocoders and talk boxes have entered the market. A YouTube program on *That Pedal Show* contains demonstrations of some of these pedals (see, https://www.youtube.com/watch?v=M9NLDvf_ZPU, accessed January 16, 2021). The program is particularly worth watching for its explanation of the complex way in which classic talk boxes function, particularly with their need of two separate amplifiers to produce sound. Some pedals that are currently available eliminate the need for the plastic tube to achieve talk box effects—using a standard microphone instead—and some have multi-function vocoder and talk box settings.

See Also: Effects Pedals

Further Reading

Anderton, Craig. 2003. "Q. What's the Difference between a Talk Box and a Vocoder?" *Sound on Sound*, October. Accessed January 18, 2021. https://www.soundonsound.com/sound-advice/q-whats-difference-between-talk-box-and-vocoder.

Wire Recording

Although not necessarily well remembered today, wire recording was for a brief period of time one of the primary ways that individuals could make live recordings, make recordings off the radio, and so on. The technology predated the widespread home use of electronic tape. As a matter of fact, the basic principles of recording sound by magnetizing metal wire and playing it back date back to the last quarter of the 19th century.

For all intents and purposes, the history of wire recording—and, indeed, recording sound on magnetic media—goes back to American inventor and electrical engineer Oberlin Smith. Smith's article (Smith 1888) shows and describes a procedure for using cotton or silk thread to which magnetic metal shavings are attached. By the late 1890s, the Danish inventor Valdemar Poulsen had developed a machine that used Smith's principles and that received U.S. Patent No. 661,619 (for Poulsen's patent application, see https://patents.google.com/patent/US661619A/en, accessed September 4, 2020). Most famously, Poulsen used his machine to make a brief recording of Austro-Hungarian emperor Franz Josef at the Paris World's Fair in 1900; this was decades before magnetic tape recording came into being.

Although the German inventor Fritz Pfleumer developed a method of using electromagnetic tape for audio recording in the late 1920s, wire recording using the principles of Smith and Poulsen enjoyed a period of popularity through the 1940s. The technologies overlapped in time; however, ultimately, tape proved to be more reliable and practical, and it had greater audio fidelity than magnetic wire. During the heyday of wire recording, several companies, including Western Chicago and Silvertone, made wire recorders for the consumer market.

Although the medium is largely forgotten today, occasionally one can find old wire recorders for sale on various internet sales sites, at antique malls, garage sales, and so on. During the 1940s, however, it was one of the few media for amateurs to make live recordings. As such, it became the stuff of legend. The most famous legend about wire recordings arose about saxophonist Dean Benedetti, who was known to have made recordings of live performances by the great bebop jazz saxophonist Charlie Parker in 1947 and 1948. The story for years was that Benedetti had recorded Parker using a wire recorder that was allegedly captured from the Nazis in World War II. However, when the missing Benedetti recordings were finally found and mastered for commercial release, it was discovered that in reality, they were all either recorded on an inexpensive Sears disc-cutting machine or on reel-to-reel tape (Harrington 1990); there was no evidence that a wire recorder had been used. According to Harrington, the legend of the missing wire recordings can be dated back to Ross Russell, a Parker biographer who wrote the novel *The Sound*. In the novel, the character Royo is based on Benedetti. Royo follows Parker around and makes numerous wire recordings at a variety of Parker's live dates. Russell's later less-fictionalized Parker biography, *Bird Lives!* continued to build on the Benedetti legend, indeed; the book opens with a scene of Benedetti allegedly making an unauthorized recording of Parker in a New York club (Russell 1973).

Despite the fact that the story of Dean Benedetti and the lost wire recordings was in large part the stuff of fiction, wire technology really was part of the recording legacy of another famous American musician: Woody Guthrie. In fact, a recording of a 1949 live performance by singer-songwriter Woody Guthrie done on a wire recorder is apparently one of only a few live recordings of Guthrie known to exist. The recording was made at the Jewish Community Center in Newark, New Jersey. Because of the educational setting of Guthrie's appearance, Marjorie Guthrie essentially interviews her husband about the songs that Woody Guthrie then performs. Paul Braverman, who made the recordings, had stored them in a closet, only to discover them at the start of the 21st century. The raw recordings were digitally cleaned up, restored, and released on compact disc as *The Live Wire: Woody Guthrie in Performance 1949*. In 2008, the Recording Academy presented the 2007 Grammy Award for Best Historical Album to compilation producers Nora Guthrie and Jorge Arévalo Mateus and mastering engineers Jamie Howarth, Steve Rosenthal, Warren Russell-Smith, and Kevin Short for *The Live Wire: Woody Guthrie in Performance 1949*.

See Also: Reel-to-Reel Tape

Further Reading

Daniel, Eric D., C. Denis Mee, and Mark H. Clark, eds. 1999. *Magnetic Recording: The First 100 Years*. New York: John Wiley & Sons/IEEE Press.

Harrington, Richard. 1990. "The Legend of Benedetti and 'Bird.'" *Washington Post*, November 18. Accessed April 1, 2020. https://www.washingtonpost.com/archive/lifestyle/style/1990/11/18/the-legend-of-benedetti-and-bird/a57e739d-2c67-42f1-946f-95982df27d38/.

Ruhlmann, William. n.d. "Woody Guthrie: *The Live Wire: Woody Guthrie in Performance 1949*." *AllMusic*. Accessed January 18, 2021. https://www.allmusic.com/album/the-live-wire-woody-guthrie-in-performance-1949-mw0000496961.

Russell, Ross. 1973. *Bird Lives!: The High Life and Hard Times of Charlie (Yardbird) Parker*. New York: Charterhouse.

Schaap, Phil. 1990. Liner notes for *The Complete Dean Benedetti Recordings of Charlie Parker*. Stamford, CT: Mosaic Records.

Smith, Oberlin. 1888. "Some Possible Forms of Phonograph." *Electrical World* 11–12 (September 1): 116–117. Accessed January 18, 2021. https://books.google.com/books?id=zYVMAAAAYAAJ&pg=RA2-PA116#v=onepage&q&f=false.

Woodwind Instruments

The modern woodwind instruments, including the flute, clarinet, oboe, bassoon, and saxophone, might not necessarily seem particularly high-tech and have undergone few substantive technological advances in decades. However, the present-day woodwinds owe a huge debt of gratitude to instrument makers, performers, and physicists, particularly in the mid-19th century, individuals who basically gave us the instruments that we use today in bands, orchestras, chamber groups, and jazz combos.

Let us first consider the flute because of the importance and influence of Theobald Boehm's work in the middle of the 19th century. As Samuel Baron wrote in his introduction to a revised edition of Dayton C. Miller's 1922 translation of Boehm's treatise on the flute and flute playing, "Theobald Boehm had a unique combination of skills. He was a master flutist, a master gold- and silversmith, and a keen student of physics and acoustics as they apply to the flute . . . All of this is known from history, and we flutists know it best of all, for we play on a Boehm flute virtually unchanged from his work bench of 1847 to the present day" (Boehm 1964, v). Indeed, Boehm's innovations in materials, in bore design, and in developing a key mechanism that enabled flutists better to play in the entire range of 12 major and 12 minor tonalities not only dramatically changed the instrument from the 18th-century flute to what it is today but it also formed the basis of the work of makers of other woodwind instruments.

The situation in the world of the clarinet was not as simple. During the 19th century, two distinct fingering systems developed: the so-called French or Boehm system designed by Hyacinthe Klosé and the German or Oehler system that developed from an earlier key and fingering system designed by Iwan Müller. Without going into the entire history of the development of the clarinet, suffice it to say that the French and German systems are the two most commonly used systems today, with the majority of clarinet players using the French system. In addition to different systems of keys and the accompanying differences of fingering notes, two instruments generally have different internal bores, which causes them to have different tone qualities. One of the interesting features of the French system, as designed by Klosé, is that the designer intentionally called his clarinet the "Boehm" system as a tribute to Theobald Boehm's pioneering work on the flute.

The middle of the 19th century also saw Adolphe Sax's invention of the saxophone. Sax filed for 14 patents in 1846 (Hart 2010); however, his woodwind instrument design work included other patent filings between 1838 and 1850. Among these are for developments that Sax made toward the instrument that bears his name, as well as for changes to the design of the bass clarinet. Perhaps one of the more interesting aspects of Sax's work on the saxophone, however, is the fact that, unlike the clarinet, which shares a similar type of single-reed mouthpiece, but which basically has a cylindrical bore, the saxophone has a conical bore. The mouthpiece, too, although similar to that of the clarinet, had some basic internal shape differences from those of clarinet mouthpieces. One of the other significant differences between the two single-reed instruments is that because of the conical bore of the saxophone, the tone holes needed to be much larger than those of clarinet. As a result, Sax had to develop a key mechanism to allow the instruments pads—as opposed to the player's fingers—to close the holes. The fingering system of the saxophone ultimately is easier than those of other woodwinds, including the flute, clarinet, and oboe. Interestingly, several of Sax's innovations dealt with the problem of the instrument's length, particularly for larger members of the family, such as the tenor, baritone, and bass saxophones. So, several of his patents dealt with the shape and upturned nature of the instruments' bells and the curve in the instruments' necks. In short, then, Sax developed an instrument that had some historical precedents but that effectively integrated the volume capabilities of brass instruments, the vibration-producing system of the reed instruments, with a fingering system that was easier to learn than other woodwind instruments of the day.

Although the design and construction principles of inventors such as Boehm, Sax, and Klosé were seminal, smaller-scale innovations in woodwind instrument technology continued to be made through the rest of the 19th century and throughout the 20th century. For example, changes to the shape and taper of instruments' bores have been made to try to improve response and intonation. To cite one example, clarinet makers have experimented with cylindrical (most of the inside of the tube is shaped like a cylinder, with little change of diameter until the bell end of the instrument is reached), largely conical (cone shaped), and polycylindrical bores.

Still a staple of jazz and concert bands today, the saxophone was the result of numerous patented design and acoustical developments made by its inventor, Adolphe Sax, in the mid-19th century. (The Metropolitan Museum of Art)

One of the other areas in which musicians have made important changes to single-reed instruments—clarinets and saxophones—is in the mouthpieces. Various designers and makers use different sizes and shapes of the bore, different tip openings (the distance between the tip of the reed and the tip of the mouthpiece), different lengths of the curve on the mouthpiece—which extends from its tip to the point at which the reed touches it without any pressure on the reed—and so on. These differences result in differing balances of overtones—which lead to tone qualities being described as dark, bright, and so on—tuning differences from register to register, how the player's embouchure functions, and response feel to the player.

More recent, and more visible, technological advances to woodwinds include flute mechanisms that avoid the use of the pins used in traditional instruments; however, one of the more interesting advances are in the use of carbon fiber bodies and magnetic, spring-less key mechanisms in the Finnish company Matit Flutes' instruments (see, http://www.matitflutes.com/index.html, accessed March 17, 2020, for information on the company, the advantages of carbon fiber bodies, and the spring-less key mechanism). And one of the more intriguing late 20th-century developments was flutist-composer Robert Dick's Glissando Headjoint for the flutist. The inventor's website includes a description of the device as well as links to several videos of flutists using the head joint, which according to Dick, was "inspired by the 'whammy bar' on the electric guitar" (see, http://robertdick.net/the-glissando-headjoint/, accessed March 5, 2020).

Another recent but more mainstream flute innovation—one that may prove to be important during flu season and during times of airborne pathogens—is the Win-D-Fender. This device, which clamps onto the head joint (mouthpiece) of the flute on the back of the blow hole, is intended to make the flute playable when wind is blowing toward the flutist and, therefore, against the player's airstream. However, throughout the 2020 COVID-19 pandemic, there were several YouTube demonstration videos and reviews that suggested that the Win-D-Fender also acts as a diffuser for the player's air, a large percentage of which does not actually go into the flute (see, e.g., https://www.youtube.com/watch?v=L5TsUiztHJg, accessed January 16, 2021). There are also videos posted of flutists playing while in-line skating and other videos of flutists playing with strong winds coming toward them using the device for its original intended purpose.

See Also: Electronic Wind Instruments (EWI)

Further Reading

Baines, Anthony. 2012. *Woodwind Instruments & Their History*. Mineola, NY: Dover.

Benade, Arthur H. 1960. *Horns, Strings, and Harmony*. Garden City, NY: Anchor Books.

Boehm, Theobald. 1964. *The Flute and Flute Playing in Acoustical, Technical, and Artistic Aspects*. New York: Dover. This is a reprint of Dayton C. Miller's 1922 translation of Boehm's seminal 19th-century treatise.

Brymer, Jack. 1990. *Clarinet*. London: Kahn & Averill.

De Lorenzo, Leonardo. 1992. *My Complete Story of the Flute: The Instrument, the Performer, the Music*. Revised and expanded ed. Lubbock: Texas Tech University Press. This is an expansion and revision of De Lorenzo's 1951 treatise on the flute.

Hart, Hugh. 2010. "June 28, 1846: Parisian Inventor Patents Saxophone." *Wired*, June 28. Accessed March 16, 2020. https://www.wired.com/2010/06/0628saxophone-patent/.

Horwood, Wally. 1983. *Adolphe Sax 1814–1894: His Life and Legacy*. Baldock, U.K.: Egon Publishers.

McBride, William. 1982. "The Early Saxophone in Patents 1838–1850 Compared." *Galpin Society Journal* 35 (March): 112–121.

Rendall, F. Geoffrey. 1971. *The Clarinet: Some Notes on Its History and Construction*. 3rd ed. New York: W.W. Norton.

Ridley, E. A. K. 1986. "Birth of the 'Boehm' Clarinet." *Galpin Society Journal* 39 (September): 68–76.

Yamaha Disklavier

In addition to being of the world's biggest manufacturers of pianos, acoustic guitars, orchestral string, woodwind, brass, and percussion instruments, Yamaha Corporation has been a leader in electronic keyboards, sound synthesis, and related areas. The company's best-known piano-based electronic instrument was the Disklavier. Yamaha had first released its Player Piano in 1982 and subsequently released the more powerful Disklavier in 1987. Yamaha has continued to incorporate new capabilities into the Disklavier, so the instrument has been in an evolutionary state since its inception.

The Disklavier was not the first hybrid example of acoustic-meets-electronic technology in the world of the piano. For example, although it was not computer based, Baldwin's Electro Piano of the 1970s essentially was an acoustic piano, with 88 keys, hammers, strings, and so on but with electronic pickups. A simple way to understand the instrument is to think of it as the piano equivalent of a hollow-bodied electric guitar. The Electro Piano might be best known through its use by jazz/pop keyboardist Dick Hyman, who composed and recorded the 1970 work *Concerto Electro*, a jazz-based piano concerto that featured the Baldwin instrument. There were other hybrid keyboard instruments, nothing that truly combined electronics and the core materials and mechanism of an acoustic piano caught on in the marketplace.

In addition to the mechanism of an acoustic piano, Yamaha's Disklavier contained sensors that could record a player's performance. The performance could then be saved as a standard MIDI file. The MIDI file could be edited, played back as it was originally recorded, imported into a music notation program for saving or for publication, and so on. So, the Disklavier was an acoustic piano, it could serve the same function as a fully electronic keyboard connected to a computer for MIDI usage, including with notation programs, and it could serve as an electronically based version of the old player piano, with the encoded paper roll replaced by computer data in MIDI format.

Because the Disklavier can play acoustic piano versions of MIDI files, it is also possible to compose pieces for the instrument using MIDI encoding without having to first have a human play the piece on the Disklavier. As a result, the Disklavier became a tool for composers, some of whom have used it to play pieces that would be impossible for a human performer to play at the piano keyboard. In this respect, the instrument can be thought of as a computer-age extension of the work that composer Conlon Nancarrow did from the late 1940s into the early 1990s. Nancarrow's compositions for player piano made use of the traditional player piano's ability to play more notes and faster and more complex rhythms than a human could play at the keyboard. Because Nancarrow's work spanned nearly 45 years, some of his later player piano compositions were written for computer-based instruments like Yamaha's Player Piano and Disklavier.

The use of the Disklavier by composers continues today and even involves competitions for composers using Yamaha's instrument. For example, Italy's Cremona Music sponsored the second edition of its Disklavier Composers Contest in September 2020 (for details about the competition, see, http://www.cremonamusica.com/en/disklavier-composers-contest/, accessed September 5, 2020). As the Disklavier has evolved, and as composers and pianists have discovered some of the capabilities offered by an instrument that integrates acoustic and MIDI-based technology, compositions for the Disklavier have moved well beyond simply using the instrument as a late 20th-century and early 21st-century version of the player piano. Some composers use the instrument's sensors as triggers, such that a live player could activate a MIDI patch by playing a particular note. Perhaps one of the easiest to understand analogies for this is to consider the well-known old Warner Bros. cartoons in which a character such as Yosemite Sam or Daffy Duck rigs up a piano or xylophone that will explode when another character (e.g., Bugs Bunny) plays a particular note in the old song "Believe Me, If All Those Endearing Young Charms" (for a compendium of several examples of Warner Bros.' use of this joke, see, for example, https://www.youtube.com/watch?v=ZLNduq2pnN0, accessed January 16, 2021). For the Disklavier, triggering technology has been used by composers both for pieces in which the Disklavier keys are played entirely by a live performer and for pieces in which a live performer plays along with pre-recorded material performed by the Disklavier.

A number of composers and pianists, including contemporary music figures such as Tod Machover, Xiao Xiao, Kyle Gann, Eve Egoyan, and others, have composed and/or included pieces in their repertoire written specifically for the Disklavier. For example, MIT professor and composer Machover has used the Disklavier as one of the so-called hyper instruments in his pieces. In Machover's work, the Disklavier takes its place among other acoustic instruments (e.g., bowed stringed instruments such as the violin and cello) that are integrated with computer technology. Pianist and composer Egoyan includes works for Disklavier played by a live pianist and prepared Disklavier (an instrument that has objects placed between the strings or on top of the strings to produce different percussion tone colors and effects) in her concert repertoire.

Because the Disklavier can so accurately reproduce a pianist's performance, it can also be used as a teaching tool. For example, a piano teacher could record a passage or an entire piece on a Disklavier and make the MIDI file available to their students. When the music is recorded, the instrument records not only the keys depressed but also pedaling and the velocity with which the keys are struck. Because of this, a student listening to the Disklavier performance essentially hears exactly what the teacher performed. Theoretically, this would give a student who then plays the passage on the Disklavier on which they just heard the music the immediate opportunity to try to match the teacher's playing, including such aspects as tempo, rubato, dynamic shadings, use of the sustain pedal, and so on.

The instrument can also be used as a virtual rehearsal pianist by vocal and instrumental soloists as well as by dancers. Of course, these performers could conceivably rehearse with an audio recording of their accompaniment, and musicians could rehearse with a computer-based virtual accompanist that follows them (e.g., SmartMusic); however, rehearsing with a Disklavier provides the opportunity to have a consistent accompaniment with the sound of an acoustic piano. If a collaborative pianist and a vocal or instrumental soloist work out details of blend, dynamic shadings, phrasing, and rubato, the pianist could record their part on a Disklavier. If their collaborator had, say, access to a Disklavier and a smart practice room that can mimic the acoustical environment of a recital hall, the instrumental or vocal soloist could conceivably more fully prepare for a performance.

> **Yamaha Corporation**
>
> Although some people might associate the Yamaha Corporation with motorcycles, quad bikes, and the like, this conglomerate evolved into the largest manufacturer of musical instruments in the world. In addition to building pianos, woodwinds, brass instruments, percussion instruments, guitars, and orchestral string instruments, the Yamaha Corporation played a prominent role in the electronic music technology of the late 20th century. Among its most noteworthy electronic contributions were the DX7 synthesizer, the Synclavier, and the Disklavier. The DX7, which Yamaha produced between 1983 and 1989, was the first popular commercial digital synthesizer and quickly became a ubiquitous part of the sound of 1980s' popular music.
>
> In addition to the more standard digital pianos, and synthesizers that Yamaha continues to produce, the company manufactures several "silent" instruments, such as a silent violin, essentially a digital violin that only produces an electrical signal and no acoustic sound. Yamaha also manufactures instruments such as the Sonogenic Keytar SHS-300 (a vaguely guitar-like portable keyboard), which includes a jam function in which an app, to which the instrument is connected via Bluetooth, adjusts the notes the Keytar player plays and changes them to match the keys and chords of the song. In other words, one can play it with absolutely no knowledge of the keyboard. There are several demonstration videos and tutorials for the Keytar on YouTube (see, e.g., https://www.youtube.com/watch?v=wFW66gBjmS8, accessed May 27, 2020).

See Also: MIDI

Further Reading
The Editors of Disklavier.com. n.d. "History of the Disklavier." Accessed June 14, 2021. https://www.disklavier.com/history-of-the-disklavier.

Yamaha DX7 Synthesizer

In part, the story of Yamaha's highly commercially successful DX7 synthesizer begins at Dartmouth College in New Hampshire. Two Dartmouth students, Cameron Jones and Sydney Alonso, took summer jobs at the College in 1972 to work on the institution's computer system. By 1975, Jones and Alonso had developed a 16-bit processor card for Dartmouth's ABLE computer. This development, and the academic interest and activity in electronic music at Dartmouth, led to Jones and Alonso increasingly working on music applications for technology. In 1979, they, along with Professor of Digital Electronics Jon Appleton, introduced the 8-bit additive sound synthesis machine, the Synclavier. The Synclavier was one of the principal synthesizers of the 1980s, especially in its later Synclavier II form. Because of its sampling capabilities, ease of programming, and the sounds that could be designed, the Synclavier was widely regarded as a tapeless studio.

The other part of the DX7 story—and the connecting thread between Yamaha and New England Digital Corporation's Synclavier—came from the other coast of the United States. Stanford University professor and composer John Chowning had developed FM (frequency modulation) synthesis that differed fundamentally from the analog synthesis systems that preceded it. After trying to interest several companies in the technology, Yamaha licensed Chowning's system. In turn, the developers of the Synclavier used Chowning's technology, as licensed through Yamaha for the first generation of their instrument, the Synclavier I.

When Yamaha introduced the DX7 in 1983, it brought out the first widely successful digital synthesizer. Although Yamaha produced the instrument only until 1989, the DX7 sold well and practically defined affordable digital sound synthesis for the period during which it was in production. To put the DX7's place in history and its relative affordability into context, at the time of its introduction, it cost approximately $2,000, which was "about a hundredth of the cost of a professional Synclavier" (Shepard 2013, 20). Shepard (2013) also estimates that at least 150,000 DX7s were sold during the synthesizer's run. The DX7 allowed for a wider frequency range than its analog and even some of its digital predecessors. This led to a brighter tone than many earlier synthesizers. Anecdotally, some users have described the tone as harsher than the analog synthesizers of the 1960s and 1970s, and some users apparently found the DX7 relatively difficult to program; however, the unusually high number of unit sales over the instrument's run suggests that it served a function in the musical world of the 1980s like perhaps no other synthesizer of the time.

Despite some of the quibbles that some users had with the DX7, it excelled in certain areas that went well beyond its price relative to other digital synthesizers of the day. The DX7 had touch-sensitive keys, so the player could change dynamic levels without having to use a volume pedal. The DX7 had 16 voices, so it was well suited to polyphonic usage. This was tempered somewhat by the fact that the instrument could only produce one tone color/patch at a time. The combination of Yamaha's chip technology and the Chowning/Stanford-owned FM synthesis technology meant that the DX7 was able to emulate acoustic instruments better than many of its competitors.

The DX7 was not the only synthesizer that Yamaha released during the mid- to late 1980s. For example, the company's DX21 was less expensive than the DX7 (by more than half), and it was polyphonic but with only eight voices. The DX21 did not have the touch-sensitive keys, so the user had to either program dynamic changes into MIDI or use a volume pedal to change dynamics. However, this 1985 release improved in a significant way over the DX7; it was multi-timbral, meaning that it could produce more than one sound/tone color (or patch) at a time.

The apparent difficulty of programming new tone colors into the DX7, combined with the instrument's popularity and the fact that it came with a variety of presets, meant that the DX7, as the first widely commercially successful digital synthesizer, helped at least in part to move sound synthesis further away than it had become by the time of the instrument's debut from user-designed tone colors/patches and toward the extensive use of presets. This in some important respects produced a fundamental change in how musicians thought of sound synthesis. The process moved from one in which users might spend as much—or possibly more—time designing tone colors by means of removing, changing the balance of, or adding overtones to, say, a basic sine wave, to a process that involved selecting the pre-programmed sounds supplied by an instrument's manufacturer.

That being said, today numerous internet sites include information about programming patches that have been used on the DX7 since the 1980s but were not part of Yamaha's presets. So, in a way, today's more advanced digital technologies have enabled users to return to some of the sonic design work that defined sound synthesis before the introduction of Yamaha's DX7. It is still a synthesizer that is emulated by some of today's programs and stand-alone synthesizers, further suggesting the iconic nature of the instrument.

See Also: Digital Synthesizers; Patches

Further Reading

The Editors of Yamaha.com. n.d. "DX7." Accessed June 14, 2021. https://www.yamaha.com/en/about/design/synapses/id_009/.

Nelson, Andrew J. 2015. *The Sound of Innovation: Stanford and the Computer Music Revolution*. Cambridge, MA: MIT Press.

Pinch, Trevor, and Frank Trocco. 2002. *Analog Days: The Invention and Impact of the Moog Synthesizer*. Cambridge, MA: Harvard University Press.

Shepard, Brian K. 2013. *Refining Sound: A Practical Guide to Synthesis and Synthesizers*. New York: Oxford University Press.

YouTube

YouTube was founded in 2005 by PayPal employees Steve Chen, Chad Hurley, and Jawed Karim. The impact of YouTube on music has been enormous; however, the platform had the humblest of beginnings. YouTube's first posting was a 2005 18-second video of cofounder Jawed Karim standing in front of the elephant enclosure at the San Diego Zoo (see, Kosoff and Leskin 2015, for a description of the event and a link to Karim's brief video). Today, YouTube's music-related offerings include countless amateur performance videos, official videos for numerous musicians, product reviews, and instructional videos for singers and instrumentalists.

Because setting up a YouTube account and posting videos are so easy, the platform has been useful for amateur musicians to get their music, performances, and videos out to the public. However, the ease of access has also caused controversy around such topics as fair use, copyright of music and videos, royalties, and associated issues. In fact, the posting of unauthorized videos and the unauthorized posting of musicians' official videos were a significant concern during the first several years YouTube was available to the public. As a result, Vivendi teamed up with YouTube for the service Vevo in 2009. According to *Business Insider*'s Megan Rose Dickey, "This was YouTube's first move to fix its relationship with music companies, which had complained about piracy and unfair licensing terms. As part of YouTube and Vevo's agreement, Vevo is free to distribute its music videos on YouTube and YouTube is able to keep showing music videos from big labels" (2013). It should be noted that the Vevo service has not completely ended violations of copyright and so on; however, Vevo has provided access to higher-quality video and higher-quality audio music and music videos from well-known artists of the late 20th century through the present than those whose legality might be more dubious.

In addition to official videos for a variety of artists, YouTube has proven to be a source for archival recordings, particularly of early 20th-century blues and country artists whose work has in some cases not been officially released on later LPs or in digital formats such as compact disc, mp3, or streaming media. Collectors of early shellac records, it seems, are eager to play their old discs and make videos of the spinning record or videos with composite photographs of the artist and so on. Before YouTube, some of these early recordings might well have been deemed lost forever, or at the very least were extremely difficult to obtain or even hear. Their accessibility on YouTube has impacted scholars and music fans alike and has widened the available repertoire for anyone with internet access around the globe.

Over the years, YouTube has seen several artists' videos go viral, and not just young phenoms such as Justin Bieber, and some of these occurrences have resulted in headlines on online news sources and have led to improvements in web-based technologies. One of the more famous examples was South Korean rapper Psy's official video for his 2012 song "Gangnam Style." As of this writing, the video for the song on Psy's YouTube channel has enjoyed over 3.8 billion views in the eight years since the song premiered (see, https://www.youtube.com/watch?v=9bZkp7q19f0, accessed October 9, 2020). In the song's initial run of popularity, at the viral stage, the official video reached 2,147,483,647 views by December 2014. It was widely reported at the time that the high viewership could no longer be counted by YouTube's 32-bit technology (see, e.g., Griggs 2014). YouTube has since upgraded its technology and can now track even more popular videos than Psy's famous work. Another well-known and even more impressive example of a viral YouTube video is Luis Fonsi's official video for "Despacito," featuring Daddy Yankee, which received over 7.1 billion views between January 13, 2017 and January 1, 2021 (see, https://www.youtube.com/watch?v=kJQP7kiw5Fk, accessed January 1, 2021).

YouTube has also made significant contributions to the musical world through the numerous tutorials on singing and playing a wide variety of instruments. It must be noted that these instructional videos vary widely. However, anecdotally, as a music educator, I have found several that are sound pedagogically and that can be helpful to developing performers. Even some of the tutorial videos that provide only a partial lesson as an encouragement to purchase full lessons or tutorial subscriptions on the poster's website can be helpful to a performer looking to learn a particular technique.

In addition to postings by individual amateur and professional vocal and instrumental teachers and coaches, some retail outlets and music-related magazines have established YouTube channels in which expert musicians explain their techniques and/or do performances that are not otherwise available online. One of numerous examples is a 2015 video of James Taylor for *Guitarist* magazine in which Taylor talks about his guitar technique, his approach to tuning, the importance of the capo in his guitar playing, and the relationship between his vocal range and his guitar playing (see, https://www.youtube.com/watch?v=dXj9D cjjWZE, accessed January 16, 2021).

Professional and amateur musicians and music merchandisers also post numerous product reviews on YouTube. Although it might seem self-evident, it is important to note that these video reviews have a clear advantage over reviews of instruments, effects pedals, and so on that one might find in a traditional print magazine: one can actually see and hear the product in many of the YouTube reviews. As is the case with tutorials, however, viewers should be advised to check out the credentials and perhaps the motives of reviewers as one considers the review's validity.

During the 2020 COVID-19 pandemic, live streaming concerts and live-streamed broadcasts on YouTube became instantly more visible, suggesting that more than just a repository of a wide range of videos, YouTube could also play a significant role in live streaming video, much as Facebook Live did during the same period. And, as some musicians were more adept than others at using software for virtual concerts, virtual worship services, and so on, the numerous YouTube videos and even extensive tutorials on using musical collaboration software proved to be very important. In fact, based on the timing of their posting, it appeared that a large number of these tutorials appeared in direct response to the fact that many musicians' needs for at least some level of technological expertise changed dramatically almost overnight.

See Also: iPads and Other Tablets; Smartphones; Social Media

Further Reading

Adib, Desiree. 2009. "Pop Star Justin Bieber Is on the Brink of Superstardom." *ABC News*, November 12. Accessed January 18, 2021. https://abcnews.go.com/GMA/Weekend/teen-pop-star-justin-bieber-discovered-youtube/story?id=9068403.

Asmelash, Leah. 2020. "The First Ever YouTube Video Was Uploaded 15 Years Ago Today. Here It Is." *CNN Business*, April 23. Accessed April 23, 2020. https://www.cnn.com/2020/04/23/tech/youtube-first-video-jawed-karim-trnd/index.html.

Dickey, Megan R. 2013. "The 22 Key Turning Points in the History of YouTube." *Business Insider*, February 15. Accessed January 18, 2021. https://www.businessinsider.com/key-turning-points-history-of-youtube-2013-2.

Griggs, Brandon. 2014. "'Gangnam Style' Breaks YouTube." CNN.com, December 3. Accessed March 17, 2020. www.cnn.com/2014/12/03/showbiz/gangnam-style-youtube/index.html.

Kosoff, Maya, and Paige Leskin. 2015. "This Is the First YouTube Video Ever Uploaded—It Was Posted 10 Years Ago Today." *Business Insider*, April 23. Accessed January 18, 2021. https://www.businessinsider.com/first-youtube-video-2015-4.

Glossary

Amplitude Modulation (AM)
In amplitude modulation, such as that used in AM radio, changes in the carrier signal are tied to changes in the wave amplitude (size). In radio, amplitude modulation has the advantage of allowing a signal to travel a greater distance than a frequency modulated (FM) signal using the same amount of power.
See also: Frequency Modulation

Analog Synthesizer
A synthesizer that uses sound-generating circuitry, generally oscillators, to produce sound.
See also: Digital Synthesizer

CD-R and CD-RW
When compact discs were one of the primary media for storing digital information, whether it be music or other types of files, it was common to see two different kinds of CDs offered for sale. CD-Rs (Compact Disc—Recordable) can be written to only once and then read any number of times. Once the user is finished writing information to the CD-R, the computer is used to finalize the disc; it cannot be further changed. The CD-RW (Compact Disc—ReWritable) can be written to multiple times; it is not finalized like a CD-R. However, CD-Rs were more compatible and stable and were more frequently used for storage of audio files such as mp3s.

Diegetic Music
Diegetic music in a film soundtrack is the music that the characters can also hear; it is part of their environment.

Digital Audio Workstation (DAW)
Digital Audio Workstation, or DAW, is a designation given to a variety of computer applications and hardware units that provide recording, playback, and audio-editing capabilities.

Digital Synthesizer
A synthesizer that uses computer-based digital information, consisting of zeros and ones, to generate sound.
See also: Analog Synthesizer

Drone
A sustained pitch, usually in a low register, that it serves as the basis for a musical composition. Perhaps the best-known instrument that produces both a drone and melodic tones is the bagpipes.

Equalization
The control of different ranges of an audio frequency. Control might range from that of treble and bass to control of narrower bands of the overall frequency range.

FM Synthesis
Originally developed by Stanford University professor and composer John Chowning in the 1970s, synthesis using frequency modulation helped to make possible digital synthesizers, such as the Synclavier and Yamaha's DX7.

Frequency Modulation
In frequency modulation, such as that used in FM synthesis or FM radio, changes in the carrier signal are tied to changes in the wave frequencies (numbers of vibrations per second). In radio, although amplitude modulation has the advantage of allowing a signal to travel a greater distance than a frequency modulated (FM) signal using the same amount of power, frequency modulation allows for better audio fidelity.
See also: Amplitude Modulation

Harmonic Series
Pitches that are part of the wave form of tones. These higher tones are often imperceptible, but their presence and relative balance play a major role in the tone quality of real or virtual instruments. For example, a clarinet produces generally only the odd-numbered tones of the harmonic series, while a flute produces both the odd- and even-numbered tones, thus accounting at least in part for the audible difference between the two instruments.

Monophonic
A musical texture in which there is a single melody line. For example, monophonic synthesizers could produce only one note at a time.

Nondiegetic Music
Nondiegetic music in a film soundtrack is the music that is added in; it is not heard by the characters.

Organum
The early European polyphonic music of the Middle Ages. Based on Gregorian chant, two distinct forms were common: (1) parallel organum, in which two parts moved in parallel motion to each other, generally at the interval of a perfect 5th, and (2) florid organum, in which a new, fast-moving melodic material in an upper voice was accompanied by slow-moving, chant-based music in a lower voice.

Ostinato
A repeating musical pattern. Ostinatos have a long history in music; however, in recent times, short repeating patterns have played roles in loops, generative music, drum machine programming, and in the material hip-hop that DJs use to construct accompaniments to rap.

Overtones
Usually imperceptible tones that are higher in pitch than the perceived pitch and that are part of the tone's harmonic series. Overtones are sometimes referred to as partials.

Peer-to-Peer (P2P) Networking
Peer-to-peer networking is a system by which individual computers can be networked together without going through a central server. This technology was at the core of millennial file-sharing sites such as Napster.

Polyphonic
A musical texture in which there are multiple simultaneous sounds. In polyphonic music, this term usually refers to multiple discreet melodies. In reference to synthesizers, this term generally refers to synthesizers that can produce more than one simultaneous sound (e.g., two, three, or more sounds or pitches) at the same time.

Semiconductor
A solid substance that has attributes in between those of an electrical conductor and an insulator. Semiconductor substances such as silicon are used in the construction of computer chips, transistors, diodes, and other post-vacuum tube electronic components.

Seventh Chord
A four-note chord that consists of a root and notes the intervals of a 3rd, 5th, and 7th above the root.

Subtractive Synthesis
A process in which various overtones are removed from a wave form to create a new wave form and sound.

Tablature
A notational system for music that is specific to a particular instrument and is based on fingerings on that instrument, as opposed to using the traditional 5-line and 4-space staff.

Tremolo
A pulsation in a musical tone that is caused by rapid period changes of volume. See also: Vibrato

Triad
A chord, arranged in thirds, consisting of three pitches.

USB
USB stands for Universal Serial Bus. It is a system for connecting various digital devices. The original USB connections, known as Type A, required that the device be oriented the correct way in order to be inserted. The newer USB-C connectors are smaller, and there is no designated up or down.

Vibrato
A pulsation in a musical tone that is caused by small changes of pitch.

Bibliography

Aamoth, Doug. 2014. "Watch Steve Jobs Unveil the iPod 13 Years Ago." *Time*, October 23. Accessed January 18, 2021. https://time.com/3533908/ipod-turns-13/.

ABC News. 2000. "Napster Has Hit a Sour Note in Court." *ABC News*, July 27. Republished January 7, 2006 as "Napster Shut Down." Accessed January 18, 2021. https://abcnews.go.com/Technology/story?id=119627&page=1.

Abdurraqib, Hanif. 2020. "Tick, Tick, Boom: In the Whimsical World of Pop Music, Sometimes Technology Has More Impact after It's Obsolete." *Smithsonian* 51, no. 4 (July/August): 22–23.

Abernathy, David. 2015. *The Prophet from Silicon Valley: The Complete Story of Sequential Circuits*. Auckland, New Zealand: AM Publishing.

Adib, Desiree. 2009. "Pop Star Justin Bieber Is on the Brink of Superstardom." *ABC News*, November 12. Accessed January 18, 2021. https://abcnews.go.com/GMA/Weekend/teen-pop-star-justin-bieber-discovered-youtube/story?id=9068403.

Ahmed, Azam. 2016. "Mexico City's Organ Grinders, Once Beloved, Feel Shunned." *New York Times*, September 12. Accessed January 18, 2021. https://www.nytimes.com/2016/09/13/world/americas/mexico-city-organ-grinders.html.

Albinsson, Staffan. 2012. "Early Music Copyrights: Did They Matter for Beethoven and Schumann?" *International Review of the Aesthetics and Sociology of Music* 43, no. 2 (December): 265–302.

Aldrich, Margret, and Michael Dregni, eds. 2011. *The Old Guitar: Making Music and Memories from Country to Jazz, Blues to Rock*. New York: Crestline.

American Federation of Musicians. n.d. "History." American Federation of Musicians. Accessed January 18, 2021. https://www.afm.org/about/history-2/.

Ames, Charles. 1987a. "Automated Composition in Retrospect: 1956–1986. *Leonardo* 20, no. 2: 169–185.

Ames, Charles. 1987b. "Reflections on James Tenney." *Perspectives of New Music* 25, nos. 1 and 2: 455–458.

Ames, Charles. 1989. "The Markov Process as a Compositional Model: A Survey and Tutorial." *Leonardo* 22, no. 2: 175–187.

Ames, Charles. 1990. "Statistics and Compositional Balance." *Perspectives of New Music* 28, no. 1: 80–111.

Anderson, Paul A. 2015. "Neo-Muzak and the Business of Mood." *Critical Inquiry* 41, no. 4 (Summer): 811–840.

Anderton, Craig. n.d. "Craig Anderton's Brief History of MIDI." MIDI Association. Accessed January 18, 2021. https://www.midi.org/midi-articles/a-brief-history-of-midi.

Anderton, Craig. 2003. "Q. What's the Difference between a Talk Box and a Vocoder?" *Sound on Sound*, October. Accessed January 18, 2021. https://www.soundonsound.com/sound-advice/q-whats-difference-between-talk-box-and-vocoder.

Arkin, Daniel. 2020. "VHS Tapes Are Back in Vogue as Everything Old Is New Again." *NBC News*, March 7. Accessed January 18, 2021. https://www.nbcnews.com/pop-culture/movies/vhs-tapes-are-back-vogue-everything-old-new-again-n1151611.

Arnold, Leo F. J. 1941. Patent granted 1942. Patent application for "Electrical Clarinet." Accessed January 18, 2021. https://patents.google.com/patent/US2301184A/en.

Asmelash, Leah. 2020. "The First Ever YouTube Video Was Uploaded 15 Years Ago Today. Here It Is." *CNN Business*, April 23. Accessed April 23, 2020. https://www.cnn.com/2020/04/23/tech/youtube-first-video-jawed-karim-trnd/index.html.

Babiuk, Andy. 2002. *Beatles Gear: All the Fab Four's Instruments, from Stage to Studio*. Revised ed. San Francisco, CA: Backbeat Books.

Barbaud, Pierre. 1966. *Initiation a la composition musicale automatique*. Paris: Dunod.

The Beatles. 2000. *The Beatles Anthology*. San Francisco: Chronicle Books.

Beauchamp, George. 1929. Patent granted 1931. Patent application for "Stringed Musical Instrument." Accessed January 18, 2021. https://patents.google.com/patent/US1808756.

Beauchamp, George. 1934. Patent granted 1937. Patent application for "Electrical Stringed Musical Instrument." Accessed January 18, 2021. https://patents.google.com/patent/US2089171?oq=patent:2089171.

Beaumont, Rachel. 2017. "The Ondes Martenot: The Eeriest Instrument Ever Invented?" Royal Opera House (U.K.). Accessed April 8, 2020. https://www.roh.org.uk/news/the-ondes-martenot-the-eeriest-instrument-ever-invented.

Becker, Stephen. 2011. "8-Track Tapes Belong in a Museum." *All Things Considered*, February 16. Accessed January 18, 2021. https://www.npr.org/sections/therecord/2011/02/17/133692586/8-track-tapes-belong-in-a-museum.

Benkler, Yochai. 2006. *The Wealth of Networks: How Social Production Transforms Markets and Freedom*. New Haven, CT: Yale University Press.

Berkowitz, Justin. 2010. "The History of Car Radios. Car Tunes: Life before Satellite Radio." *Car and Driver*, October 25. Accessed January 18, 2021. https://www.caranddriver.com/features/a15128476/the-history-of-car-radios/.

Beuttler, Bill. 2020. "Christian Scott aTunde Adjuah: Don't Stream; Stretch." *JazzTimes*, March 24. Accessed January 18, 2021. https://jazztimes.com/features/profiles/christian-scott-atunde-adjuah-dont-stream-stretch/.

Bickerton, R. C., and G. S. Barr. 1987. "The Origin of the Tuning Fork." *Journal of the Royal Society of Medicine* 80 (December): 771–773.

Bicknell, Jeanette. 2015. *A Philosophy of Song and Singing: An Introduction*. New York: Routledge.

Boehm, Theobald. 1964. *The Flute and Flute Playing in Acoustical, Technical, and Artistic Aspects*. New York: Dover. This is a reprint of Dayton C. Miller's 1922 translation of Boehm's seminal 19th-century treatise.

Bonds, Ray. 2003. *The Illustrated Directory of Guitars*. St. Paul, MN: Voyageur Press.

Bowers, Q. David. *Encyclopedia of Automatic Musical Instruments*. Lanham, MD: Vestal Press.

Boynton, Brad. 2017. "A Brief History of the Cajon." *Drum Magazine*, December 8. Accessed January 18, 2021. https://drummagazine.com/a-brief-history-of-the-cajon/.

Brend, Mark. 2005. *Strange Sounds: Offbeat Instruments and Sonic Experiments in Pop*. San Francisco, CA: Backbeat Books.

Brend, Mark. 2012. *The Sound of Tomorrow: How Electronic Music Was Smuggles Into the Mainstream*. New York and London: Bloomsbury.

Brenner, David. 2018. "Performing Rebellion: Karaoke as a Lens into Political Violence." *International Political Sociology* 12, no. 4 (December): 401–417.

Brock, Pope. 2008. *Charlatan: America's Most Dangerous Huckster, the Man Who Pursued Him, and the Age of Flimflam*. New York: Three Rivers Press.

Brymer, Jack. 1990. *Clarinet*. London: Kahn & Averill.

Bughin, Jacques, Jeongmin Seong, James Manyika, Michael Chui, and Raoul Joshi. 2018. "Notes from the AI Frontier: Modeling the Impact of AI on the World Economy." The McKinsey Global Institute, September 8. Accessed January 18, 2021. https://www.mckinsey.com/featured-insights/artificial-intelligence/notes-from-the-ai-frontier-modeling-the-impact-of-ai-on-the-world-economy#.

Bush, Douglas E., and Richard Kassel, eds. 2006. *The Organ: An Encyclopedia*. New York: Routledge.

Business Wire. 2005. "Beatnik's Audio Engine Reaches a Quarter of a Billion Mobile Phones; One of the Fastest Adoptions of Embedded Software Technology in the Handset Industry." *Business Wire*, February 14. Accessed June 1, 2020. https://www.tmcnet.com/usubmit/2005/Feb/1116860.htm.

Bustillos, Maria. 2020. "José Andrés Is Feeding the World." *AARP: The Magazine* 63, no. 5B (August/September): 52–55, 82.

Calore, Michael. 2009. "May 12, 1967: Pink Floyd Astounds with 'Sound in the Round.'" *Wired*, May 12. Accessed January 18, 2021. https://www.wired.com/2009/05/dayintech-0512/.

The Canton Palace Theatre. n.d. "A Brief History of the Kilgen Organ." Accessed January 18, 2021. https://cantonpalacetheatre.org/about-us/brief-history-of-the-kilgen-organ/.

Caulfield, Keith. 2020. "Paul McCartney's *McCartney III* Debuts at No. 1 on *Billboard*'s Top Album Sales Chart." *Billboard*, December 28. Accessed

January 7, 2021. https://www.billboard.com/articles/business/chart-beat/9504857/mccartney-iii-debuts-top-album-sales-chart/.

Chang, Jeff. 2005. *Can't Stop Won't Stop: A History of the Hip-Hop Culture*. New York: Picador.

Chodosh, Sara. 2018. "You Should Be Listening to Video Game Soundtracks at Work." *Popular Science*, January 26. Accessed January 18, 2021. https://www.popsci.com/work-productivity-listening-music/.

Clark, Paul W., and Laurence A. Lyons. 2014. *George Owen Squier: U.S. Army Major General, Inventor, Aviation Pioneer, Founder of Muzak*. Jefferson, NC: McFarland.

Clifford, Stephanie. 2007. "Pandora's Long Strange Trip: Online Radio That's Cool, Addictive, Free, and—Just Maybe—a Lasting Business." *Inc.*, October 1. Accessed May 11, 2020. https://www.inc.com/magazine/20071001/pandoras-long-strange-trip.html.

CNN Business. 2020. "Music's Biggest Stars Are Performing Online during the Pandemic." *CNN Business*, March 17. Accessed March 18, 2020. https://www.cnn.com/videos/business/2020/03/17/coronavirus-musicians-social-media-orig.cnn-business.

Coleman, Mark. 2004. *Playback: From the Victrola to MP3, 100 Years of Music, Machines, and Money*. Cambridge, MA: Da Capo Press.

Collins, Karen. 2008. *Game Sound: An Introduction to the History, Theory, and Practice of Video Game Music and Sound Design*. Cambridge, MA: The MIT Press.

Collins, Mike. 2015. *In the Box Music Production: Advanced Tools and Techniques for Pro Tools*. Burlington, MA: Focal Press.

Cornish, Audie. 2018. "20 Years of Cher's 'Believe' and Its Auto-Tune Legacy." NPR, October 22. Accessed January 18, 2021. https://www.npr.org/transcripts/659611154.

Coryat, Karl. 2005. *Guerrilla Home Recording: How to Get Great Sound from Any Studio (No Matter How Weird or Cheap Your Gear Is)*. San Francisco, CA: Backbeat Books.

Couture, François. n.d. "Emerson, Lake & Palmer: 'Lucky Man.'" *AllMusic*. Accessed January 18, 2021. https://www.allmusic.com/song/lucky-man-mt0053497938.

Crow, Michael, and Barry Bozeman. 1998. *Limited by Design: R&D Laboratories in the U.S. National Innovation System*. New York: Columbia University Press.

D'Amico, Tony. 2017. "Introducing the Audience to the Pit." *International Musician*, May 1. Accessed January 18, 2021. https://internationalmusician.org/introducing-audience-pit/.

Daniel, Eric D., C. Denis Mee, and Mark H. Clark, eds. 1999. *Magnetic Recording: The First 100 Years*. New York: John Wiley & Sons/IEEE Press.

Davis, Stephen. 2001. *Old Gods Almost Dead: The 40-Year Odyssey of the Rolling Stones*. New York: Broadway Books.

DeBord, Matthew. 2016. "We Tried 4 of the Best Audio Systems in Cars—Here's How They Stacked Up." *Business Insider*, November 22. Accessed

January 18, 2021. https://www.businessinsider.com/best-car-audio-systems-compared-2016-11.

De Lorenzo, Leonardo. 1992. *My Complete Story of the Flute: The Instrument, the Performer, the Music*. Revised and expanded ed. Lubbock: Texas Tech University Press. This is an expansion and revision of De Lorenzo's 1951 treatise on the flute.

DeMauro, Thomas A. 2021. "1967 Cadillac Fleetwood Eldorado." *Hemmings Classic Car* 17, no. 6 (March): 36–40, 42.

Dickey, Megan R. 2013. "The 22 Key Turning Points in the History of YouTube." *Business Insider*, February 15. Accessed January 18, 2021. https://www.businessinsider.com/key-turning-points-history-of-youtube-2013-2.

Dietmeier, Bob. 1955. "Can't Hardly See No End to Artists, Song Hits, Developments: Music, Juke Box Industries Rich in All Categories, Says Operator Poll." *Billboard* 67, no. 13 (March 26): 1, 147.

DJ Booma. 2017. *How to Be a DJ in 10 Easy Lessons: Learn to Spin, Scratch and Produce Your Own Mixes*. London: QED Publishing.

Dolby, Thomas. 2016. *The Speed of Sound: Breaking the Barrier between Music and Technology*. New York: Flatiron Books.

Dopyera, Rudolph. 1933. Patent application for "Musical Instrument." Accessed January 18, 2021. https://patents.google.com/patent/US1896484.

Douglas, Adam. 2019. "The Synths behind 8 Classic TV Theme Songs: *Miami Vice*, *Doctor Who*, and More." *Reverb*, June 11. Accessed January 18, 2021. https://reverb.com/news/the-synths-behind-8-classic-tv-theme-songs.

Drew, Rob. 2004. "'Scenes' Dimension of Karaoke in the United States." In Andy Bennett and Richard A. Peterson, eds., *Music Scenes: Local, Translocal and Virtual*, 64–79. Nashville, TN: Vanderbilt University Press.

Du Gay, Paul, Stuart Hall, Linda Janes, Anders K. Madsen, Hugh MacKay, and Keith Negus. 2013. *Doing Cultural Studies: The Story of the Sony Walkman*. 2nd ed. London: Sage Publications.

Duggan, Mary K. 1992. *Italian Music Incunabula*. Berkeley and Los Angeles: University of California Press.

Duke, Dan. 2015. "Jimmy Ray Dunn Starts His Own Radio Station—on the Internet." *Virginian-Pilot*, November 5. Accessed January 18, 2021. https://www.pilotonline.com/entertainment/music/article_6cef89c1-e04b-56d0-b12a-3936e362bdeb.html.

Ebert, Roger. 1995. "Theremin: An Electronic Odyssey." Rogerebert.com, December 15. Accessed January 18, 2021. https://www.rogerebert.com/reviews/theremin-an-electronic-odyssey-1995.

Eder, Bruce. n.d. "Wendy Carlos: *Switched-On Bach*." *AllMusic*. Accessed January 18, 2021. https://www.allmusic.com/album/switched-on-bach-mw0000976916.

The Editors of *Billboard*. 1965a. "Lear Cartridge-Equipped Fords Getting a Fast Start." *Billboard* 77, no. 39 (September 25): 1, 10.

The Editors of *Billboard*. 1965b. "RCA Fires 175-Title Burst with Release of Stereo 8 Cartridges." *Billboard* 77, no. 39 (September 25): 3.

The Editors of Bose.com. n.d. "Dream + Reach: The First 50 Years of Bose." Bose.com. Accessed April 29, 2020. https://www.bose.com/en_us/better_with_bose/dream_and_reach.html.

The Editors of *Consumer Reports*. 2020. "Noise-Cancelling Headphones." *Consumer Reports Buying Guide 2020* 84, no. 13: 103.

The Editors of *Consumer Reports*. 2021. "Face-Off: New (and Improved) Boom Boxes." *Consumer Reports* 86, no. 2 (February): 16.

The Editors of *Encyclopædia Britannica*. 2020a. "Music Box." *Encyclopædia Britannica*. Accessed May 8, 2020. https://www.britannica.com/art/music-box.

The Editors of *Encyclopædia Britannica*. 2020b. "Player Piano." *Encyclopædia Britannica*. Accessed March 10, 2020. https://www.britannica.com/art/player-piano.

The Editors of *International Musician*. 2020. "Cool Tools: Guitar Synthesizer." *International Musician* 118, no. 7 (July): 24.

The Editors of *Ohio History Central*. "Voice of America." *Ohio History Central*. Accessed January 18, 2021. https://ohiohistorycentral.org/w/Voice_of_America.

The Editors of Pandora.com. n.d. "About the Music Genome Project." Pandora.com. Accessed May 11, 2020. https://www.pandora.com/about/mgp.

The Editors of *Reverb*. 2020. "11 Tools for Collaborating on Music Remotely." *Reverb*, March 24. Accessed May 15, 2020. https://reverb.com/news/ways-to-collaborate-on-music-remotely.

The Editors of the *San Diego Union-Tribune*. 2018. "From the Archives: September 20, 1919: 50,000 Hear President Wilson." *San Diego Union-Tribune*, September 20. Accessed January 18, 2021. https://www.sandiegouniontribune.com/news/150-years/sd-me-150-years-september-20-htmlstory.html.

Emerson, Ken. 1998. *Doo-Dah! Stephen Foster and the Rise of American Popular Culture*. New York: Da Capo.

Epstein, Dan. 2015. "Coolio's 'Gangsta's Paradise': The Oral History of 1995's Pop-Rap Smash." *Rolling Stone*, August 7. Accessed December 14, 2020. https://www.rollingstone.com/music/music-news/coolios-gangstas-paradise-the-oral-history-of-1995s-pop-rap-smash-50357/.

Eschner, Kat. 2017. "John Philip Sousa Feared 'The Menace of Mechanical Music': Wonder What He'd Say about Spotify." *Smithsonian Magazine*, November 6. Accessed January 18, 2021. https://www.smithsonianmag.com/smart-news/john-philip-sousa-feared-menace-mechanical-music-180967063/.

Faragher, Scott. 2011. *The Hammond Organ: An Introduction to the Instrument and the Players Who Made It Famous*. Milwaukee, WI: Hal Leonard Books.

Farkas, Remy. 2017. "Quiet Ann Arbor Looks to Limit Piped Music." *Michigan Daily*, December 5. Accessed May 26, 2020. https://www.michigandaily.com/section/ann-arbor/quiet-ann-arbor-looks-limit-pipped-music.

Feldman, Heidi C. 2006. *Black Rhythms of Peru: Reviving African Musical Heritage in the Black Pacific*. Middletown, CT: Wesleyan University Press.

Ferris, Robert. 2016. "How Amar Bose Used Research to Build Better Speakers." CNBC.com, March 24. Accessed January 18, 2021. https://www.cnbc.com/2016/03/24/how-amar-bose-used-research-to-build-better-speakers.html.

Fine, Thomas. 2008. "The Dawn of Commercial Digital Recording. *ARSC Journal* 39, no. 1 (Spring). Accessed June 12, 2020. http://www.aes.org/aeshc/pdf/fine_dawn-of-digital.pdf.

Finley, Larry. 1965. "Tape Cartridge Tips." *Billboard* 77, no. 39 (September 25): 10.

Flanagan, Andrew. 2017. "Pandora Co-Founder and CEO Tim Westergren to Step Down." *NPR*, June 27. Accessed January 18, 2021. https://www.npr.org/sections/therecord/2017/06/27/534542303/pandora-co-founder-and-ceo-tim-westergren-to-step-down.

Floros, Constantin. 2011. *The Origins of Western Notation*. Revised and translated by Neil Moran. Frankfurt am Main, Germany: Peter Lang AG.

France-Presse, Agence. 2016. "First Recording of Computer-Generated Music—Created by Alan Turing—Restored." *Guardian*, September 26. Accessed July 22, 2020. https://www.theguardian.com/science/2016/sep/26/first-recording-computer-generated-music-created-alan-turing-restored-enigma-code.

Franzen, Carl. 2014. "The History of the Walkman: 35 Years of Iconic Music Players." *Verge*, July 1. Accessed June 10, 2020. https://www.theverge.com/2014/7/1/5861062/sony-walkman-at-35.

Free, John R. 1974. "Look and Listen: News, Comment, and Opinion from the World of Home-Entertainment Electronics." *Popular Science* 204, no. 2 (February): 50.

Fuller, David. 1986. "Automatic Instrument." In Don M. Randel, ed., *The New Harvard Dictionary of Music*, 61–63. Cambridge, MA: The Belknap Press of Harvard University Press.

Furseth, Jessica. 2017. "Noise-Cancelling Headphones: The Secret Survival Tool for Modern Life." *Guardian*, March 16. Accessed January 18, 2021. https://www.theguardian.com/technology/2017/mar/16/noise-cancelling-headphones-sound-modern-life.

Gagliardi, Tino. 2020. "From the Theatre, Touring, and Booking Division—Safety Protocols for a Return to Work for Pit Musicians." *International Musician* 118, no. 8 (August): 8–9.

Gagniuc, Paul A. 2017. *Markov Chains: From Theory to Implementation and Experimentation*. Hoboken, NJ: John Wiley & Sons.

Garner, Louis E., Jr. 1967. "For That Different Sound, Music a la Theremin." *Popular Electronics* 27, no. 5 (November): 29–33.

Garvey, Marianne. 2020. "Garth Brooks and Trisha Yearwood's Emotional Home Concert Crashes Facebook Live." *CNN*, March 24. Accessed March 24, 2020. https://www.cnn.com/2020/03/24/entertainment/garth-brooks-trisha-yearwood-concert-trnd/index.html.

George, Nelson. 2012. "Hip-Hop's Founding Fathers Speak the Truth." In Murray Forman and Mark A. Neal, eds., *That's the Joint!: The Hip-Hop Studies Reader*, 2nd ed., 44–55. New York: Routledge.

Germaine, Thomas. 2020. "Best Noise-Canceling Headphones of 2020." *Consumer Reports*, January 26. Accessed May 1, 2020. https://www.consumerreports.org/noise-canceling-headphones/best-noise-canceling-headphones-of-the-year/.

Giuliano, Geoffrey, and Vrnda Devi. 1999. *Glass Onion: The Beatles in Their Own Words*. Cambridge, MA: Da Capo Press.

Glass, Louis, and William S. Arnold. 1890. Patent for Coin-Operated Phonograph. Accessed January 18, 2021. https://patents.google.com/patent/US428750A/en.

Glinsky, Albert. 2000. *Theremin: Ether Music and Espionage*. Urbana and Chicago: University of Illinois Press.

Gordon, Doug. 2018. "Musician Thomas Dolby Blinded Us with Ringtones." Wisconsin Public Radio (WPR), June 16. Accessed January 18, 2021. https://www.wpr.org/musician-thomas-dolby-blinded-us-ringtones.

Gracyk, Theodore. 1996. *Rhythm and Noise: An Aesthetics of Rock*. Durham, NC: Duke University Press.

Graham, Jefferson. 2016. "App Makes It a Cinch to Add Backup Band." *USA Today*, January 21, Money section: 3b.

Gray, Elisha. 1875. "Electric Telegraph for Transmitting Musical Tones." U.S. Patent 166095 granted July 27. Accessed November 30, 2020. https://patents.google.com/patent/US166095.

Griggs, Brandon. 2014. "'Gangnam Style' Breaks YouTube." CNN.com, December 3. Accessed March 17, 2020. www.cnn.com/2014/12/03/showbiz/gangnam-style-youtube/index.html.

Guerrieri, Matthew. 2013. "With the '70s-era Lyricon, Woodwind Met Synthesizer." *Boston Globe*, July 6. Accessed January 18, 2021. https://www.bostonglobe.com/arts/music/2013/07/06/non-event-experimental-music-concert-include-lyricon-recalling-the-lyricon-era-woodwind-like-synthesizer/kYw3FB20rj7AzFGkMMPWHI/story.html.

Haire, Meaghan. 2009. "A Brief History of the Walkman." *Time*, July 1. Accessed December 8, 2020. http://content.time.com/time/nation/article/0,8599,1907884,00.html.

Hammar, Nickolai, and Colin Marshall. 2020. "Playing Music Together Online Is Not as Simple as It Seems." NPR.com, July 15. Accessed July 16, 2020. https://www.npr.org/2020/07/14/891091995/playing-music-together-online-is-not-as-simple-as-it-seems.

Harrington, Richard. 1990. "The Legend of Benedetti and 'Bird.'" *Washington Post*, November 18. Accessed April 1, 2020. https://www.washingtonpost.com/archive/lifestyle/style/1990/11/18/the-legend-of-benedetti-and-bird/a57e739d-2c67-42f1-946f-95982df27d38/.

Hart, Hugh. 2010. "June 28, 1846: Parisian Inventor Patents Saxophone." *Wired*, June 28. Accessed March 16, 2020. https://www.wired.com/2010/06/0628saxophone-patent/.

Harvell, Ben. 2012. *Make Music with Your iPad*. Indianapolis, IN: John Wiley & Sons.

Hasty, Katie. 2007. "Swift's Un-Swift Climb." *Billboard* 119, August 4: 9.

The Henry Ford. 2020. "Boomboxes at The Henry Ford." The Henry Ford, February 25. Accessed June 26, 2020. https://www.thehenryford.org/explore/blog/boomboxes-at-the-henry-ford.

Heylin, Clinton. 1994. *Bootleg: The Secret History of the Other Recording Industry*. New York: St. Martin's.

Hiller, Lejaren A., Charles Ames, and Robert Franki. 1985. "Automated Composition: An Installation at the 1985 International Exposition in Tsukuba, Japan." *Perspectives of New Music* 23, no. 2: 196–215.

Hiller, Lejaren A., and Leonard M. Isaacson. 1959. *Experimental Music: Composition with an Electronic Computer*. New York: McGraw-Hill. Reprinted 1979. Westport, CT: Greenwood Press.

Hischak, Thomas S. 2013. "Musical Film." In Charles H. Garrett, ed., *The Grove Dictionary of American Music*, 2nd ed., vol. 5, 617–620. New York: Oxford University Press.

Holland, Oscar. 2020. "Designing the World's First Home Computers." *CNN*, May 3. Accessed May 4, 2020. https://www.cnn.com/style/article/home-computers-design-history/index.html.

Horwood, Wally. 1983. *Adolphe Sax 1814–1894: His Life and Legacy*. Baldock, U.K.: Egon Publishers.

Huber, David M. 2018. *Modern Recording Techniques*. 9th ed. New York: Routledge.

Hunter, Dave. 2013. *Guitar Effects Pedals: The Practical Handbook*. San Francisco, CA: Backbeat Books.

Isaacson, Leonard, and Lejaren Hiller. 1993. "Musical Composition with a High-Speed Digital Computer." In *Machine Models of Music*, 9–21. Cambridge, MA: MIT Press. This is a reprint of an article originally published in the *Journal of the Audio Engineering Society* in 1958.

JackTrip.org. 2020. "JackTrip: A System for High Quality Audio Network Performance over the Internet." JackTrip.org, May 11. Accessed July 16, 2020. https://ccrma.stanford.edu/software/jacktrip/.

Jacobson, Adam. 2019. "Behind iHeart's Product Strategy Pitch." *Radio + Television Business Report*, April 3. Accessed January 18, 2021. https://www.rbr.com/behind-ihearts-product-strategy-pitch/.

Jam, Jimmy, and Terry Lewis, producers. 1986. Liner notes for *Crash* by the Human League, LP, A&M Records SP-5129.

Jarnow, Jesse. 2020. "The Endless Potential of the Pedal Steel Guitar, an Odd Duck by Any Measure." NPR, January 7. Accessed January 18, 2021. https://www.npr.org/2020/01/07/793989801/the-endless-potential-of-the-pedal-steel-guitar-an-odd-duck-by-any-measure.

Jasen, David A. 2003. *Tin Pan Alley: An Encyclopedia of the Golden Age of American Song*. New York: Routledge.

Jones, Andy. 2018. "Brian Eno and Peter Chilvers Bloom: 10 Worlds Review." *MusicTech*, December 28. Accessed October 30, 2020. https://www.musictech.net/reviews/brian-eno-and-peter-chilvers-bloom-10-worlds-review/.

June-Friesen, Katy. 2015. "For a Brief Time in the 1930s, Radio Station WLW in Ohio Became America's One and Only 'Super Station.'" *Humanities: The Magazine of the National Endowment for the Humanities* 36, no. 3 (May/June). Accessed January 18, 2021. https://www.neh.gov/humanities/2015/mayjune/feature/in-the-1930s-radio-station-wlw-in-ohio-was-americas-one-and-only-sup.

Kaufmann, Thomas, Juliette Sterkens, and John M. Woodgate. 2015. "Hearing Loops: The Preferred Assistive Listening Technology." *Journal of the Audio Engineering Society* 63, no. 4 (April): 298–302.

Kelley, Caitlin. 2019. "VMAs Ratings Plunge to All-Time Low This Year." *Forbes*, August 30. Accessed August 7, 2020. https://www.forbes.com/sites/caitlinkelley/2019/08/30/vmas-ratings-plunge-to-all-time-low-this-year.

Kelley, Frannie. 2009. "A Eulogy for the Boombox." NPR, April 22. Accessed January 18, 2021. https://www.npr.org/2009/04/22/103363836/a-eulogy-for-the-boombox.

Kellman, Andy. n.d. "Bow Wow Wow." *AllMusic*. Accessed January 18, 2021. https://www.allmusic.com/artist/bow-wow-wow-mn0000094126/biography.

Kempema, Jocelyn. 2008. "Imitation Is the Sincerest Form of . . . Infringement? Guitar Tabs, Fair Use, and the Internet." *William & Mary Law Review* 49, no. 6: 2265–2307. Accessed online January 18, 2021. https://scholarship.law.wm.edu/cgi/viewcontent.cgi?referer=https://www.google.com/&httpsredir=1&article=1166&context=wmlr.

Kennelly, Arthur E. 1938. *Biographical Memoir of George Owen Squier, 1865–1934*. Washington, DC: National Academy of Sciences of the United States of America.

Kim, Michelle. 2019. "How the Record Industry Is Trying to Make Vinyl More Environmentally Friendly." *Pitchfork*, May 9. Accessed January 18, 2021. https://pitchfork.com/thepitch/how-the-record-industry-is-trying-to-make-vinyl-more-environmentally-friendly/.

King, Brad. 2002. "The Day the Napster Died." *Wired*, May 15. Accessed January 18, 2021. https://www.wired.com/2002/05/the-day-the-napster-died/.

Kittler, Friedrich. 1999. *Gramophone, Film, Typewriter*. Translated by Geoffrey Winthrop-Young and Michael Wurtz. Stanford, CA: Stanford University Press.

Klopus, Joe. 1991. Rare 'Bird.'" *Chicago Tribune*, May 2. Accessed April 1, 2020. https://www.chicagotribune.com/news/ct-xpm-1991-05-02-9102080914-story.html.

Knauf, Ken. 1955. "Juke Box Bonanza: World Market Boom." *Billboard*, 67, no. 13 (March 26): 52.

Knight, Roderic. 2019. "Elisha Gray and the Musical Telegraph." *Galpin Society Journal* 72 (March): 205–231, 248, 250–251.

Korg, Inc. n.d. "About KORG." Korg.com. Accessed May 27, 2020. https://www.korg.com/us/corporate/.

Kosoff, Maya, and Paige Leskin. 2015. "This Is the First YouTube Video Ever Uploaded—It Was Posted 10 Years Ago Today." *Business Insider*, April 23.

Accessed January 18, 2021. https://www.businessinsider.com/first-youtube-video-2015-4.

Krummel, D. W. 1992. *The Literature of Music Bibliography: An Account of the Writings on the History of Music Printing & Publishing.* Metuchen, NJ: Scarecrow Press.

Krummel, D. W., and Stanley Sadie, eds. 1990. *Music Printing and Publishing.* New York: W.W. Norton.

Labbe, Mark. 2018. "Music Composed by AI Breathes New Life into Video Games." *SeachEnterpriseAI*, September 5. Accessed January 18, 2021. https://searchenterpriseai.techtarget.com/feature/Music-composed-by-AI-breathes-new-life-into-video-games.

Lahiff, Keris. 2020. "Two 'Surprising' Stocks to Sell—and One to Buy—Heading into the New Year, according to Trader." CNBC, December 24. Accessed December 24, 2020. https://www.cnbc.com/2020/12/24/sell-zoom-beyond-shares-into-the-new-year-according-to-trader.html.

Lamb, C., J. E. Burns, J. Scaffidi, and J. Murdock. 1994. "Karaoke: Research with a Two Drink Minimum." Paper presented at the annual convention of the Association for Education in Journalism and Mass Communication, Washington, DC. Accessed January 18, 2021. http://www2.southeastern.edu/Academics/Faculty/jeburns/karaoke.html.

Lamont, Tom. 2013. "Napster: The Day the Music Was Set Free." *Guardian*, February 23. Accessed January 18, 2021. https://www.theguardian.com/music/2013/feb/24/napster-music-free-file-sharing.

Lanza, Joseph. 1994. *Elevator Music: A Surreal History of Muzak, Easy-Listening, and Other Moodsong.* New York: St. Martin's Press.

Lanza, Joseph. 2004. *Elevator Music: A Surreal History of Muzak, Easy-Listening, and Other Moodsong.* Revised and expanded edition. Ann Arbor: University of Michigan Press.

Lapin, Andrew. 2016. "The Bizarre History of a Bogus Doctor Who Prescribed Goat Gonads." *National Geographic*, July 15. Accessed May 19, 2020. https://www.nationalgeographic.com/news/2016/07/documentary-interview-medicine-science/#close.

Lawing, Scott. 2018. "How Does a Guitar Pickup Really Work?" *Guitar World*, February 23. Accessed September 21, 2020. https://www.guitarworld.com/gear/how-does-a-guitar-pickup-really-work.

Layton Turner, Marcia. 2020. "Why Bose's Move to Close Stores Should Actually Boost Profits." *Forbes*, January 24. Accessed May 1, 2020. https://www.forbes.com/sites/marciaturner/2020/01/24/bose-closing-of-brick-and-mortar-stores-is-expected-to-boost-profits.

Lazarus, David. 2017. "Whatever Happened to Muzak? It's Now Mood, and It's Not Elevator Music." *Los Angeles Times*, July 7. Accessed January 18, 2021. https://www.latimes.com/business/lazarus/la-fi-lazarus-store-music-20170707-story.html.

Ledbetter, James. 2019. "How Tim Westergren Steered Pandora from 'the Brink of Shutting Down' to a $3.5 Billion Exit." *Inc.*, June 6. Accessed

January 18, 2021. https://www.inc.com/james-ledbetter/pandora-founder-tim-westergren-fast-growth-tour-san-francisco.html.

Lee, R. Alton. 2002. *The Bizarre Careers of John R. Brinkley*. Lexington: University Press of Kentucky.

Lefcowitz, Eric. 1985. *The Monkees Tale*. San Francisco, CA: The Last Gasp.

Leland, John. 2017. "I Didn't Move Here to Avoid Chaos." *New York Times* 166, no. 57,576 (April 23), Entertainment: 1, 6.

Lemley, Brad. 2004. "*Discover* Dialogue: Amar G. Bose." *Discover*, October 1. Accessed January 18, 2021. https://www.discovermagazine.com/technology/discover-dialogue-amar-g-bose.

Lendino, Jamie. 2019. "Apple *GarageBand* (for Mac) Review." *PC Magazine*, September 26. Accessed October 29, 2020. https://www.pcmag.com/reviews/apple-garageband-for-mac.

The Leonard Bernstein Office. n.d. "About: Educator: Young People's Concerts." The Leonard Bernstein Office. Accessed January 18, 2021. https://leonardbernstein.com/about/educator/young-peoples-concerts.

Leslie, Donald J. 1945. "Rotatable Tremulant Sound Producer." U.S. Patent application, July 9. Accessed January 18, 2021. https://patents.google.com/patent/US2489653.

LeVitus, Bob, Edward C. Bair, and Bryan Chaffin. 2018. *iPad for Dummies*. 10th ed. Hoboken, NJ: John Wiley & Sons.

Levy, Kenneth. 1987. "On the Origin of Neumes." *Early Music History* 7: 59–90.

Lewis, Tom. 1991. *Empire of the Air: The Men Who Made Radio*. New York: HarperCollins.

Lewisohn, Mark. 1988. *The Beatles Recording Sessions: The Official Abbey Road Studio Session Notes 1962–1970*. New York: Harmony Books.

The Library of Congress. n.d. "Emile Berliner and the Birth of the Recording Industry." The Library of Congress. Accessed March 30, 2020. https://www.loc.gov/collections/emile-berliner/about-this-collection/.

Light, Enoch, producer. 1967. Liner notes for *Kites Are Fun* by the Free Design. LP, Project 3 PR5019SD.

The Lightning Arranger Company. 1931. *The Lightning Instructor for The Lightning Arranger*. Allentown, PA: Lightning Arranger Company.

Limina, Dave. 2002. *Hammond Organ Complete*. Boston: Berklee Press Publications.

Mackey, Bob. 2016. "Day of the Tentacle Composer Peter McConnell on Communicating Cartooniness." *USGamer*, March 8. Accessed January 18, 2021. https://www.usgamer.net/articles/day-of-the-tentacle-composer-peter-mcconnell-on-communicating-cartooniness.

MacMahon, Bernard, director. 2017. *American Epic: The First Time America Heard Itself*. Documentary film. New York: Touchstone.

MacMahon, Bernard, and Allison McGourty, with Elijah Wald. 2017. *American Epic: The Companion Book to the PBS Series*. New York: Touchstone.

Madrigal, Alexis C. 2013. "Someone Had to Invent Karaoke—This Guy Did." *Atlantic*, December 18. Accessed January 18, 2021. https://www.theatlantic.com/technology/archive/2013/12/someone-had-to-invent-karaoke-this-guy-did/282491/.

Majewski, Lori, and Jonathan Bernstein. 2014. *Mad World: An Oral History of New Wave Artists and Songs That Defined the 1980s*. New York: Abrams Image.

Maloof, Rich. 2004. *Jim Marshall—The Father of Loud: The Story of the Man behind the World's Most Famous Guitar Amplifiers*. San Francisco, CA: Backbeat Books.

Margotin, Philippe, and Jean-Michel Guesdon. 2016. *The Rolling Stones: All the Songs: The Story Behind Every Track*. New York: Black Dog & Leventhal Publishers.

Marr, Bernard. 2019. "The Amazing Ways Artificial Intelligence Is Transforming the Music Industry." *Forbes*, July 5. Accessed July 22, 2020. https://www.forbes.com/sites/bernardmarr/2019/07/05/the-amazing-ways-artificial-intelligence-is-transforming-the-music-industry/.

Martin, Steven M, director. 1993. *Theremin: An Electronic Odyssey*. VHS Video. Orion Classics. Released on DVD MGM/Fox Video, 2003.

Massingill, Randi L. 1997. *Total Control: The Michael Nesmith Story*. Mesa, AZ: FLEXquarters.

McBride, William. 1982. "The Early Saxophone in Patents 1838–1850 Compared." *Galpin Society Journal* 35 (March): 112–121.

McDonald, Glenn, and Ken Mingis. 2019. "The Evolution of the iPad." *Computerworld*, October 1. Accessed January 18, 2021. https://www.computerworld.com/article/3269331/the-evolution-of-the-ipad.html.

McIntyre, Hugh. 2018. "The Piracy Sites That Nearly Destroyed the Music Industry: What Happened to Napster?" *Forbes*, March 21. Accessed January 18, 2021. https://www.forbes.com/sites/hughmcintyre/2018/03/21/what-happened-to-the-piracy-sites-that-nearly-destroyed-the-music-industry-part-1-napster.

McKee, Ruth, and Jamie Grierson. 2017. "Roland Founder and Music Pioneer Ikutaro Kakehashi Dies Aged 87." *Guardian*, April 2. Accessed January 18, 2021. https://www.theguardian.com/music/2017/apr/02/roland-founder-and-music-pioneer-ikutaro-kakehashi-dies-aged-87.

McLerran, Marina. 2017. "A Brief History of the Music Box." *McLerran Journal*, June 4. Accessed January 18, 2021. https://www.mclerranjournal.com/technology-1/2017/6/4/a-short-history-of-the-music-box.

McNairn, Bonnie, with James Wilson. 1993. "A Simple Theremin from Schematic to Performance." *Experimental Musical Instruments* 8, no. 3 (March): 26–27.

McNamee, David. 2009. "Hey, What's That Sound: Linn LM-1 Drum Computer and the Oberheim DMX." *Guardian*, June 22. Accessed January 18, 2021. https://www.theguardian.com/music/2009/jun/22/linn-oberheim-drum-machines.

Michel, Norbert J. 2006. "The Impact of Digital File Sharing on the Music Industry: An Empirical Analysis." *Topics in Economic Analysis & Policy* 6, no. 1. This article is available for download from the Recording Industry Association of America (RIAA) at https://www.riaa.com/wp-content/uploads/2004/01/art-the-impact-of-digital-file-sharing-on-the-music-industry-michel-2006.pdf.

The MIDI Association. n.d.a. "Details about MIDI 2.0TM, MIDID-CI, Profiles and Property Exchange." MIDI.org. Accessed November 20, 2020. https://www.midi.org/midi-articles/details-about-midi-2-0-midi-ci-profiles-and-property-exchange.

The MIDI Association. n.d.b. "The History of MIDI." MIDI.org. Accessed January 18, 2021. https://www.midi.org/midi-articles/the-history-of-midi.

Miller, Judith. 2018. *Collectibles Handbook & Price Guide 2019–2020*. London: Miller's.

Miller, Mary K. 2002. "It's a Wurlitzer: The Giant of the Musical Instrument Collection Makes Tunes Rootin 'Tootin' or Romantic." *Smithsonian Magazine* (April). Accessed January 18, 2021. https://www.smithsonianmag.com/history/its-a-wurlitzer-61398212/.

Miller, Michael. 2020. *My iPad for Seniors*. 7th ed. Carmel, IN: Que Publishing.

Milliman, Ronald E. 1982. "Using Background Music to Affect the Behavior of Supermarket Shoppers." *Journal of Marketing* 46, no. 3 (Summer): 86–91.

Mobile Fidelity Sound Lab. n.d.a. "History." MoFi.com. Accessed May 29, 2020. https://mofi.com/pages/about-us.

Mobile Fidelity Sound Lab. n.d.b. "Limited Editions Form the Original Masters." Advertising liner note supplement.

Moog, Robert A. 1961. "A Transistorized Theremin." *Electronics World* 65 (January): 29–32, 125.

Moog Music, Inc. 2011a. *A Brief History of the Minimoog*, Part 1. Accessed January 18, 2021. https://www.youtube.com/watch?v=sLx_x5Fuzp4.

Moog Music, Inc. 2011b. *A Brief History of the Minimoog*, Part 2. Accessed January 18, 2021. https://www.youtube.com/watch?v=xh4Ok0ex2vU.

Moore, Thurston, ed. 2005. *Mix Tape: The Art of Cassette Culture*. New York: Universe Publishing.

Moreno, Ashley. 2013. "It's Still All about Peer-to-Peer Connectivity: Alex Winter Raps about Napster with Shawn Fanning and Sean Parker." *Austin Chronicle*, March 12. Accessed January 18, 2021. https://www.austinchronicle.com/daily/sxsw/2013-03-12/its-still-all-about-peer-to-peer-connectivity/.

Morgan, Robert P. 1972. Liner notes for *Edgard Varèse: Offrandes, Intégrales, Octandre, Ecuatorial*. Reissued on CD as Elektra/Nonesuch 9 71269–2.

Moskovciak, Matthew. 2006. "Bose Acoustic Wave Music System II Review." CNET.com, Accessed January 18, 2021. https://www.cnet.com/reviews/bose-acoustic-wave-music-system-2-review/.

Murnane, Kevin. 2017. "'Sgt. Pepper's' Was a Perfect Storm of Musical and Recording Creativity." *Forbes*, June 3. Accessed January 18, 2021. https://www.forbes.com/sites/kevinmurnane/2017/06/03/sgt-peppers-was-a-perfect-storm-of-musical-and-recording-creativity.

The Music House Museum. 1994. *The Music House*. 2nd ed. Traverse City, MI: The Music House.

Nathan, John. 1999. *Sony: A Private Life*. New York: Houghton Mifflin.

National Association of Music Merchandisers (NAMM). 2005. "Irv Kratka." NAMM Oral History Program, September 13. Accessed January 18, 2021. https://www.namm.org/library/oral-history/irv-kratka.

National Museums Liverpool. n.d. "Emergence of Multi-Track Recording." Accessed August 4, 2020. https://www.liverpoolmuseums.org.uk/emergence-of-multitrack-recording.

National Public Radio (NPR) Staff. 2013. "'Something Being Born': On Making a Classic Album with a Boombox." *Weekend Edition Saturday*, August 10. Accessed January 18, 2021. https://www.npr.org/sections/therecord/2013/08/11/210524003/something-being-born-on-making-a-classic-album-with-a-boombox.

Naylor, Tim. 2015. "The *Record Collector* Guide to 8 Track Cartridge Stereo." *Record Collector*, February 23. Accessed January 18, 2021. https://recordcollectormag.com/articles/8-track-cartridge-stereo.

Nixon, Marni, and Stephen Cole. 2006. *I Could Have Sung All Night: My Story*. New York: Billboard Books.

NPR Music. 2009. "The History of the Boombox." NPR Music. Accessed January 18, 2021. https://www.youtube.com/watch?v=e84hf5aUmNA.

Odell, Jennifer. 2019. "The History of Play-A-Long (and More)." *JazzTimes*, April 25. Accessed January 18, 2021. https://jazztimes.com/features/profiles/history-of-play-a-long/.

O'Neill, John J. *Prodigal Genius: The Extraordinary Life of Nicola Tesla*. Paperback ed. Valley Cottage, NY: Discovery Publisher.

Ord-Hume, Arthur W. J. G. 1973. *Clockwork Music: An Illustrated History of Mechanical Musical Instruments from the Musical Box to the Pianola, from Automaton Lady Virginal Players to Orchestrion*. New York: Crown Publishers.

Ord-Hume, Arthur W. J. G. 1995. *The Musical Box: A Guide for Collectors*. Atglen, PA: Schiffer Publishing.

Owsinski, Bobby. 2017. *The Mixing Engineer's Handbook*. 4th ed. Burbank, CA: Bobby Owsinski Media Group.

Palmer, Robert. 1997. Liner notes for *Kind of Blue* by Miles Davis. CD reissue. Columbia Legacy CK 64935.

Pantsios, Anastasia. 2012. "Canton Morning Team Settles Lawsuit against Former Station." *Cleveland Scene*, July 31. Accessed June 9, 2020. https://www.clevescene.com/scene-and-heard/archives/2012/07/31/canton-morning-team-settles-lawsuit-against-former-station.

Paris, Steve. 2020. "DaVinci Resolve 17 Review." *TechRadar*, November 11. Accessed November 23, 2020. https://www.techradar.com/reviews/davinci-resolve-17.

Parkinson, Justin. 2016. "What Is Shop Music Doing to Your Brain?" *BBC News Magazine*, June 1. Accessed January 18, 2021. https://www.bbc.com/news/magazine-36424854.

Payne, Christine. 2019. "MuseNet." *OpenAI*, April 25. Accessed August 20, 2020. https://openai.com/blog/musenet/.

Peoples, Glenn. 2020. "How SiriusXM Is Beating the Subscriber Paradox & Scoring a Better Return Than Streaming Services." *Billboard*, February 7. Accessed December 14, 2020. https://www.billboard.com/articles/business/8550404/siriusxm-listener-subscriber-growth-data-analysis-streaming-radio/.

Perone, James E. 1988. *Pluralistic Strategies in Musical Analysis: A Study of Selected Works of William Albright.* Ph.D. dissertation. Buffalo, NY: State University of New York at Buffalo.

Perone, James E. 2017. *The Words and Music of Taylor Swift.* Santa Barbara, CA: Praeger Publishers.

Perry, Peter. 2006. "*InForm*: A Music Analysis System." *Music Educators Journal* 93, no. 1 (September): 23.

Pinch, Trevor, and Frank Trocco. 2002. *Analog Days: The Invention and Impact of the Moog Synthesizer.* Cambridge, MA: Harvard University Press.

Pino, David. 1980. *The Clarinet and Clarinet Playing.* Mineola, NY: Dover.

Port, Ian. 2019. *The Birth of Loud: Leo Fender, Les Paul, and the Guitar-Pioneering Rivalry That Shaped Rock 'n' Roll.* New York: Scribner.

Prendergast, Mark. 2003. *The Ambient Century: From Mahler to Moby—The Evolution of Sound in the Electronic Age.* New York and London: Bloomsbury.

QY Research. 2020. "Global Karaoke Market Research Report 2020." Accessed November 22, 2021. https://www.marketstudyreport.com/reports/global-karaoke-market-research-report-2020.

Rendall, F. Geoffrey. 1971. *The Clarinet: Some Notes on Its History and Construction.* 3rd ed. New York: W.W. Norton.

Rex, Nicholas. 2014. *Close up Tight: The Life and Music of Bill Chase.* Master's Thesis, M.A. in Jazz History & Research. Newark, NJ: Rutgers University.

Ridley, E. A. K. 1986. "Birth of the 'Boehm' Clarinet." *Galpin Society Journal* 39 (September): 68–76.

Robles, Sonia. 2019. *Mexican Waves: Radio Broadcasting along Mexico's Northern Border, 1930–1950.* Tucson: University of Arizona Press.

Rogers, Scott. 2014. *Level Up! The Guide to Great Video Game Design.* 2nd ed. Chichester, U.K.: John Wiley & Sons.

Rogers, Vince, director and producer. 2011. *Vox Pop: How Dartford Powered the British Beat Boom.* Video documentary. London: BBC Productions.

Rose, Joel, and Jacob Ganz. 2011. "The MP3: A History of Innovation and Betrayal." *NPR*, March 23. Accessed January 18, 2021. https://www.npr.org/sections/therecord/2011/03/23/134622940/the-mp3-a-history-of-innovation-and-betrayal.

Ross, Michael. 2015. "Pedal to the Metal: A Short History of the Pedal Steel Guitar." *Premier Guitar*, February 17. Accessed January 18, 2021. https://www.premierguitar.com/articles/22152-pedal-to-the-metal-a-short-history-of-the-pedal-steel-guitar.

Ruhlmann, William. n.d. "Woody Guthrie: *The Live Wire: Woody Guthrie in Performance 1949*." *AllMusic*. Accessed January 18, 2021. https://www.allmusic.com/album/the-live-wire-woody-guthrie-in-performance-1949-mw0000496961.

Rushent, Martin, and the Human League, producers. 1981. Liner notes for *Dare* by the Human League, LP, A&M Records SP-4892.

Russell, Ross. 1973. *Bird Lives! The High Life and Hard Times of Charlie (Yardbird) Parker.* New York: Charterhouse.

Samuel, Harold E. 1986. "Printing of Music." In Don Randel, ed., *The New Harvard Dictionary of Music*, 655–656. Cambridge, MA: The Belknap Press of Harvard University Press.

Sandoval, Andrew. 2001. "The True Story of the Monkees" (liner booklet for the *The Monkees Music Box*). Four CD set. Rhino Records R2 76706.

Schaap, Phil. 1990. Liner notes for *The Complete Dean Benedetti Recordings of Charlie Parker*. Stamford, CT: Mosaic Records.

Schaffer, Stan. 1984. "Math Music Add up for Whitehall Inventor." *Morning Call*, March 22. Accessed January 18, 2021. https://www.mcall.com/news/mc-xpm-1984-03-22-2400593-story.html.

Scheinman, Ted. "Status Cymbals: A Selection of Top Answers to the Centuries-Old Musical Question, How Do You Get By without an Actual Drummer." *Smithsonian* 51, no. 4 (July/August): 23.

Schloss, Joseph. 2012. "Sampling Ethics." In Murray Forman and Mark A. Neal, eds., *That's the Joint! The Hip-Hop Studies Reader*, 2nd ed., 610–630. New York: Routledge.

Schubin, Mark. 2011. "Headphones, History, & Hysteria." *SVG News*, February 11. Accessed January 18, 2021. https://www.sportsvideo.org/2011/02/11/headphones-history-hysteria/.

Scott, Jonathan. 2019. *The Vinyl Frontier: The Story of the Voyager Golden Record*. London: Bloomsbury Sigma.

Seagrave, Shane. 2007. "Organ Figures." *Carousel Organ* no. 33 (October): 15–22.

Seifer, Marc. 2016. *Wizard: The Life and Times of Nikola Tesla: Biography of a Genius*. New York: Kensington Publishing Corp.

Sewell, Amanda. 2020. *Wendy Carlos: A Biography*. New York: Oxford University Press.

Shah, Haleema. 2019. "How the Hawaiian Steel Guitar Changed American Music." *Smithsonian Magazine*, April 25. Accessed January 18, 2021. https://www.smithsonianmag.com/smithsonian-institution/how-hawaiian-steel-guitar-changed-american-music-180972028/.

Shannon, John R. 2009. *Understanding the Pipe Organ: A Guide for Students, Teachers and Lovers of the Instrument*. Jefferson, NC: McFarland & Company.

Shatavsky, Sam. 1969. "The Best Tape System for You: Reel, Cassette, or Cartridge." *Popular Science* 194, no. 2 (February): 126–129.

Shepard, Brian K. 2013. *Refining Sound: A Practical Guide to Synthesis and Synthesizers*. New York: Oxford University Press.

Simpson, Dave. 2016. "How We Made Laurie Anderson's 'O Superman.' " *Guardian*, April 19. Accessed January 18, 2021. https://www.theguardian.com/culture/2016/apr/19/how-we-made-laurie-anderson-o-superman.

Slonimsky, Nicolas, Laura Kuhn, and Dennis McIntire. 2001. "Nancarrow, Conlon." In Nicolas Slonimsky and Laura Kuhn, eds., *Baker's Biographical Dictionary of Musicians*, vol. 4, 2565. New York: Schirmer Books.

Smith, Ernie. 2016. "The Story of the Bose Wave, the Stereo System Built for the Infomercial Era." *Vice*, December 23. Accessed January 18, 2021. https://www.vice.com/en_us/article/4xap3d/the-story-of-the-bose-wave-the-stereo-system-built-for-the-infomercial-era.

Smith, Kathleen E. R. 2003. *God Bless America: Tin Pan Alley Goes to War*. Lexington: University of Kentucky Press.

Smith, Oberlin. 1888. "Some Possible Forms of Phonograph." *Electrical World* 11–12 (September 1): 116–117. Accessed January 18, 2021. https://books.google.com/books?id=zYVMAAAAYAAJ&pg=RA2-PA116#v=onepage&q&f=false.

Sony Corporation. n.d. "History." Sony.com. Accessed May 19, 2020. https://www.sony.net/SonyInfo/CorporateInfo/History/.

The Staff of the *Buffalo News*. 1991. "Pipes & Bells & Whistles for a Feast of Pull-out-the-Stops Music, There's Nothing Like Listening to Buffalo's Classic Theater Organs." *Buffalo News*, September 20. Accessed December 17, 2020. https://buffalonews.com/news/pipes-bells-whistles-for-a-feast-of-pull-out-the-stops-music-theres-nothing-like/article_90d400bd-ade1-5e96-865d-b49037ff7fc0.html.

Stamp, Jimmy. 2013. "A Partial History of Headphones." *Smithsonian Magazine*, March 19. Accessed January 18, 2021. https://www.smithsonianmag.com/arts-culture/a-partial-history-of-headphones-4693742/.

Stassen, Murray. 2020. "Can AI-Driven A&R Transform the Music Business?" *Music Business Worldwide*, February 27. Accessed January 18, 2021. https://www.musicbusinessworldwide.com/can-ai-driven-ar-transform-the-music-business/.

Steffen, David J. 2005. *From Edison to Marconi: The First Thirty Years of Recorded Music*. Jefferson, NC: McFarland.

Sterling, Christopher H. n.d. "Mutual Broadcasting System." *Encyclopædia Britannica*. Accessed January 18, 2021. https://www.britannica.com/topic/Mutual-Broadcasting-System.

Stimson, Arthur J. 1933. Patent granted 1934. Patent application for "Electrophonic Stringed Musical Instrument." Accessed January 18, 2021. https://patents.google.com/patent/US1962919?oq=Us1962919.

Stross, Randall. 2007. *The Wizard of Menlo Park: How Thomas Alva Edison Invented the Modern World*. New York: Three Rivers Press.

Stross, Randall. 2010. "The Incredible Talking Machine." *Time*, June 23. Accessed January 18, 2021. http://content.time.com/time/specials/packages/article/0,28804,1999143_1999210,00.html.

Swallow, Matthew J. 2016. *MIDI Electronic Wind Instrument: A Study of the Instrument and Selected Works*. Doctoral dissertation. Morgantown: West Virginia University. Accessed January 18, 2021. https://researchrepository.wvu.edu/etd/6750.

Sweet, Michael. 2015. *Writing Interactive Music for Video Games*. New York: Addison-Wesley.

Sweetwater Sound. 2004. "Virtual Analog Synths vs. Analog Synths." Sweetwater.com, April 22. Accessed January 18, 2021. https://www.sweetwater.com/insync/virtual-analog-synths-vs-analog-synths/.

Swift, Taylor. 2014. Spoken introductions to bonus tracks. *1989*. CD. Big Machine Records BMRBD0550A.

Teagle, John, and John Sprung. 1995. *Fender Amps: The First Fifty Years*. Milwaukee, WI: Hal Leonard.

Teal, Larry. 1963. *The Art of Saxophone Playing.* Miami, FL: Summy-Birchard.

Techmoan. 2016. "Boom Box Time Capsule." Accessed January 18, 2021. https://www.youtube.com/watch?v=2QGWyxI-ZPs. A YouTube reflection back on the popular boomboxes of the 1980s and 1990s.

Thomas, Helen, Claudine King-Dabbs, and Dan Walker, producers. 2007. *James May's 20th Century.* London: BBC Productions. Issued on DVD in 2012 as Athena Learning AMP-8788.

Thompson, Dave. 2020. "Cassette Tapes Are Making a Comeback." *Goldmine*, May 5. Accessed January 18, 2021. https://www.goldminemag.com/collector-resources/cassette-tapes-are-making-a-comeback.

Ugrešić, Dubravka. 2011. *Karaoke Culture.* English translation by David Williams. Rochester, NY: Open Letter.

Vail, Mark. 2002. *The Hammond Organ: Beauty in the B.* San Francisco, CA: Backbeat Books.

Velasco, Carl. 2017. "The MP3 Is Dead: Here's a Brief History of MP3." *Tech Times*, May 13. Accessed January 18, 2021. https://www.techtimes.com/articles/207213/20170513/the-mp3-is-dead-heres-a-brief-history-of-mp3/.

Vincent, James. 2016. "This AI-Written Pop Song Is Almost Certainly a Dire Warning for Humanity: Let's Not Rule It Out, Anyway." *Verge*, September 26. Accessed January 18, 2021. https://www.theverge.com/2016/9/26/13055938/ai-pop-song-daddys-car-sony.

Vogel, Peter, and James Gardner. 2012. "Interview: Peter Vogel." Radio New Zealand (RNZ), May. Accessed January 18, 2021. https://www.rnz.co.nz/concert/programmes/hopefulmachines/20131119. This is an update and correction of an interview transcript originally published on May 8, 2010. The exact May 2012 date of republication is not given.

Walker, Rob. 2009. "The Song Decoders." *New York Times Magazine*, October 14. Accessed November 30, 2020. https://www.nytimes.com/2009/10/18/magazine/18Pandora-t.html.

Warfield, Patrick. 2009. "John Philip Sousa and 'The Menace of Mechanical Music.'" *Journal of the Society for American Music* 3, no. 4 (November): 431–463.

Warren, Rich. 1991. "Bose Improves Remarkable Acoustic Wave Music System." *Chicago Tribune*, December 20. Accessed January 18, 2021. https://www.chicagotribune.com/news/ct-xpm-1991-12-20-9104240238-story.html.

Warwick, Josh. 2014. "Happy Birthday to the World's Best-Known Ringtone." *Telegraph*, April 4. Accessed January 18, 2021. https://www.telegraph.co.uk/technology/nokia/10734730/Happy-birthday-to-the-worlds-best-known-ringtone.html.

Watts, Janet. 1999. "No Thank You for the Muzak." *Guardian*, December 1. Accessed January 18, 2021. https://www.theguardian.com/society/1999/dec/01/guardiansocietysupplement2.

Wegman, Jesse. 2000. "The Story behind 'Purple Haze.'" National Public Radio (NPR), September 18. Accessed January 18, 2021. https://www.npr.org/2000/09/18/1088122/jimi-hendrix-purple-haze.

Weiss, Brett. 2010. "Rock on with Vintage Jukeboxes." *AntiqueWeek*, October 15. Accessed January 18, 2021. http://www.antiqueweek.com/Article.asp?newsid=1796.

Wenger Corporation. n.d. "What Is VAE® Technology?" Wenger Corporation. Accessed January 18, 2021. https://www.wengercorp.com/Lit/Wenger_VAE_BRO_LT0373.pdf.

Werner, Karen, ed. 2012. *MIM: Highlights from the Musical Instrument Museum*. Phoenix, AZ: Musical Instrument Museum.

White, Forrest. 1994. *Fender: The Inside Story*. San Francisco, CA: Miller Freeman Books.

White, Marla. 2006. "Guerrilla Home Recording." *Music Educators Journal* 93, no. 1 (September): 23–24.

Wickman, Jim. 1955. "New Types of Equipment, New Types of Locations." *Billboard* 67, no. 13 (March 26): 50–51.

Wilford, John N. 1971. "Bill Lear Thinks He'll Have the Last Laugh." *New York Times*, April 4, Section A: 11.

Williams, Cameron. 2017. "How MTV Changed the World with Its Industry of Cool." Special Broadcast Services (SBS) Australia, February 13. Accessed January 18, 2021. https://www.sbs.com.au/guide/article/2017/02/13/how-mtv-changed-world-its-industry-cool.

Williams, David B., and Peter R. Webster. 2008. *Experiencing Music Technology*. Updated 3rd ed. Boston: Schirmer.

Williams, Stephen. 2011. "For Car Cassette Decks, Play Time Is Over." *New York Times*, February 4. Accessed January 18, 2021. https://www.nytimes.com/2011/02/06/automobiles/06AUDIO.html.

Wiltshire, Alex. 2020. *Home Computers: 100 Icons That Defined a Digital Generation*. Cambridge, MA: The MIT Press.

Winter, Alex, director. 2013. *Downloaded*. Documentary film. New York: VH1

Wolff, Christoph, ed. 1983. *The New Grove Bach Family*. New York: W.W. Norton.

Woolley, Scott. 2016. *The Network: The Battle for the Airwaves and the Birth of the Communications Age*. New York: Ecco Press.

Xenakis, Iannis. 1971. *Formalized Music: Thought and Mathematics in Composition*. Translated by Christopher Butchers, G. W. Hopkins, and Mr. and Mrs. John Challifour. Bloomington: Indiana Press.

Xun, Zhou, and Francesca Tarocco. 2007. *Karaoke: A Global Phenomenon*. London: Reaktion Books.

Zak, Albin J., III. 2001. *The Poetics of Rock: Cutting Tracks, Making Records*. Berkeley: University of California Press.

Zappa, Frank, with Peter Occhiogrosso. 1989. *The Real Frank Zappa Book*. New York: Poseidon Press.

Zarczynski, Andrea. 2020. "Record High 34.9 Million Paid Subscribers Marks SiriusXM Milestone Year." *Forbes*, January 7. Accessed December 14, 2020. https://www.forbes.com/sites/andreazarczynski/2020/01/07/record-high-349-million-paid-subscribers-marks-siriusxm-milestone-year.

Zurcher, Neil. 2005. *Strange Tales from Ohio*. Cleveland, OH: Gray & Company.

Zwonitzer, Mark, with Charles Hirschberg. 2002. *Will You Miss Me When I'm Gone: The Carter Family and Their Legacy in American Music*. New York: Simon & Schuster.

Index

Note: Page numbers in **boldface** indicate main entries in the Encyclopedia.

ABBA, 30
Abbey Road Studios, 96, 146, 157, 244 (sidebar)
Ableton Live, 51, 220
Acapella, xxii, xxxi, 27–28, 228
Activision, 40
Advanced Audio Coding (AAC), 110, 151
Aebersold, Jamey, 259–60
Aeolian piano rolls, 193
Akai, xxi, 81, 157. *See also* Electronic Wind Instruments (EWI)
Al-Jazari, Ismail, 53–54
Albright, William, *Sphaera*, 36
Alesis drum machines, 56
Alfred Music Publishing, *Tone 'N' Tempo Changer*, 260
Alonso, Sydney, 277–78. *See also* Synclavier; Yamaha Corporation, DX7 synthesizer
Altec Lansing, Voice of the Theatre speaker systems, 181
Amati family, 174
Amazon, 134; Amazon Music, 109, 151, 237; Echo Show, 63–64
American Bandstand, xvii
American Graffiti, 113
American Federation of Musicians, xvi, 54, 56–57 (sidebar), 221–22
American Roots (streaming audio service), 227, 237–38
American Society of Composers, Authors and Publishers (ASCAP), 216
Ampex, 64
AmpKit, 4–5
amplification, **1–5**
Analog Synthesizers, **5–7**
Anderson, Laurie, "O Superman," 153, 267
Antares Audio Technologies, xxx, 10. *See also Auto-Tune*

Appcompanist, 261
Apple Computer, xxx, 39, 162; Apple Music, 237; *GarageBand*, xxi, xxxi, 11, 87–89, 104, 122, 142, 157, 220, 263–64; *iMovie*, 28, 104; iPads, 103–5; iPods, 105–8; *iTunes*, xxx, 32, 103, 107–10, 151; *Logic Pro X*, 88, 263
Appleton, Jon, 128, 278. *See also* Synclavier; Yamaha Corporation, DX7 synthesizer
Argent, Rod, 73
Armstrong, Edwin, 204
Arnold, Leo F. J., 79–80
Arnold, William S., 111
ARP String Ensemble, 6
Arrau, Claudio, xxx, 30
Ars Nova (software company), 101
Artificial Double Tracking, 10, 244 (sidebar)
Artificial intelligence, **7–10**
Artuia, drum machines, 56; synthesizers, 7
Artusi, Inc., 99
Association for Technology in Music Instruction (ATMI), 100
Atlanta Symphony and Chorus, 43
Atlantic Records, 156
aTunde Adjuah, Christian Scott, 261
Audacity, 11, 104, 142, 157, 213, 243, 263–64
Audio induction systems, 82
Audio Virtual Reality Engine (*AVRe*), 37–38, 254–55
Audition, 243
Auto-Tune, xxx, **10–12**, 117

Babcock, Alpheus, xxiv
Bach, Johann Sebastian, xxiv

Bagdasarian, Ross. *See* Seville, Alan
Baldwin Electro-Piano, 74
Baldwin, Nathaniel, 95
Bally Manufacturing Corporation, 253
Band Box. *See* Thomas Organs
Band-in-a-Box. *See* PG Music
BandLab, xxxi, 28–29
Barbaud, Pierre, 34
Barlow, John Perry, 190
Bashful Cousin Oswald, 211
The Beach Boys, 120; "Good Vibrations," 249
The Beatles, xxviii, 35, 85, 96; *Abbey Road*, xx, xxviii, 6, 145; "Being for the Benefit of Mr. Kite," 243; *A Hard Day's Night*, 152; "Hey Bulldog," 152; "I'm Only Sleeping," 243; "Penny Lane," 152; "Rain," 243;"Revolution," 152; *Sgt. Pepper's Lonely Hearts Club Band*, 156, 235–36; "Strawberry Fields Forever," 131, 152, 235–36, 243; "Tomorrow Never Knows," 10, 120; use of guitar volume pedal, 62; use of Leslie speaker, 10; use of Moog synthesizer, xx, xxviii, 145–46
Beatnik Audio Engine, 229 (sidebar)
Beauchamp, George D., xvii–xviii, xxvi, 67–68, 209–10
Beaver, Paul, 144
Beck, Jeff, 62
Beck, Zane, 187
Beethoven, Ludwig van, 132, 223
Behringer, drum machines, 56; synthesizers, 7
Bell, Alexander Graham, 59, 123, 135
Bell Laboratories, xxvii, 71, 250, 266
Benedetti, Dean, 269–70
Berlin, Irving, 203, 224
Berliner, Emile, xxv, 59, 90–91, 220. *See also* gramophones and victrolas
Bernardi, Bill, 80. *See also* Lyricon
Bernstein, Leonard, xvi–xvii
Berry, Chuck, 69; "School Days," 113
Bethany Station (Voice of America), 203
Bieber, Justin, xxi–xxii, 280
The Black-Eyed Peas, "Boom Boom Pow," 11
The Black Keys, 26 (sidebar)
Blackmer, David E., 53. *See also* dbx noise reduction system
Blake, Eubie, 193
Blombach, Ann, 99
Blondie, *Autoamerican*, 81
Bloom, 89, 90, 122. *See also* Generative Music

Bluetooth technology, 13, 16, 97, 113, 143, 277 (sidebar)
Boehm, Theobald, xiii, xxiv, 271
Booker T. and the M.G.'s, "Green Onions," 178
Boomboxes, xxviii, **13–16**, 24
Bose, Amar G., and the Bose Corporation, **16–19**. *See also* Bose Corporation
Bose Corporation, 16–19, 125; Acoustic Wave Music System, xxx, 17, 125; automotive speaker systems, 17–18; Bose 2201 speaker system, 16–17; Noise Canceling Headphones, 18; QuietComfort headphones, 18; QuietControl headphones, 18
Boulez, Pierre, 241
Bow Wow Wow, "C·30 C·60 C·90 Go!," 24
Bowie, David, "Heroes," 89; "Space Oddity," 7, 132
Braille, Louis, 169
Brandenburg, Karlheinz, 150. *See also* MP3s
brass instruments, **19–22**
Braven BRV-XXL/2 boombox, 16
Braverman, Paul, 270
Brecker, Michael, 80
Brenston, Jackie and His Delta Cats, "Rocket 88," 47–48
Brinkley, John R., 202, 205–6 (sidebar). *See also* XERA (radio station)
Brooks, Garth, 230
Brown, James, "Give It Up, Turn It Loose," 51
The Buggles, "Video Killed the Radio Star," xxix, 6, 152
Eric Burden and the Animals, "Sky Pilot," 84
Bush, Kate, 137
The Byrds, xx; "Goin' Back," xxviii, 5, 144; *The Notorious Byrd Brothers*, 144

C. G. Conn (musical instrument company): Sousaphone, 21; Stroboconn, 76–77
Caceres, Juan Pablo, 28
Cadenza Live Accompanist, 261
Cage, John, 242; *HPSCHD*, 36; *ORGAN2/ ASLSP—As SLow aS Possible*, 179–80
Cahill, Thaddeus, 170
Carlos, Wendy, xx; *A Clockwork Orange*, 145; *Switched-On Bach*, xx, xxviii, 6, 144–45
Carnes, Kim, 7
Carpenter, Pete, *The Rockford Files*, 146

Carré, Benoît, 9, 34–35
The Carter Family, xvi, 71, 203
Carvin, 4, 70
Cash, Johnny, *At Folsom Prison*, 31
Casio Computer Co., Ltd., 46–47 (sidebar); Casiotone CTS-200, 47 (sidebar); VL-1, 44–45; VL-TONE, 44
cassette tapes, **23–27**
CDs, xx, xxix–xxx, 30–33
Celestion speakers, 3
Chafe, Chris, 28
Chamberlain, Harry, xxvii, 130; Chamberlain (musical instrument), xxvii, 54, 130; Rhythmate, xxvii, 54, 130
Chandler, Chas, 141
Charles, Ray, 73; "What'd I Say," 75, 83–84
Chase, Bill, 21
Chen, Steve, 279. *See also* YouTube
Cher, "Believe," xxx, 10
Chilvers, Peter, 35, 89, 122. *See also* Generative Music
Chowning, John, 45, 278. *See also* Synclavier; Yamaha Corporation, DX7 synthesizer
Christian, Charlie, 69
Chrysler Corporation, 64
Cicala, Roy, 243
Clapton, Eric, 3, 48, 62
Clark, Melville, 193
Clark, Steve, 175. *See also* Luis and Clark
Clavioline, 5, 75
Cleartune, 78, 227
Cleveland Orchestra, 43
A Clockwork Orange, 52
Collaboration software, **27–30**
Columbia Broadcasting System (CBS), 203
Columbia Records, xxviii, 71
Command Records, 149, 257
Compact discs (CDs), **30–33**
Computer-composed music, **33–35**
Computer-generated music, **36–38**
Computone. *See* Lyricon
Confrey, Zez, *Kitten on the Keys*, 166
Conn, C. G. *See* C. G. Conn
Conniff, Ray, 171
Connorized piano rolls, 193
Conrad, Frank, 202
Cooder, Ry, *Bop 'Til You Drop*, xxix, 43
Coolio, "Gansta's Paradise," 216–17
Corea, Chick, 73
Cortez, Dave, "Baby," "Happy Organ," 179
Costello, Elvis, *My Aim Is True*, 31
Cowell, Henry, xxvi, 54
CRAVE synthesizers, 7

Creating Music, 34
Creedence Clearwater Revival, "45 Revolutions per Minute," 242
Cristofori, Bartolomeo, xxiv
Crosley, Powel, Jr., 202. *See also* Bethany Station; WLW
CSL Research Laboratory, 8–9. *See also* Sony Corporation
Ctesibius of Alexandria, xxiii
Culture Club, 153

Daddy Yankee, 280
"Daddy's Car," 9, 34–35
Dale, Dick, 68
Danelectro, 70
Darnielle, John. *See* The Mountain Goats
Dartmouth College, 277–78
Davinci Resolve, 28, 142
Davies, Dave, 69. *See also* The Kinks
Davis, Miles, *Bitches Brew*, 21; *Kind of Blue*, 31
The Day the Earth Stood Still, xviii, xxvi, 249
dbx noise reduction system, 53
DeArmond Trem Trol 800, 61
Decca Records, 42
del Rosario, Roberto, 115
The Delicate Delinquent, 249
DeLuca, Pat, 102
Denney, Dick, 2
Denon, 42–43
Depeche Mode, 7
Desktop and laptop computers, **39–40**
Devo, 147, 153
DeWine, Mike, xxii
Dick, Robert, 273
Diddley, Bo, 68
DiFranco, Charlotte, 102
Digital Audio Tape (DAT), **41–42**
Digital Audio Workstations (DAW), 29
digital recording, **42–44**
digital recording designations, 32 (sidebar)
digital synthesizers, **44–46**
distortion, **47–49**
DJ-ing, xxix, **49–51**
DJ Kool Herc, xxix, 50, 122, 216
DMX Krew, 56
Dobro Corporation, 68, 210
Dolby Laboratories, noise reduction systems, xxviii, 23–24, 51–53; surround-sound system, 52, 149, 198
Dolby noise reduction, **51–53**
Dolby, Ray, xxviii, 51. *See also* Dolby Laboratories

Dolby, Thomas, xx, 7, 37, 228–29 (sidebar), 254–55; "She Blinded Me with Science," 153, 215. *See also* Audio Virtual Reality Engine (*AVRe*)
Dolenz, Micky, 144. *See also* The Monkees
Donkey Kong, 253
The Doors, "Light My Fire," 84; *Strange Days*, 144
Dopyera, John, xvii–xviii, xxvi, 68, 186, 209–10. *See also* Dobro Corporation; National String Instrument Corporation
Dopyera, Rudolph, 186, 210. *See also* Dobro Corporation
Douglas, Jerry, 211
Dowd, Tom, 156
Drake, "God's Plan," 55
Dropbox, 28, 29
drum machines, **53–57**
DTS surround sound system, 198–99
Dudley, Homer, 266. *See also* vocoders and talk boxes
Dunn, Jimmy Ray, 102

E.M.I., 96, 234
EarBeater, 100
Eastman Wind Ensemble, xxix, 43
Earth, Wind & Fire, 14; "Let's Groove Tonight," 267
The Ed Sullivan Show, xvi, xxviii
Edison, Thomas, xxv, 59, 71, 95, 220. *See also* phonograph (Edison cylinder player)
Edison phonograph, **59–61**
effects pedals, 48, **61–64**
Egoyan, Eve, 276
8-track tapes, xix, 23, **64–67**
electric guitars, **67–70**
Electric Light Orchestra, *A New World Record*, 267
electrical recording, xxv, **70–73**, 92
Electro Harmonix, effects pedals, 63; SYNTH9 synthesizer emulater, 63
Electronic Courseware Systems, 100
electronic dance music, 253
electronic keyboard instruments, **73–76**
electronic tuners, **76–79**
Electronic Valve Instruments (EVI), 80
electronic wind instruments (EWI), **79–82**
Ellis, Don, *Man Belongs to Earth* soundtrack, 149, 198
Ellison, Steve, 37, 254–55. *See also* Audio Virtual Reality Engine (AVRe)
Emerson, Keith. *See* Emerson, Lake & Palmer
Emerson, Lake & Palmer, "Lucky Man," xxviii, 6, 146
Emmons, Buddy, 187. *See also* Sho-Bud

Eno, Brian, 89–90, 122, 142–43; *Ambient 1: Music for Airports*, 89. *See also* Generative Music
eStand, 103–4
Eurythmics, 7, 153

Facebook, 230; Facebook Live, xxii, xxxi, 29, 228, 230–31
Fairlight CMI, xxix, 44, 45, 122, 215
Fanning, Shawn, 151, 188, 190. *See also* Napster
Fantasia, 149
Farfisa electronic organs, 179
Federal Communications Commission (FCC), 204
Fender Electric Instrument Company: amplifiers, 1–2, 213; electric bass guitars, 69–70; electric guitars, xxvii, 68–69
Fender, Leo, xxvii, 1–2, 213. *See also* Fender Electric Instrument Company
Fennell, Frederick, xxix, 43
Ferguson, Graeme, 198
Finale, xxi, xxx, 14, 139, 162
Finale PrintMusic, 104
Fitzgerald, Ella, 25–26
Flow Machines, 8–9, 34–35
Fogel, Lawrence Jerome, 18
Foley, Red, 203
Fonsi, Luis, "Despacito," 280
Ford, Mary, 156
Ford Motor Company, xix, 64, 65; Lincoln-Mercury Division, xxx, 31
forScore, 103–4, 169
45-rpm records, xix, xxvii, **83–85**
Foster, Stephen, "Oh! Susanna," 223–24
Fostex, xxi, 157
Frampton, Peter, "Do You Feel Like We Do," 267; "Show Me the Way," 267
Franz, Frederick, 134
Franz Manufacturing Company, 134
Fraunhofer-Gesellschaft, 106, 150
Freed, Alan, 49
Fripp, Robert, 89
Frogger, 253
Fullerton, George, 2 (sidebar)

Gabrieli, Giovanni, *Sonata pian'e e forte*, 166, 168
Galvin Manufacturing Corporation, xxvi, 205. *See also* Motorola
Galvin, Paul, 205. *See also* Galvin Manufacturing Corporation
Gann, Kyle, 276
Gannon, Peter, 260. *See also* PG Music

GarageBand, **87–89**. *See also* Apple Computer
Gaye, Marvin, "Sexual Healing," 55
General Electric, 123
General Motors, xix, 64
Generative Music, **89–90**
Gershwin, George, 193, 224
Gibson Brands, Clavioline, 75; electric guitars, xxvii, 68, 69, 70
Gilliland, Ezra, 95
Glass, Louis, 111
Glasser Bows, 175
Gnutella, 189
Goodman, Benny, *Let's Dance* (radio program), 203
Google Drive, 28
gramophones and victrolas, **90–93**
Grampian Transport Museum, 128
Grand Master Flash and the Furious Five, 217 (sidebar)
Grand Ole Opry, xvi
Graves, Josh, 211
Gray, Elisha, xviii–xix, xxv, 201
Green Acres, 62
Greenwood, Jonny, xviii, 173. *See also* Radiohead, "How to Disappear Completely"
Guarneri family, 174
Guitar Hero, 69, 255
guitars: electric guitar, xviii, 67–70; Hawai'ian guitar, xviii, 211; resonator guitar, xvii–xviii, xxvi, 209–12. *See also* effects pedals; pedal steel guitars
Gutenberg, Johannes, 164
Guthrie, Marjorie, 270
Guthrie, Nora, 270
Guthrie, Woody, 270

Hal Leonard Corporation, *Amazing Slow Downer*, 260
Half-speed-mastered recordings, 72 (sidebar)
Hammacher Schlemmer, 12
Hammond organs, 119–20, 178
Han, Ulrich, xxiii, 165
Hancock, Herbie, 73
Happy Days, 113
Harris, Charles K., "After the Ball," 224
Harrison, George, 9, 62, 120; *Thirty Three & ⅓*, 81. *See also* The Beatles
Harrison, Henry C., 71. *See also* Bell Laboratories
Harvey, Will, 39–40
Hayman, Dick, *Concerto Electro*, 74
headphones, **95–97**

Heap, Imogen, "Hide and Seek," 267
hearing loops, 182
Hee Haw, xvii
Heil, Bob, 266, 267. *See also* Heil Talk Box
Heil Talk Box, 266–68
Hendrix, Jimi, 3, 48, 120; "Purple Haze," 141, 235
Herschell Carrousel Museum, 129
Hertz, Heinrich, 201
Hill, Brad, 72 (sidebar), 257
Hiller, Lejaren, 33–34; *HPSCHD*, 36; "Illiac Suite," 33–34
Hindemith, Paul, 133, 193
Höfner guitars, 70
Hohner, Clavinet, 75; Electra Melodica, 80; Pianet, 75
Holly, Buddy, 69, 212
The Hollywood Revue of 1929, 148
Hoopi'i, Sol, 186
Hootenanny, xvii
Hope-Jones, Robert, xxv, 245. *See also* Rudolph Wurlitzer Company
House, Son, 210–11
House on the Rock, 128, 160
Houston, Whitney, "I Wanna Dance with Somebody," 55
Howarth, Jamie, 270
Hullabaloo, xvii
The Human League, *Crash*, 219; *Dare*, 45, 55
Hurdy-Gurdy, 129–30 (sidebar)
Hurley, Chad, 279. *See also* YouTube

IBM, 39
Ibuka, Masaru, 233 (sidebar). *See also* Sony Corporation
iHeart Radio, 102, 172, 227
IMAX films, 149, 198
iMUSE, 37, 254, 265
Incredible Bongo Band, "Apache," 51
Indiana University, 189
InForm, 100–101
Inoue, Daisuke, 115. *See also* karaoke
Instagram, xxii, 230
instructional software, **99–101**
internet radio, **101–3**. *See also* radio
Isaacs, Bud, 187
Isaacson, Leonard, 33–34; "Illiac Suite," 33–34
iMovie. *See* Apple Computer
iPads and other tablets, **103–5**
iPods and other portable digital music players, **105–8**
iTunes, **108–10**. *See also* Apple Computer

J. P. Seeburg, 112
Jackson, Harold, 187. *See also* Sho-Bud
Jackson, Michael, "Beat It," 153; "Billie Jean," 81
JackTrip, 28
James, Rick, "Mary Jane," 6
JamKazam, xxii, 28
The Jazz Singer, 147–48, 246
Jennings, Tom, 2, 5. *See also* Vox
Jensen, Peter, xxv, 1, 123, 181
Jobs, Steve, 106
Joel, Billy, 30
John, Elton, "Someone Saved My Life Tonight," 6
Johnson, Eldridge, 91. *See also* Victor Talking Machine Company
Johnson, James P., 193
Jolson, Al, 148. *See also The Jazz Singer*
Jones, Cameron, 277–78. *See also* Synclavier; Yamaha DX7 synthesizer
Jones, Brian, 69, 131. *See also* The Rolling Stones
Joplin, Scott, 193; *Maple Leaf Rag*, 166, 193
Juke 8 karaoke machine, 115
jukeboxes, **111–13**
JVC boomboxes, 14

Kakehashi, Ikutaro, 76 (sidebar). *See also* Roland Corporation
karaoke, **115–18**; and the Kachin rebellion in Myanmar, 116; as a sociological phenomenon, 116
Karaoke app, 117
Kardashian, Kim, 230
Karim, Jawred, xxxi, 279. *See also* YouTube
Kauffman, Clayton "Doc," 2 (sidebar)
Kazaa, 189
KDKA (radio station), xxv, 202
Keb Mo, 211
Kekuku, Joseph, 185–86
Kellogg, Edward W., 123
Kenny G, 81
Keuffel & Esser, 161
Kilgen organs, 246
King Musical Instruments, 21
King Crimson, 132
The Kinks, 47–48
Klosé, Hyacinthe, xxiv, 271
Konami, 253
Koninklijke Philips N.V., xxviii, xxix–xxx, 23, 30, 157–58 (sidebar)

Korg, Inc., 79 (sidebar); drum machines, 56; electronic tuners, 77–78; synthesizers, 7, 46
Koss, John, 96
Kraftwerk, 147
Kramer, Eddie, 141
Kratka, Irv, 259. *See also* Music Minus One
Krause, Bernie, 144
Kubrick, Stanley, 145
Kurzweil, Ray, 73–74 (sidebar)
Kurzweil Music Systems, 74 (sidebar)

Lake, Greg. *See* Emerson, Lake & Palmer
Land, Michael, 37, 254. *See also* LucasArts
Lear, Bill, 64, 205. *See also* 8-track tapes
Leguia, Luis, 175. *See also* Luis and Clark
Lennon, John, 9; "Meat City," 243; "Revolution No. 9," 242. *See also* The Beatles
Leslie, Donald, 119. *See also* Leslie speaker
Leslie speaker, **119–21**, 125
Lexicon Pro, 262
Light, Enoch, 149, 257
Lightning Arranger, 225–26 (sidebar)
Limewire, 189
Linn LM-1 drum machine, xxix, 55–56
Lipps, Inc., "Funkytown," 267
The Living Daylights (film), 14
Lizzo, 230
LL Cool J, "I Can't Live without My Radio," 14
Lo-Fi audio recording and culture, 15, 26
Lodge, Oliver, 123
Logic Pro X. *See* Apple Computer
loops, **121–23**
The Lost Weekend, xviii, xxvi, 249
loudspeakers, **123–26**. *See also* Bose Corporation; Leslie speaker
Louisiana Hayride, xvi, xvii
Love, Mike. *See* The Beach Boys
Lowrey organs, 178
Lucas, George, 198. *See also* LucasArts
LucasArts, 37, 254
Luis and Clark, 175
Lynn, Alison, 176. *See also* Moxie Strings
Lynne, Jeff, "Mr. Blue Sky," 267. *See also* Electric Light Orchestra
Lyricon, 80–81

Madonna, 137, 153
Maestro FZ-1 effects pedal, 62

Maceo, Teo, 21
MacGAMUT, 99
Machover, Tod, 276
Mackay, Andy, 81
Maelzel, Johann, xiv, xxiv, 132
Magenta (Google songwriting software), 9, 34
MAGIX, 263
Magnavox, xxv, 1
MakeMusic, Inc., *New SmartMusic*, 101, 261; *SmartMusic*, 101, 261. *See also* Finale; Finale PrintMusic
Making Music, 34, 100
Manchester, Melissa, 26
Mangione, Chuck, 26
Mantovani and His Orchestra, 171
Manzarek, Ray, 73; "Riders on the Storm," 75. *See also* The Doors
Marantz, 157
Marching 110 (Ohio University Marching Band), xxii
Marconi, Guglielmo, 201
Marr & Colton theater organs, 246
Marshall amplifiers, 2, 3
Marshall, Jim, 3
Martenot, Maurice, xviii, xxvi, 173. *See also* Ondes Martenots
Martin, Chris, 230
Martin, Constant, 5
Martin, George, 156, 235–36, 243, 244 (sidebar)
Mateus, Jorge Arévalo, 270
Matit Flutes, 273
Maxell, 26
Maxfield, Joseph P., 71. *See also* Bell Laboratories
Maxwell, James Clerk, 201
May, Brian, 3
McCartney, Paul, 9, 70, 146; *McCartney III*, 258. *See also* The Beatles
McConnell, Peter, 37, 254. *See also* LucasArts
Mechanical organs, **127–30**
Mechanical rights, 194–95 (sidebar)
Mellotrons, xxviii, 54, 121, **130–32**
Melodrive, 38, 255
Memorex, 25
Memphis Jug Band, 71
Mercadier, Ernest, 95
Mercedes-Benz, xxx, 31
Messiaen, Olivier, xviii; *Trois petites liturgies de la presence divine*, 173; *Turangalîla-symphonie*, 173
Metallica, 189

Metheny, Pat, 7
metronomes, **132–35**, 168–69
microphones, **135–38**, 174–75
MIDI, xx–xxi, xxix, 81, **138–40**, 161, 163, 169, 179, 265, 275
Midsomer Murders, xxvi
Midwestern Hayride, xvi, xvii, 202
Miller, Mitch, 115
The Milton Berle Show, xvi
Mixing boards, **140–43**
Miyamoto, Shigeru, 253
Mobile Fidelity Sound Lab, 72 (sidebar), 257
The Monkees, xx; "Daily Nightly," xxviii, 5, 144; "Star Collector," 144
The Monkees (television program), 152
Montez, Chris, "Let's Dance," 179–80
Montovani, 171
Mood Media, 171
The Moody Blues, 131
Moog synthesizers, xix–xx, xxviii, 6, 7, **143–47**
Moog, Robert, xviii, xix–xx, xxvi, xxviii, 5, 143
Moore, Scotty, 69
Mortier Dance Organ, 128
Morton, Jelly Roll, 193
Motion picture soundtracks, **147–50**
Motion Pictures Experts Group, 106, 150. *See also* MP3s
Motorola, xxvi, 64
The Mountain Goats, 15, 26 (sidebar)
Moxie Strings, 176
MP3 players, 105–8
MP3s, 40, 105–8, **150–52**
MPEG-4, 110, 151
"Mr. Shadow," 9
MTI RehearScore, 261
MTV and music videos, xvii, xxix, **152–55**
Müller, Iwan, 271
multi-track recording, 140–41, **155–58**
Muntz, Earl, 64
MuseNet, 9, 36–37. *See also* OpenAI
Music boxes, xv, **158–61**
Music Construction Set, 39–40
Music Genome Project. *See* Pandora and the Music Genome Project
The Music House Museum, 160
Music Memos, 226
Music Minus One, 259
music notation, xxiii
music notation software, **161–64**
music printing with movable type, xiv, **164–67**
Music Recorder, 226

The Music Studio, 40
music videos, 152–55
musical Instrument Museum, 160
musical notation, **167–70**
MusiciansKit, 134, 227
MusicMatch Jukebox, 151
Musitronics Mu-Tron III effects pedal, 62–63
Mutual Broadcasting System, xxvi, 203
Muzak, **170–72**
MyPianist, 261
MySpace, 229–30

Nancarrow, Conlon, 194
National Broadcasting Company (NBC), 203
National String Instrument Corporation, 210
Neo-Bechstein electric piano, 74
Napster, xxx, xxxi, 40, 106, 188–90, 237
Nesmith, Michael, 144, 152. See also The Monkees
Neumatic notation, xiv, xxiii
New England Digital Corporation, 45; Synclavier, 45; Synclavier II, 45, 278
New Order, "Blue Monday," 56
New York Philharmonic, xvi–xvii
Nintendo, 253
Nishikado, Tomohiro, 253
Nixon, Marni, 148
Noble, Roger, 80. See also Lyricon
Nokia, 228 (sidebar)
Nord drum machines, 56
Norelco 22RL962, 13. See also Koninklijke Philips N.V.
Notion, 104, 164
Novation synthesizers, 7
NS Design electric string instruments, 176

Oberheim DMX drum machine, 56
Okeh Records, 71
Ondes Martenots, xviii, xxvi, 5, **173–74**
OneDrive, 29
Ono, Yoko, 242
OpenAI, 9, 36–37
Optical Music Recognition (OMR) software, 164
Orbison, Roy, 3
The Orchard, 8. See also Sony Corporation
orchestral string instruments, **174–76**
organs, xxiii–xxiv, **177–80**; electronic organs, 178; mechanical organs, 127–30; pipe organs, 177–78, 179; theater organs, 244–47; water organs, xxiii, 177

Owens, Buck, 144

PA systems, **181–82**
Pacific Arts, 152
Paderewski, Ignacy, 193
Page, Jimmy, xviii, 62, 249
Palace Theatre (Canton, Ohio), 246
Panasonic, 14, 43; Panapet transistor radio, 251; Toot-a-Loop transistor radio, 251–52
Pandora, 172, 218, 227, 237
Pandora and the Music Genome Project, 37, **182–84**
Parker, Charlie, 269–70
Parker, Sean, 151, 188, 190. See also Napster
patches, **184–85**, 279
Paul, Les, xxvii, 68, 156, 242
Paul Revere and the Raiders, 179
pedal steel guitars, **185–87**
Peer, Ralph, 71
peer-to-peer file sharing, 40, 106, **188–90**
Pepper, J. W., 20
percussion instruments, **191–92**
Peterson electronic tuners, 77
Petrucci, Ottaviano, xxiii, 165–66
Pfleumer, Fritz, 269
PG Music, *Band-in-a-Box*, 260–61; *RealBand*, 261
Philips. See Koninklijke Philips N.V.
Phillips, Sam, 212
phonograph (Edison cylinder player), xxv, 59–61
PhotoScore, 139, 164
pianos, xxiv; electric pianos, 73–75; player pianos, xv, xxv, 192–95, 275–76
PiBox, 27
Pickett, Wilson, 235
Pierson, Kate, 179
Pinder, Mike. See The Moody Blues
Pink Floyd, 120
Plant, Robert, xviii, 249
player pianos, xv, xxv, **192–95**, 275–76
playlist mentality, 109–10
PlayScore, 139, 164
Pollard Industries, xxix, 191–92
Polyphon Musikwerke, xv, 159–60
The Porter Wagoner Show, xvii
Post, Mike, *The Rockford Files*, 146
Poulsen, Valdemar, 269
Practica Musica, 101
Presley, Elvis, xvi, xxvii; "Jailhouse Rock," 152
PreSonus, AudioBox, 104; *Studio One Professional*, 263

Preston, Billy, 75
The Pretenders, 153
Price, Alan, 179. *See also* The Animals
Pridham, Edwin, xxv, 1, 123, 181. *See also* Magnavox
Prince, 55, 153
Pro Tools, 263
Project 3 Records, 257
Prophet-5 synthesizer, xxx, 7
Psy, "Gangnam Style," 280
Puff Daddy, "I'll Be Missing You," 217
Pyke, Magnus, 153, 215
Pythagoras, xiii–xiv, xxiii

QRS Music Company, 193, 194
quadraphonic and surround sound, **197–99**

Rachmaninoff, Sergei, 193
radio, xv–xvi, xxv, xxvii, **201–6**; AM radio, 85, 204, 251; FM radio, 24, 203; internet radio, 101–3; satellite radio, 217–19; transitor radios, xix, 250–52
Radiohead, "How to Disappear Completely," xviii
Rapaport, Samuel M. "Skip," 225 (sidebar)
RCA, 202, 248. *See also* RCA Victor
RCA Victor, xxvii, 64, 91
RealBand. *See* PG Music
Realistic Concertmate MG-1 synthesizer, 147
Recording Industry Association of America (RIAA), 42
reel-to-reel tape, **206–9**
Regency TR-1 transistor radio, 250
Reis, Johann Philipp, 123
resonator guitars, **209–12**
reverb, **212–13**
Revere, Paul, 179. *See also* Paul Revere and the Raiders
Rey, Alvino, 187
Rhodes electric pianos, 74–75
Rhodes, Harold Burroughs, 74. *See also* Rhodes electric pianos
Rhythmate. *See* Chamberlain, Harry
Rhythmicon, xxvi, 54
Ribbon controller, 249 (sidebar)
Rice, Chester W., 123
Richards, Keith. *See* The Rolling Stones
Richland Carousel Park (Mansfield, Ohio), 129
Rickenbacker, Adolph, xviii, 67–68
ringtones, 228–29 (sidebar)
Riviera Theatre (North Tonawanda, New York), 245
Robinson, Sylvia, 217 (sidebar)

Rock-Ola jukeboxes, 113
Rodgers, Jimmie, xvi, 71, 203
Roland Corporation, 7, 56, 76 (sidebar), 81; BOSS DB-11 Music Conductor, 78; BOSS DR-55 (Dr. Rhythm), 55; Jupiter-6 synthesizer, xxx, 138; MC-8 MicroComposer, xxix, 122–23, 219; TR-808 drum machine, xxix, 54–55; TR-909 drum machine, xxx
The Rolling Stones, "Dandelion," 131; "(I Can't Get No) Satisfaction," 62; "She's a Rainbow," 131; "We Love You," 131
Roni Music, 260
Roosevelt, Theodore, 201
Rosenthal, Steve, 270
Roxy Music, 89, 142–43
Rudolph Wurlitzer Company, xxv, 54, 75, 127, 129, 245
Russell, Ross, 270
Russell-Smith, Warren, 270
Rykodisc, 31

sampling, xx, **215–17**
Santana, Carlos, 48
Sanyo, 53
Sarnoff, David, 202
satellite radio, **217–19**
Satie, Erik, 170
Sax, Adolphe, xiv, xxiv, 20, 271–72
Say Anthing (film), 14
Schaeffer, Pierre, 241
Schlick, Arnolt, xxiii
Schumann, Robert, 223
SCORE, 162
Scott, James, 193
Scott, Tom, 81
Seeburg. *See* J. P. Seeburg
Sega, 253
Seiko, 77–78
Selmer Corporation, Clavioline, 5, 75
Sennheiser Communications, 96
sequencers, **219–20**
Sequential Circuits, xxix, 138; Prophet 600 synthesizer, 138
78-rpm records, **220–22**
Seville, Alan, "The Chipmunk Song (Christmas Don't Be Late)," 242–43; "Witch Doctor," 242
The Shadows, 68
Shannon, Del, "Runaway," 5, 75
sheet music, **223–26**
Shepp, Archie, *On Green Dolphin Street*, xxix, 42–43
Shindig!, xvii
Sho-Bud, 187

Shore, John, xxiv
Short, Kevin, 270
Short, William, 125
Shure Vagabond, 88, 137
Sibelius, xxi, 104, 139, 162
Side Man. *See* Rudolph Wurlitzer Company
Siemans, Werner von, 123
Silvertone, 269
Simmons Company, xxix, 192
Sing Along System karaoke machine, 115
Sing Along with Mitch, 115–16
Singular Sound drum machines, 56
Sirius, xxxi, 217–18. *See also* SiriusXM
SiriusXM, 218
Skype, 27
SmartMusic. *See* MakeMusic, Inc.
smartphones, **226–29**
SmartScore, 139, 164
Smash Mouth, 249
Smith, Dave, xxix, 138
Smith, Jimmy, 120
Smith, Kate, *The Kate Smith Hour*, 203
Smith, Leland C., 162
Smith, Mike, 179
Smith, Oberlin, 269
Snodgrass, Jennifer Sterling, 100
social media, **229–31**
Solina String Ensemble, 6
Sonic Foundry, *Acid pH1*, 51, 87, 220, 263; *Sound Forge*, 87, 263
Sony Corporation, xxix–xxx, 31, 41, 42, 233 (sidebar); Car Discman, 31–32, 232; Discman, xxx, 31, 231–34; TR-63 transistor radio, 250; Walkman, xix, xxix, 231–34; Sony Digital Pictures, 87
Sony Walkman and Discman, **231–34**
Soul Train, xvii
Sound Blaster computer audio cards, 254, 265
Soundbrenner Core, 78, 134
SoundCloud, 29
SoundStorming, xxii
Soundstream, 42
Soundtrap, xxii
Sousa, John Philip, xv, 20, 166
Space Invaders, 253
The Specials, 153
Spellbound, xviii, xxvi, 249
Spielberg, Stephen, *Jurassic Park*, 198
Spillman Engineering, 129
Spotify, xxxi, 172, 218, 237. *See also* Streaming audio services
Squier, George Owen, 170. *See also* Muzak
Springsteen, Bruce, *Nebraska*, 157

Stage Show, xvi
Stanford University, 28, 45, 162
A Star Is Born, 52
State University of New York at Buffalo, 33
Steely Dan, "Peg," 81
Steiner, Max, *Gone with the Wind*, 148
Steiner, Nyle, 21, 80. *See also* Electronic Wind Instruments (EWI)
Stereophonic sound, xxvii, 124, 140–41, **234–36**
Stimson, Arthur J., xviii, xxvi, 68
Sting, "Every Breath You Take," 217
Stockhausen, Karlheinz, 241
Story & Clark Pianos, 193
Stowasser, Ignaz, 20
Stradivari, Antonio, 174
Stravinsky, Igor, 193
streaming audio services, **236–38**
Stretch Music, 261
Studio One Professional. *See* PreSonus
Stylophone synthesizer, 7
Subotnick, Morton, 34, 100
The Sugarhill Gang, "Rapper's Delight," 122, 216
Sun Studios, 212
Swift, Taylor, 226, 229–30
Synclavier, 45–46, 278. *See also* New England Digital Corporation

T-Pain, "Buy U a Drank (Shawty Snappin')," 11
tablature systems, xxiii, **239–41**
tablet computers, 103–5
Taito, 253
Talk Box. *See* Heil Talk Box
tape manipulation, **241–44**
Tárrega, Francisco, *Gran Vals*, 228 (sidebar)
TASCAM, xxi; Digital Portastudio, 157; Portastudio (Teac 144), xxix, 24–25, 157
Taylor, James, 281
TDK, 42
Tedesco, Tommy, 62
TEFview, 241
Telarc, xxix, 43
Tesh, John, 49
Tesla, Nicola, 201
Theater Organs, **244–47**
Thee Milkshakes, 26 (sidebar)
Theremin, Leon, xviii, xxvi, 54, 247. *See also* Rhythmicon, Theremins
Theremins, xviii, xxvi, 5, **247–50**; use in movie and television soundtracks, xxvi, 248–49

Thingamagig multi-effects processor, 63–64
33–1/3-rpm records, xix, xxvii
Thomas Organs, 54
3M, 42
THX sound certification, 149, 198–99
Tidal, 237
TikTok, 155
Tokyo Tsushin Kogyo K. K. *See* Sony Corporation
The Tornados, "Telstar," 5
Townshend, Pete, 3
Townsend, Ken, 244 (sidebar)
Trackd, 27
Transistor radios, **250–52**
Tumblr, 230
tuning devices, xxiv
tuning and temperament systems, xxiv
Turing, Alan, xxvii
Tuscora Park (New Philadelphia, Ohio), 129
Twitter, 230

Ultimate Ears Hyperboom boombox, 16
Unicord Corporation, Univox, 5, 75; Univox Comac-Piano, 75
University of Illinois, 33
University of Southern California, 189
Urban, Keith, 230
U.S. Marine Band, xv

Vallée, Rudy, 181
Varèse, Edgard, xviii, 241–42; *Ecuatorial*, 173; *Poème électronique*, xxviii, 197, 242
The Ventures, 68
Vestex, 157
Vevo, xxxi, 155, 280
VH-1, xvii, 154. *See also* music videos
Victor Talking Machine Company, 61, 71, 73, 91, 220 21. *See also* RCA Victor, Victrola, 59, 73, 90–93
Victrolas. *See* Victor Talking Machine Company
video games, 37, **253–56**
vinyl albums, **256–59**
virtual accompanists, **259–63**
virtual studio software, **263–64**
virtual synthesizers, **264–66**
Vivendi, xxxi
Vocalstyle piano rolls, 193
vocoders and talk boxes, **266–68**
Vogel, Peter, 215. *See also* Fairlight CMI
Vogue Records, 222

Voice Memos, 226
Voice Record, 226
Votey, Edwin S., xxv, 193
Vox, amplifiers, 2, 3; effects pedals, 62, 63; musical instruments, 179

Wagner, Richard, 148
Walker, T-Bone, 69
Waller, Fats, 193
Walsh, Joe, "Rocky Mountain Way," 267
Warner Music Group, 8
Wenger Corporation, 262 V-Room, 262; Virtual Acoustical Environment technology, 262
Wente, Edward C., 71
West, Kanye, 230
West Side Story, 148
Westergren, Tim, 182. *See also* Pandora and the Music Genome Project
Western Chicago, 269
Western Electric, 71
Westinghouse, 202
WGN (radio station), 203
Williams, John, *Star Wars* trilogy, 148–49
Wilson, Woodrow, xxv, 1, 135
Win-D-Fender, 273
Winkel, Dietrich Nikolas, xiv, xxiv, 132
Winter, Alex, 190
Winter, Edgar, 75
wire recording, **269–70**
Wireless Specialty Apparatus Co., 95
Withey, Conrad, 8
WLW (radio station), xxvi, 202–3
WolframTones, 38, 255
Wonder, Stevie, xx, 55, 75, 267; "Higher Ground," 63; "Pastime Paradise," 216–17
Wood, Chet, 138
Wood, Mark, 176. *See also* Wood Violins
Wood, Natalie, 148
Wood Violins, 176
Woodstock Festival, 181
Woodwind instruments, **270–73**
WOR (radio station), xxvi, 203
Wurlitzer. *See* Rudolph Wurlitzer Company
Wurlitzer, Rudolph, 245. *See also* Rudolph Wurlitzer Company
WXYZ (radio station), xxvi, 203

Xenakis, Iannis, 34, 241
XERA (radio station), xxvi–xxvii, 202–3

Xiao Xiao, 276
XM Radio, xxx, 217–18. *See also* SiriusXM

Yale University, 189
Yamaha Corporation, xxi, 45–46, 157, 277 (sidebar); Disklavier, xxx, 275–77; DX7 synthesizer, xxx, 45–46, 277–79; DX21 synthesizer, 278; keyboards, 46–47, 275; SILENT instrument series, 21–22, 176; Sonogenic Keytar, 277
Yamaha Disklavier, **275–77**
Yamaha DX7 Synthesizer, **277–79**

Yearwood, Trisha, 230
Yokee, 117
Young Chang, 74 (sidebar)
YouTube, xxi–xxii, xxxi, 155, **279–81**; YouTube Live, xxxi, 228; YouTube Music, 237

Zappa, Frank, 62, 156
Zoom (electronics company), 44; G1X FOUR multi-effects pedal, 63
Zoom (virtual meeting software), xxii, xxxi, 27

About the Author

James E. Perone earned degrees in music education, clarinet performance, and music theory from Capital University and the State University of New York at Buffalo. Jim is currently professor emeritus of music at the University of Mount Union in Alliance, Ohio. He has been active as an author since the early 1990s. After researching and writing several music theory-related reference volumes and bio-bibliographies of American composers ranging from Howard Hanson and Louis Moreau Gottschalk to Paul Simon and Carole King, he focused his research and critical analysis more squarely on popular music. Jim's previous publications include *Paul Simon: A Bio-Bibliography* (Greenwood, 2000), *Songs of the Vietnam Conflict* (Greenwood, 2001), *Music of the Counterculture Era* (Greenwood, 2004), *The Words and Music of Carole King* (Praeger, 2006), *Mods, Rockers, and the Music of the British Invasion* (Praeger, 2008), *The Words and Music of Elvis Costello* (Praeger, 2015), *Smash Hits: The 100 Songs That Defined America* (Greenwood, 2015), *The Words and Music of James Taylor* (Praeger, 2017), *Listen to New Wave Rock!* (Greenwood, 2018), and *Listen to the Blues!* (Greenwood, 2019). Jim serves as the editor of Greenwood's "Exploring Musical Genres" series and previously served as the editor of Praeger Publishers' "Praeger Singer-Songwriter Collection."

www.ingramcontent.com/pod-product-compliance
Lightning Source LLC
Chambersburg PA
CBHW082027300426
44117CB00015B/2374